Symbol	Meaning
p	Sample proportion
H_0	Null hypothesis
H_a	Alternate hypothesis
σ_p	Sample standard deviation for proportions
$t_{0.025}$	Student's t distribution at 5% critical value
α	Probability of Type-I error
χ^2	Tests independence of 2 variables
E	Expected frequency
O	Observed frequency
ANOVA	Analysis of variance
SS	Sum of squares
MS	Mean square
F	F distribution
$\chi^2_{0.05}$	Chi square at 5% significance level
μ_R	Mean number of runs
σ_R	Standard deviation of the number of runs
$=$	Equal to
$<$	Less than
$\leq$	Less than or equal to
$>$	Greater than
$\geq$	Greater than or equal to
$\approx$	Approximately equal to
z	Standard score
df	Degree of freedom

Statistics
and Probability
in Modern Life

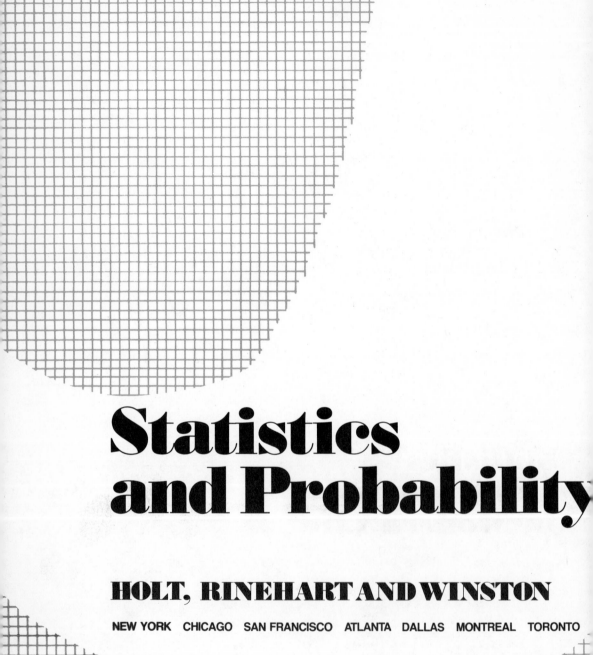

Statistics
and Probability

HOLT, RINEHART AND WINSTON

NEW YORK   CHICAGO   SAN FRANCISCO   ATLANTA   DALLAS   MONTREAL   TORONTO

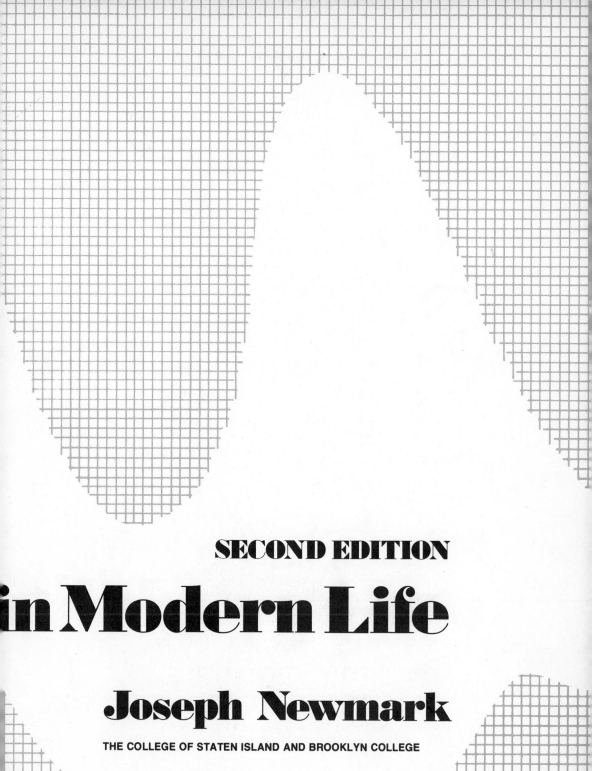

SECOND EDITION

in Modern Life

Joseph Newmark

THE COLLEGE OF STATEN ISLAND AND BROOKLYN COLLEGE

To Trudy, Sharon, Rochelle, and Stephen

Preface

Statistics and Probability in Modern Life, Second Edition, reflects the increasing range of applications of statistics. Each chapter begins with newspaper or magazine clippings which show in applied context the ideas discussed in the chapter. Hopefully these clippings will motivate the reader by showing how ideas of probability and statistics are used in everyday situations. Also many applied problems have been added in the body of the text and in the exercises.

Numerous learning aids have been added. Each chapter contains a section called *Preview* which specifically outlines the main ideas of the chapter. At the end of each chapter are a *Self-Study Guide* and a set of *Mastery Tests.* The Self-Study Guide summarizes chapter material and presents for review formulas and terms discussed in the chapter. The Mastery Tests are designed to measure comprehension of material presented in the chapter and to present challenging problems to the more demanding student. The questions in Form *A* of the Mastery Tests require short answers, whereas those in Form *B* require longer, more detailed answers. Questions in Form *B* are more challenging and supplement exercise sets.

Most examples and almost all exercises have been updated or changed completely. New material has been added and many existing sections have been expanded to include historical material. Answers to most exercises and to questions in Mastery Tests are provided. This edition like the first is made readable through the use of simple, understandable language.

PREFACE

I would like to thank the many teachers and students who received the first edition of this book so warmly. I am grateful to those who took the time to send comments, suggestions, and corrections. I am also grateful to those who reviewed this edition of the book: Ralph D'Agostino, Boston University; Gail Davis, Los Angeles Pierce College, Woodland Hills, California; Sammie Deek, Richland College, Dallas; David Levine, Baruch College, New York City; Robert Winkler, Indiana University, Bloomington; Bennie Zinn, San Antonio College, Texas.

My thanks also go to the authors and publishers who granted permission to use the statistical tables so necessary for this work. Among others, I am indebted to the Literary Executor of the late Sir Ronald A. Fisher, F.R.S., and to Oliver and Boyd, Ltd., Edinburgh, for their permission to reprint tables from *Statistical Methods for Biological, Agricultural, and Medical Research*.

I also wish to thank the staff of Holt, Rinehart and Winston for their enthusiastic interest in this project.

Finally, and most importantly, I wish to thank my wife Trudy and our children Sharon, Rochelle, and Stephen for their understanding and patience as they endured the enormous strain associated with completing this project. Without their encouragement it could not have been undertaken or completed.

Joseph Newmark

To the Instructor

In recent years more and more schools are offering courses on Introductory Statistics and Probability on an elementary level. This text is designed to serve as a general introduction to modern statistics and probability for students in all academic areas. This text presents to the college student the basic statistical ideas which are needed in such areas as sociology, business, economics, ecology, education, medicine, psychology, mathematics, and so forth. Students in such fields must frequently demonstrate a knowledge of the language and methods of statistics. Accordingly, exercises and examples have been chosen to interest such students.

This text evolved from classroom notes based upon my years of teaching at Staten Island Community College and Brooklyn College. No mathematical background, other than simple arithmetic, is needed as a prerequisite. A special effort is made to communicate with students who are not prepared for elaborate symbolism or complex arithmetic. The mathematical content is complete and correct, yet the language is elementary and readily understandable. Words, phrases, and any mode of expression that students find difficult to comprehend are avoided. The introduction of new terminology is held to a minimum. Points which students commonly misunderstand or completely miss are carefully explained in *Comments*.

Each chapter starts with an introductory section called *Preview* which explains the scope, basic concepts, and goals of that chapter. At the end of each chapter is a *Self-Study Guide,* which can be used to reinforce the ideas covered in the chapter.

Exercises and examples are abundant and are related directly to the student's experience.

An instructor should find no difficulty in selecting material for a one- or two-semester course. The following outlines indicate how this text can be used.

ONE-SEMESTER COURSE (MEETS 40 TIMES PER SEMESTER, 40 MINUTES PER SESSION)

Text Material	Approximate Amount of Time	Prerequisite Needed for Each Chapter
Chapter 1	1 lesson	none
Chapter 2	4 lessons	none
Chapter 3	5 lessons	none
Chapter 4	2 lessons	Chapter 2, Section 2.1 or the equivalent
Chapter 5	4 lessons	none
Chapter 6 (skip §6.4)	4 lessons	Chapter 5
Chapter 7	4 lessons	Chapter 6
Chapter 8	4 lessons	Chapter 2
Chapter 10	4 lessons	Chapter 8
	32*	

*The remaining meetings can be devoted to exams and review.

TWO-SEMESTER COURSE

| | SEMESTER 1 | |
Text Material	Approximate Amount of Time	Prerequisite Needed for Each Chapter
Chapter 1	1 lesson	none
Chapter 2	4 lessons	none
Chapter 3	5 lessons	none
Chapter 4	2 lessons	Chapter 2, Section 2.1, or equivalent
Chapter 5	4 lessons	none
Chapter 6	4 lessons	Chapter 5
Chapter 7	4 lessons	Chapter 6
Chapter 8	4 lessons	Chapter 2
Chapter 9	4 lessons	none
	32*	

Text Material	SEMESTER 2 Approximate Amount of Time	Prerequisite Needed for Each Chapter
Chapter 10	4 lessons	Chapter 8
Chapter 11	5 lessons	Chapter 8
Chapter 12	7 lessons	Chapter 8
Chapter 13	4 lessons	none
Chapter 14	4 lessons	Chapter 12
Chapter 15	4 lessons	Chapters 10, 12
	28*	

*The remaining meetings can be devoted to exams and review.

An instructor's manual which provides detailed solutions for exercises and suggests teaching techniques for each chapter is available from the publisher upon request.

Joseph Newmark

To the Student

This is a nonmathematical text on elementary probability and statistics. If you are afraid that your mathematical background is rusty, don't worry. The only math background needed to satisfactorily use this book is arithmetic. For those, however, who are fairly knowledgeable in math, this text will not be distracting. Instead you will see how statistics can be used in interesting and challenging problems.

The examples are plentiful and are chosen from everyday situations. Also, the ideas of statistics are applied to a variety of subject areas. This variety indicates the general applicability of statistical methods. In frequent *Comments* appropriate explanations of statistical theory are given and the "why's" of statistical methods answered.

Occasionally a section or exercise is starred. This means that it is slightly more difficult and may require some time and thought.

Each chapter concludes with a *Self-Study Guide* which includes a summary and a review of the formulas and major terms and ideas introduced in the chapter. In addition there are *Mastery Tests* which will help in preparing for exams. Also, answers to practically all exercises and mastery-test questions are given at the end of the book.

I hope that you will find reading and using this book an enjoyable and rewarding experience since a basic knowledge of statistics is essential. Good luck!

Joseph Newmark

Contents

14 ANALYSIS OF VARIANCE   409

15 NONPARAMETRIC STATISTICS   427

APPENDIX: STATISTICAL TABLES   452

ANSWERS TO SELECTED EXERCISES AND MASTERY TEST QUESTIONS   479

INDEX   513

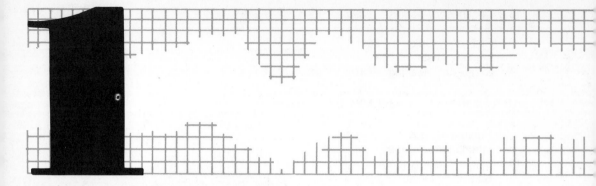

THE NATURE OF STATISTICS

What is Statistics?

Family of 4 Can Live a Little for 15.5G

Washington, April 10 (AP)—The typical urban family of four requires $15,500 a year to maintain a moderate standard of living, the Labor Department said today. This, because of inflation, is $1,200 more than in the previous year.

The same family can live at an austere level for $9,800 a year, or at a level allowing some luxuries for $22,500 a year, the government said in its annual analysis of hypothetical family budgets.

The costs, calculated for fall 1975, rose 7% for the low budget, 8% for the moderate budget and 8.2% for the higher budget over the previous year.

By comparison, the size of such budgets a year earlier rose between 12.4% and 14.2%. Last year's smaller increase, though still high by post-World War II standards, reflected the slowing of inflation.

The consumer price index, the best measure of the impact of inflation on consumers, increased at a rate of 12.2% in 1974 and 7.2% in 1975. Consumer prices have risen another 1.5% since the fall.

The annual survey attempts to calculate costs of three different levels of living for a hypothetical urban family consisting of a 38-year-old husband employed full time; his wife, who doesn't work outside the home; a 13-year-old son and an 8-year-old daughter. The couple is assumed to have been married about 15 years and to be "settled in the community."

The budgets are not based on how families actually spent their money but reflect assumptions about the manner of living. Low-budget families live in rented housing, use public transportation or drive a used car and do most of their own cooking and washing.

At the moderate level, families are assumed to have purchased their own home six years ago, drive a late model car, buy more meat at the market and dine out occasionally. The higher budget family buys a new car every four years and can afford more household goods and services.

Source: Daily News Sunday, April 11, 1976

THE SHRINKING AMERICAN WORKWEEK
Average Hours Worked Per Week
in Private Nonfarm Jobs

| 1910 | 1930 | 1950 | 1960 | 1970 | 1975 |
| 51.3 hours | 42.1 hours | 39.8 hours | 38.6 hours | 37.1 hours | 36.1 hours |

THIS YEAR some big labor unions are renewing proposals for even shorter workweeks, without cuts in pay. By 1985, if present trends continue, the average workweek will decline to 34 hours.

Note: Figures for 1910 and 1930 are for workers in manufacturing.

Source: U.S. Dept. of Labor

U.S. News & World Report, March 15, 1976. Reprinted from *U.S. News & World Report*, Copyright © 1976 U.S. News & World Report, Inc.

Statistics, such as the ones found in these newspaper articles, are frequently released by the United States Department of Labor. They provide statistical data on inflation, cost of living, and other issues of concern to us.

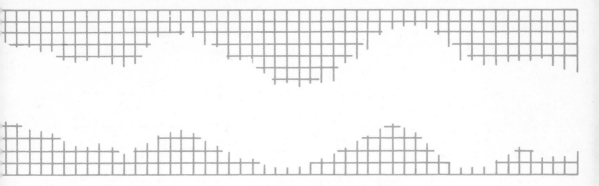

Preview

In this chapter we will discuss the following topics:

1. *The Nature of Statistics*
A discussion of statistics and numerous examples of how it is used.

2. *Descriptive Statistics and Statistical Inference*
An analysis of the two major areas of statistics — descriptive statistics and statistical inference. Descriptive statistics involves collecting data and tabulating it meaningfully. Statistical inference involves making predictions based upon the sample data.

3. *Population and Sample*
We distinguish between part of a group called a sample as opposed to the whole group called the population.

4. *Historical Discussion*
A brief discussion of the historical development of statistics and probability.

Introduction

What is **statistics?** Most of us think of statistics as having something to do with charts or tables of numbers. While this idea is not wrong, mathematicians and statisticians use the word statistics in a more general sense. Roughly speaking, the term statistics, as used by the mathematician, involves collecting numerical information called **data,** analyzing it, and making meaningful decisions based upon the data.

The role played by statistics in our daily activities is constantly increasing. As a matter of fact, the nineteenth century prophet H. G. Wells predicted that "statistical thinking will one day be as necessary for efficient citizenship as the ability to read and write." The following examples on the uses of statistics indicate that a knowledge of statistics today is quickly becoming an important tool, even for the layman.

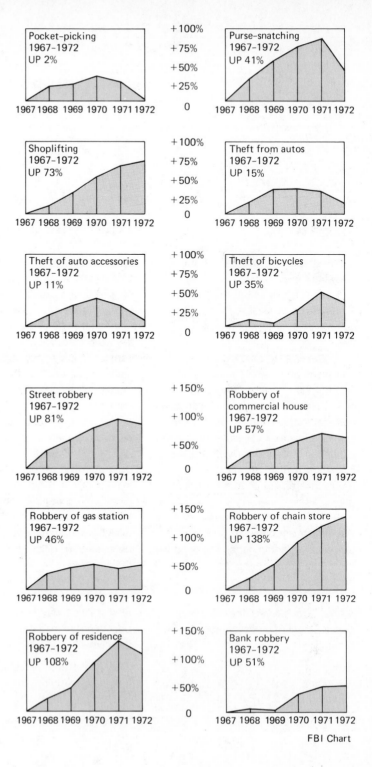

Figure 1.1

FBI crime charts. (*Crime in the U.S.*,
1972, pgs. 17 and 24)

1. The latest Department of Commerce statistics show that the cost of living in a large midwestern city rose by 8.0% in one year. The newspaper article on page 2 illustrates this use of statistics.

2. Statistics show that the use of drugs by pregnant women can be dangerous to the unborn child.

3. Statistics collected by the Metropolis Insurance Company show that today people on the average are living longer than did their parents.

4. The FBI crime charts given in Figure 1.1 show that crime increased at an alarming rate during the years 1967–1972. The same is not true for the years 1974–1975. The latest FBI statistics indicate that while major crime in the United States increased in 1975 by 9% over 1974, the increase was considerably smaller than in previous years (see Figure 1.2).

SLOWDOWN IN CRIME SPIRAL

The FBI's latest crime report, released March 25, offers hope that the nation's crime epidemic may be abating.

While the figures show that crime increased in 1975 by 9 per cent over 1974, that is only half the 18 per cent increase from 1973 to 1974. Even more encouraging, the increase was smaller each quarter, from an 18 per cent increase in the first quarter down to a 4 per cent increase in the last quarter.

Change in Number
of Major Crimes
From 1974 to 1975:

Murder	Down	1%
Forcible rape	Up	1%
Robbery	Up	5%
Aggravated assault	Up	5%
Burglary	Up	7%
Larceny-theft	Up	12%
Motor-vehicle theft	Up	2%

Figure 1.2

FBI crime chart. (*U.S. News & World Report,* April 5, 1976. Reprinted from *U.S. News & World Report,* Copyright © 1976 U.S. News & World Report, Inc.)

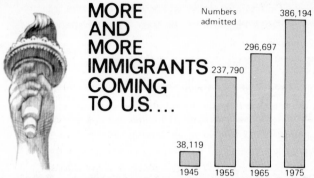

MORE AND MORE IMMIGRANTS COMING TO U.S. ...

Numbers admitted

38,119 — 1945
237,790 — 1955
296,697 — 1965
386,194 — 1975

...WITH A STRIKING SHIFT IN THEIR ORIGINS

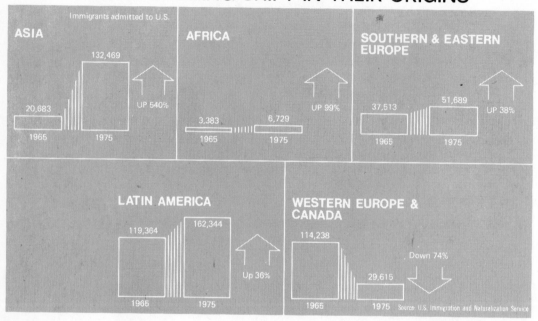

Immigrants admitted to U.S.

ASIA
20,683 (1965)
132,469 (1975)
UP 540%

AFRICA
3,383 (1965)
6,729 (1975)
UP 99%

SOUTHERN & EASTERN EUROPE
37,513 (1965)
51,689 (1975)
UP 38%

LATIN AMERICA
119,364 (1965)
162,344 (1975)
Up 36%

WESTERN EUROPE & CANADA
114,238 (1965)
29,615 (1975)
Down 74%

Source: U.S. Immigration and Naturalization Service

Figure 1.3

Immigration charts. (*U.S. News & World Report,* April 5, 1976. Reprinted from *U.S. News & World Report,* Copyright © 1976 U.S. News & World Report, Inc.)

5. Statistics show that if the air we breathe is excessively polluted, then un-doubtedly some people will become ill.

6. Statistics show that the world's population is growing at a faster rate than the availability of food.

7. Statistics show that the United States is facing a serious energy shortage and that some other source of energy must be found to meet the ever increasing demand.

8. Statistics show that any student, no matter what their high school background, will succeed in college if properly motivated.

9. Both the Gallup and Harris Polls use statistics in determining public opinion on controversial issues.

10. The statistics given in Figure 1.3 show that more and more immigrants are coming to the United States. Furthermore, Western Europe and Canada which used to be the source of the majority of our immigrants are no longer in the foreground. Figure 1.3 shows that Asia and Latin America are now the largest source of new immigrants. Such statistics have important implications for sociologists and economists.

Very little formal mathematical knowledge is needed to collect and tabulate data. However, the interpretation of the data in a meaningful way requires careful analysis. If not done by a statistician or mathematician who has been trained to interpret data, statistics can be misused. The following examples indicate how statistics can be misused.

In the city of Tunaville the occurrence of polio, thought to be nonexistent, increased by 100% from the year 1975 to 1976. Such a statistic would horrify any parent. However, upon careful analysis it was found that in 1975 there were 2 reported cases of polio out of a population of five million and in 1976 there were 4 cases.

As another example, consider the television commercial in which a car manufacturer claims that "9 out of 10 of our cars sold in the United States during the past 11 years are still on the road." The recorded sales of this car over the last 11 years are shown in Figure 1.4.

Are you convinced that the manufacturer is telling the truth or is there something misleading about this claim?

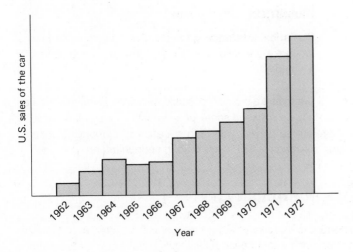

Figure 1.4

Recorded car sales.

1.1 CHOICE OF ACTIONS SUGGESTED BY STATISTICAL STUDIES

Since the word statistics is often used by many people in different ways, mathematicians have divided the field of statistics into two major areas called **descriptive statistics** and **statistical inference.**

Descriptive statistics involves collecting data and tabulating results using, for example, tables, charts, or graphs to make the data more manageable and meaningful. Very often, certain numerical computations are made which enable us to analyze data more intelligently. Most of us are concerned with this branch of statistics. For example, even if we know the income of every family in California, we are still unable to analyze the figures because there are too many to consider. These figures, somehow, must be condensed so that meaningful statements can be made.

Statistical inference, on the other hand, is much more complicated. To understand what is involved, imagine that we are interested in the average height of students at the University of California. Since there are so many students attending the university, it would require an enormous amount of work to interview each student and gather all the data. Furthermore, the procedure would undoubtedly be costly and could take too much time. Possibly we can obtain the necessary information from a sample of sufficient size that would be accurate for our needs. We could then use the data based upon this sample to make predictions about the entire student body, called the **population.** This is exactly what statistical inference involves. We have the following definitions.

DEFINITION 1.1

A **sample** is any small group of individuals selected to represent the entire group called the **population.**

DEFINITION 1.2

Statistical inference is the study of procedures by which we make predictions about a population on the basis of a sample.

Of course, we would like to make the best possible decisions about the population. To do this successfully we will need some ideas from the theory of probability. Statisticians must therefore be familiar with both statistics and probability. Exactly how such decisions are made will be discussed in later chapters.

1.2 STATISTICS IN MODERN LIFE

Statistics is so important to our way of living that many of us often use statistical analysis in making decisions without even realizing it. Today statistics is used not only by the mathematician but also by the nonmathematician in such areas as psychology, ecology, sports, insurance, education, biology, business, agriculture, music, and sociology, to name but a few. The fields of study to which statistics and probability are being applied is constantly increasing. In particular, its usefulness in the fields of biology, economics, and psychology is so enormous that the subjects of biometrics, econometrics, and psychometrics have come into being. These subjects involve the application of the ideas and techniques of both statistics and probability to biology, economics, and psychology.

Although statistics is one of the oldest branches of mathematics, it was not until the twentieth century that its use became widespread. Originally, it involved summarizing data by means of charts and tables. Historically, the use of statistics can be traced back to the ancient Egyptians and Chinese who used statistics for keeping state records. The Chinese under the Chou Dynasty, 2000 B.C., maintained extensive lists of revenue collection and government expenditures. They also maintained records on the availability of warriors.

The study of statistics was really begun by an Englishman John Graunt (1620–1674). In 1662 he published his book, *Natural and Political Observations Upon the Bills of Mortality.* Graunt studied the causes of death in different cities and noticed that the percentage of deaths from different causes was about the same and did not change considerably from year to year. For example, deaths from suicide, accidents, and certain diseases not only occurred with surprising regularity but with approximately the same percentage from year to year. Furthermore, Graunt's statistical analysis led him to discover that there were more male than female

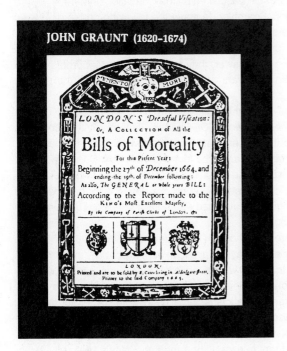

Illustration "Bills of Mortality" redrawn from *Devils, Drugs, and Doctors* by Howard W. Haggard, M.D. Copyright 1929 by Harper and Row, Publishers, Inc.; renewed 1957 by Howard W. Haggard. Reprinted by permission of the publisher.

births. But, since men were more subject to death from occupational hazards, diseases, and war, it turned out that at marriageable age the number of men and women was about equal. Graunt believed that this was nature's way of assuring monogamy.

After Graunt published his *Bills of Mortality,* many other mathematicians became interested in statistics and made important contributions. Pierre-Simon Laplace (1749–1827), Abraham De Moivre (1667–1754), and Carl Friedrich Gauss (1777–1855) studied and applied the **normal distribution** (see page 223). Karl Pearson (1857–1936) and Sir Francis Galton (1822–1911) studied the **correlation coefficient** (see page 255). These are but a few of the many mathematicians who made valuable contributions to statistical theory. In later chapters we will further discuss their works.

Although a great deal of modern statistical theory was known before 1930, it was not commonly used, simply because the accumulation and analysis of statistical data involved time consuming, complicated computations. However, things changed with the invention of the computer and its ability to perform long and difficult calculations in a relatively short period of time. Statistics soon began to be used for **inference,** that is, in making generalizations on the basis of samples. Also, probability theory was soon applied to the statistical analysis of data. The use of statistics for inference resulted in the discovery of new techniques for treating data.

Interestingly enough, the principles of the theory of probability were developed in a series of correspondences between Blaise Pascal (1623–1662) and Pierre Fermat (1602–1665). Pascal was asked by the Chevalier de Méré, a French mathematician and professional gambler, to solve the following problem: In what proportion should two players of equal skill divide the stakes remaining on the gambling table if they are forced to stop before finishing the game? Although Pascal and Fermat agreed on the answer, they both gave different proofs. It is in these correspondences during the year 1654 that they established the modern theory of probability.

A century earlier the Italian mathematician and gambler Girolomo Cardan (1501–1576) wrote *The Book On Games Of Chance.* This is really a complete textbook for gamblers since it contains many tips on how to cheat successfully. The origins of the study of probability are to be found in this book. Cardan was also an astrologer. According to legend, he predicted his own death astrologically and to guarantee its accuracy he committed suicide on that day. (Of course, that is the most convincing way to be right!) He also had a temper and is said to have cut off his son's ears in a fit of rage.

As statistics developed, probability began to assume more importance because of its wide range of applications. Today the application of probability in gambling is but one of its minor uses. In recent years, statistics has even been applied to determine the total population of various species of living things. In particular, by using very simple procedures, statisticians have been able to predict the total population of such endangered species as the whooping crane and various fish. In each case a number of birds or fish are caught, tagged with a label or some other form of identification, and then released for breeding. When they are recap-

tured or sighted at a later date, the proportion of tagged fish or birds out of the total catch is calculated and used to predict the total population of the species.

In the future many new and interesting applications of statistics and probability are likely to be found.

OVERVIEW OF TEXT

In preceding sections we mentioned several uses of statistics to convince you that the development of statistics and probability is not static. It is constantly changing. Who knows what a beginning course in statistics will be like by the year 2000? Undoubtedly, different things will be stressed and new applications for statistics and probability will be found. Yet certain basic ideas of probability and its uses in statistical studies will not be changed. Such ideas are too fundamental. It is with these ideas that we will concern ourselves in this text.

In the following chapters we will develop the techniques used in all applications of statistics and discuss the role played by probability in these applications.

EXERCISES

1. Analyze each of the following situations to determine if you should believe the claim made. In each case explain your answer.
   a. A television commercial claims that "People who use Brand X toothpaste have 50% fewer cavities." Should you use Brand X?
   b. A "statistician" made the following comparison. Between 1960 and 1976, salaries of firemen in a large city rose by the same rate as did the number

of children per family. He concluded that the more money a fireman earns in a community, the larger the size of his family. Do you agree?

c. A large oil company claims that by adding the active ingredient TLG to its gasoline, the average driver should now get 10% more miles to the gallon. Do you agree with this commercial claiming statistical proof of product superiority?

d. The following ad appeared in a magazine: "Use Gezunt's one-a-day multiple vitamins. Last year 90% of our users were not sick a single day." Should you use Gezunt's vitamins?

2. During 1976 two popular rock singing groups, the Pebblechips and the Stoneheads, each performed at three concerts. The Pebblechips attracted 79,000, 47,000, and 81,000 people, whereas the Stoneheads attracted 58,000, 71,000, and 67,000 people. Which of the following conclusions require statistical inference and which require only descriptive statistics?

a. On the average, the Pebblechips attracted more people than did the Stoneheads.

b. There is a considerable difference between total attendance at the Pebblechips' concerts and the Stoneheads' concerts.

c. The Pebblechips are likely to attract a larger audience than the Stoneheads in any future concert.

3. In a recent study it was found that 99% of the criminals who committed sex crimes drank milk on a regular basis for a number of years when they were young. Can one conclude that if you drank milk when you were young, you are likely to commit sex crimes? Explain.

4. Examine a daily newspaper carefully for its use of statistics. Classify the statistics as descriptive or inferential.

5. Listen to radio and television newscasts. The word statistics is often used by newscasters. Classify the uses as descriptive or inferential.

6. Read *Statistics: A Guide to the Unknown* edited by Judith M. Tanur for a number of interesting and wide ranging applications of statistics. (See the list of suggested reading at the end of this chapter.)

7. Can you find examples of how you use statistics in your daily experiences?

8. Imagine that you are interested in determining the number of people in San Francisco who write exclusively with their left hand. How could you go about doing this, that is, how would you select a sample?

9. Describe a procedure which the administration of the City University of New York could use to determine the number of students, out of an approximate enrollment of 220,000, who work part-time or full-time.

10. Suppose you are interested in determining the percentage of boys who are at least as tall as their fathers. Can you obtain this percentage by selecting all your friends and relatives as a random sample?

11. Consider the television commercial which states "4 out of 5 dentists recommend sugarless gum for their patients who chew gum". If you select 5 dentists at random, would you expect 4 of them to recommend sugarless gum? Explain.

In this chapter we discussed the nature of statistics and how it is used. We distinguished between descriptive statistics and statistical inference. A brief discussion on the origins of statistics and probability was given. Finally, it was pointed out how statistics and probability are constantly gaining in importance because of their ever increasing wide range of applications.

The following is a beginning list of terms basic to the vocabulary of statistics. As a review, you should be able to demonstrate your knowledge of them by giving definitions, descriptions, or specific examples. Page references are given for each term so that you can easily check your answer.

Statistics (page 3)   Sample (page 8)
Data (page 3)   Population (page 8)
Descriptive statistics (page 8)   Statistical inference (page 8)

Mastery tests appear at the end of each chapter. You will probably find these tests more useful if you take them after you have solved most of the exercises given in the chapter.

MASTERY TESTS

Form *A*

Determine whether each of the following statements are true or false.

Use the following information for questions 1–3: During the month of July three families who live on Main Street accumulated 470, 220, and 330 pounds of newspapers for recycling. Three families who live on South Street accumulated 360, 340, and 350 pounds of newspapers for recycling.

1. The conclusion "The average amount of newspapers collected on Main Street is less than the average amount collected on South Street" involves descriptive statistics only.   *T*
2. The conclusion "For the month of August the families of South Street will contribute more pounds of newspapers for recycling than the families on Main Street" involves statistical inference.   *T*
3. The conclusion "The amount of newspapers collected from the families on Main Street is generally more consistent than the amount collected from the families on South Street" involves statistical inference.   *F*
4. To obtain a random sample, we can open a telephone book and select every tenth name listed.   *F*
5. Statistics show that in the 1976 Democratic Presidential Primary in New York State only 20% of the registered Democrats actually voted. Thus, we can conclude that 80% of the registered Democrats did not care who was nominated.   *F*

Form *B*

1. In order to avoid any layoffs, Governor Pothole asked all the workers of the State Highway Maintenance Department to agree to a 25% cut in their yearly $20,000 salary. Reluctantly the workers agreed. A year later each worker received a 25% increase in his yearly salary. Are the workers now making as much money as they were before? Explain your answer. *Depends on the inflation diff.*

2. After analyzing the police department records of Sun Valley, a young and inexperienced statistician comes to the following conclusion. "Since 2/3 of all the rape and murder victims on the town's college campus were relatives or former friends of their assailants, a female college student is much safer going out at night with strangers than remaining in her room." Do you agree with this conclusion? Explain. *No*

3. The American Cancer society claims that statistics show that smoking is dangerous to your health. Yet, the tobacco industry says that the statistical claims of these experiments are not conclusive. Explain how both sides in this controversy interpret these statistics. *The Amer. G. society smokes the freq of certain illnesses in smokers. Tob. Indy refers to there being non-smokers.*

4. *analysis using* The Cuban National Commission for the Propaganda and Defense of Havana Tobacco once noted that the human life span has doubled since the tobacco plant was discovered. Is it true that increased use of tobacco will result in a longer life? Explain your answer. *no, life span has increased due to other factors.*

5. The alumni of U.F.O. University recently concluded a survey of its graduates showing that the average annual salary of a graduate 10 years after graduation was $27,412. Can we conclude that the education received by the university's students is so superior that 10 years after graduation the students will have a higher paying job than students who attend other universities? Explain your answer. *no you would need comparison studies on other universities.*

6. In one issue, January 2, 1967, *Newsweek* reported on page 10 that Mao-Tse Tung cut the salaries of certain Chinese government officials by 300%. Is it possible for a salary to be cut by 300%? Explain your answer. *no*

7. In 1976, 234 accidents involving drunken drivers and 15,763 accidents involving drunken pedestrians were reported in Danville. Can we conclude that in Danville it is safer to be a drunken driver than a drunken pedestrian? Explain your answer. *No, as drunken driver is maneuvering a machine rather than just himself.*

8. Trudy has been giving her daughter Sharon 5¢ each time she takes out the garbage. Trudy decides to give her daughter a 20% increase. By how much will Sharon's earnings be increased? *1¢*

9. In its annual report for 1975 the Aviation Council reported that there were 2 separate incidents in which a plane was struck by lightning during a blinding rainstorm. Furthermore, on clear sunny days there were 24 incidents in which engine failure resulted from birds sucked into plane's engines. In view of these findings, the council concluded that it is less dangerous to fly an airplane during a blinding rainstorm than it is on a clear sunny day when birds are present. Do you agree with this conclusion? Explain your answer.

No, as I would assume any incident involving birds in the engine could not be as serious as one involving lightning.

10. Read the book *How to Lie with Statistics* by Darrell Huff for a number of interesting examples on the misuses of statistics. (See the list of suggested reading at the end of this chapter.)

11. Read the article "Statistics-Watching: A Guide for the Perplexed" in the *New York Times* (November 21, 1975, p. 45). This article discusses how economic statistics can often be confusing.

SUGGESTED READING

Campbell, Stephen K. *Flaws and Fallacies in Statistical Thinking.* Englewood Cliffs, New Jersey: Prentice-Hall, 1974.

Dale, Edwin L. Jr. "Statistics-Watching: A Guide for the Perplexed" *The New York Times,* November 21, 1975, p. 45.

Galton, Francis. "Classification of Men According to Their Natural Gifts" in *The World of Mathematics* edited by James R. Newman. New York: Simon & Schuster, 1956. Vol. 2, part VI, chap. 2.

Graunt, John. "Foundations of Vital Statistics" in *The World of Mathematics* edited by James R. Newman. New York: Simon & Schuster, 1956. Vol. 3, part VIII, chap. 1.

Huff, Darrell. *How to Lie with Statistics.* New York: Norton, 1954.

Martin, Thomas L. Jr. *Malice in Blunderland.* New York: McGraw-Hill, 1973.

Newmark, Joseph and Lake, Frances. *Mathematics as a Second Language* 2nd ed. Reading, Mass: Addison-Wesley Publishing Co., 1977.

Tanur, Judith M., et al. *Statistics: A Guide to the Unknown.* San Francisco, Calif.: Holden-Day, 1972.

Wallis, W. Allen and Roberts, Harry V. *Statistics: A New Approach.* Glencoe, Ill.: The Free Press, 1956.

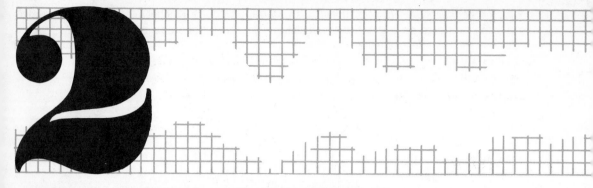

THE DESCRIPTION OF SAMPLE DATA

U.S. News & World Report, March 1, 1976. Reprinted from *U.S. News & World Report,* Copyright © 1976 U.S. News & World Report, Inc.

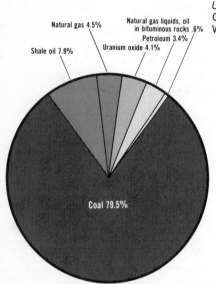

Natural gas 4.5%

Natural gas liquids, oil in bituminous rocks .6%

Petroleum 3.4%

Uranium oxide 4.1%

Shale oil 7.9%

Coal 79.5%

U.S. fossil fuel reserves economically recoverable with present technology. Data from Dept. of Interior, AEC, American Petroleum Institute, American Gas Association.

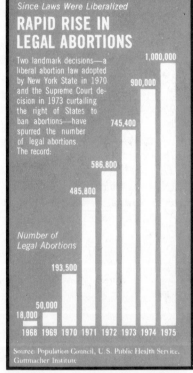

Since Laws Were Liberalized

RAPID RISE IN LEGAL ABORTIONS

Two landmark decisions—a liberal abortion law adopted by New York State in 1970 and the Supreme Court decision in 1973 curtailing the right of States to ban abortions—have spurred the number of legal abortions. The record:

Number of Legal Abortions

1,000,000
900,000
745,400
586,800
485,800
193,500
50,000
18,000

1968 1969 1970 1971 1972 1973 1974 1975

Source: Population Council, U.S. Public Health Service, Guttmacher Institute

Statistics often force government officials and prospective politicians to take a definite stand on controversial issues. The information contained in the graph on the right is one such example.

Now consider the **circle graph.** Statistics show that America has been using its natural resources at an increasing rate. As a result, government officials are looking for alternate sources of energy. The circle graph indicates some possible sources.

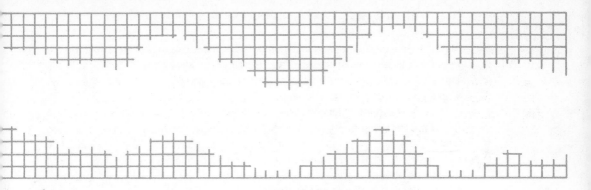

Preview

In this chapter we will discuss the following topics:

1. *Frequency Distribution*
A convenient way of grouping data so that meaningful patterns can be found.

2. *Histogram*
A representation of the graph of a frequency distribution.

3. *Circle Graphs, also called Pie Charts*
A graphical way of presenting data often used when discussing distributions of money.

4. *Frequency Polygons and Bar Graphs*
Variations of the histogram used when we wish to emphasize changes in frequency, for example, in business and economic situations.

5. *Normal Distribution*
A frequency distribution the graph of which resembles a normal curve, perhaps the most important distribution in statistics. We will analyze it in great detail.

6. *Vocabulary*
The following words are introduced: random sample, frequency, class mark, relative frequency area under a curve, and pictograph.

Introduction

Suppose that John, a student in a sociology class, is interested in determining the average age at which women in New York City marry for the first time. He could go to the Marriage License Bureau and obtain necessary data. (We assume that all marriages are reported to the Bureau.) Since there are literally thousands of numbers to analyze, he would want to organize and condense the data so that meaningful interpretations can be drawn from them. How should he proceed?

One of the first things to do when given a mass of data is to group it in some meaningful way and then construct **frequency distributions** of the data. After this is done, one can use various forms or **graphs** of the distribution so that different distributions can be discussed and compared.

In this chapter we will discuss some of the common methods of graphically describing data. In the next chapter we will discuss the numerical methods for analyzing data. In each case we will use examples from everyday situations so that you can see how and where statistics is applied.

2.1 FREQUENCY DISTRIBUTIONS

Rather than analyze thousands of numbers, John decides to take a sample. He will examine the records of the Marriage Bureau on a given day and then use the information obtained from this sample to make inferences about the ages of *all* women who marry for the first time.

The most important thing in taking a sample of this sort is the requirement that it be a **random sample.** This means that each individual of the population, a woman who applies for a marriage license in our case, must have an equally likely chance of being selected. If this requirement is not satisfied, one cannot make meaningful inferences about the population based upon the sample. Since we will have more to say about random samples in Chapter 10, we will not analyze here whether or not John's sampling procedure is random.

John has obtained the ages of 150 women for a given day from the Marriage Bureau. Table 2.1 lists the ages in the order in which they were obtained from the Bureau records.

TABLE 2.1

AGE AT WHICH 150 WOMEN IN NEW YORK CITY MARRIED FOR THE FIRST TIME

20	17	24	18	16	23	21	17	21	27	23	17	16	19	14
23	26	38	33	20	33	26	26	26	33	26	35	28	35	24
21	18	21	23	22	21	19	23	19	18	19	15	23	21	18
27	40	27	26	27	25	27	25	25	25	30	19	26	32	22
24	19	22	19	24	15	20	22	23	26	23	18	21	20	21
34	41	35	20	29	20	29	27	29	32	29	29	32	28	31
19	22	23	25	23	23	18	19	24	24	21	20	24	22	20
30	31	39	44	38	37	30	37	33	30	36	34	36	32	26
25	23	17	24	18	24	24	21	16	20	18	22	25	24	17
28	29	34	31	25	34	25	36	28	27	31	27	28	30	28

The only thing we can say for sure is that most women were in their 20s and that one was as young as 14 while another woman was as old as 44 years when they married. However, since the ages are not arranged in any particular order, it is somewhat difficult to conclude anything else. Clearly, the data must be reorganized. We will use a frequency distribution to do so.

DEFINITION 2.1

A **frequency distribution** is a convenient way of grouping data so that meaningful patterns can be found. The word *frequency* will mean how often some number occurs.

A frequency distribution is easy to construct. We first make a list of numbers starting at 14 and going up to 44 to indicate the age of each woman. This list will be the first column. In the second column we indicate tally marks for each age, that is, we go through the list of original numbers and put a mark in the appropriate space for each age. Finally, in the third column we enter the total number of tally marks for each particular age. The sum for each age gives us the frequency column. When applied to our example we get the distribution shown in Table 2.2.

TABLE 2.2

CONSTRUCTION OF A FREQUENCY DISTRIBUTION FOR AGE AT WHICH 150 WOMEN IN NEW YORK CITY MARRIED FOR THE FIRST TIME

COLUMN 1	COLUMN 2	COLUMN 3
Age	Tally	Frequency
14	|	1
15	||	2
16	|||	3
17							5
18										8
19											9
20											9
21												10
22												10
23														12
24												10
25											9
26										8
27									7
28								6
29								6
30							5
31	||||	4
32	||||	4
33	||||	4
34	||||	4
35	|||	3
36	|||	3
37	||	2
38	||	2
39	|	1
40	|	1
41	|	1
42		0
43		0
44	|	1

ungrouped data

A number of interesting things should be immediately obvious from Table 2.2. Since 23 occurred most often, this is the age at which most women married. Also, the frequency began to decrease as age increased beyond the age of 23.

The data in Table 2.2 can also be arranged into a more compact form as shown in Table 2.3. In this table we have arranged the data by age groups, also called classes. We select a group size of 3 years since this will result in 10 different groups. Although any number of groups could be used, it is more convenient to work with 10. Classes are usually of equal size. Generally speaking, if we subtract the smallest age from the largest age and divide the results by 10, we will get a number, rounded off if necessary, which can be used as the size of each group. In our case, we have

$$\frac{44 - 14}{10} = \frac{30}{10} = 3$$

Notice the last age group 41–44 is of size 4. Why?

TABLE 2.3

AGE AT WHICH 150 WOMEN IN NEW YORK CITY MARRIED FOR THE FIRST TIME

CLASS NUMBER	AGES	CLASS MARK	TALLY	CLASS FREQUENCY	RELATIVE FREQUENCY
1	14–16	15		6	6/150
2	17–19	18		22	22/150
3	20–22	21		29	29/150
4	23–25	24		31	31/150
5	26–28	27		21	21/150
6	29–31	30		15	15/150
7	32–34	33		12	12/150
8	35–37	36		8	8/150
9	38–40	39		4	4/150
10	41–44	42.5		2	2/150
			Total Frequency =	150	

We have labeled the first column of Table 2.3 **class number** because we will need to refer to the various age groups in our later discussions. Thus, we will refer to age group 26–28 as class 5. Notice also that each class has an *upper limit,* the oldest age, and a *lower limit,* the youngest age. A **class mark** represents the point which is midway between the limits of a class, that is, it is the midpoint of a class. Thus, 18 years is the class mark of class 2. We have indicated the class mark for each class in the third column of Table 2.3.

Comment The class mark need not be a whole number.

Comment Although we mentioned that equal class intervals are usually used, this does not necessarily include the first and last classes.

EH, WHAT'S IT TO YOU, SONNY?

In the fourth column of Table 2.3 we have the **class frequency,** that is, the total tally for each class. Finally, the last column gives the relative frequency of each class. Formally stated, we have the following definition:

DEFINITION 2.2

The **relative frequency** of a class is defined as the frequency of that class divided by the total number of measurements (also called total frequency).
Symbolically, we let f_i denote the frequency of class i where i represents any of the classes, and we let n represent the total number of measurements. Then

$$\text{Relative frequency} = \frac{f_i}{n} \text{ for class } i$$

Since in our example the total frequency is 150, the relative frequency of class 6, for example, is 15 divided by 150, or 15/150. Here $i = 6$ and $f_i = 15$. Similarly, the relative frequency of class 7 where $i = 7$ and $f_i = 12$ is 12/150. The relative frequency of class 10 is 2/150.

Once we have a frequency table, we can present the information it contains in the form of a graph called a **histogram.** To do this we first draw 2 lines, one horizontal, that is, across, and one vertical, that is, up-down. We mark the class boundaries along the horizontal line and indicate frequencies along the vertical line. We draw rectangles over each interval, with the height of each rectangle equal to the frequency of that class.

The histogram for the data of Table 2.3 is shown in Figure 2.1. The area under the histogram for any particular rectangle or combination of rectangles is proportional to the relative frequency. Thus the rectangle for class 4 will contain 31/150 of the total area under the histogram. The rectangle for class 7 will contain 12/150 of

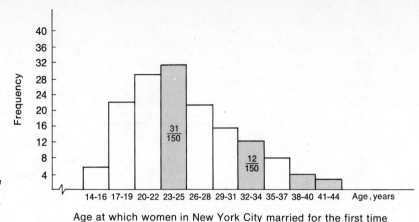

Figure 2.1

Histogram for data on age
at which 150 women in
New York City married for
the first time.

Age at which women in New York City married for the first time

the total area under the histogram. The rectangles for classes 9 and 10 together will contain 6/150 of the total area under the histogram since

$$\frac{4}{150} + \frac{2}{150} = \frac{4+2}{150} = \frac{6}{150}$$

The area under a histogram is important in statistical inference and we will discuss it in detail in later chapters.

Comment By changing the number of intervals used, we can change the appearance of a histogram and hence the information it gives. (See Exercise 7 of this section.)

To further illustrate the concept of a frequency distribution and its histogram, let us consider the following example:

EXAMPLE 1

Manya, the receptionist in the Student Counseling Office at Fantasyland University, keeps a daily list of the number of students who appear for counseling. During the first 10 weeks of the fall semester, the following number of students appeared for some form of counseling:

30	28	16	23	22	10	8	15	16	23
29	30	22	15	24	9	5	4	15	24
21	24	16	14	21	13	6	2	23	22
24	21	27	13	20	8	4	3	24	20
22	18	23	23	17	7	6	1	16	17

Construct a frequency distribution for the preceding data and then draw its histogram.

SOLUTION

We will use 10 classes. The largest number is 30 and the smallest
number is 1. To determine the class size, we subtract 1 from 30
and divide the result by 10 and get:

$$\frac{30 - 1}{10} = \frac{29}{10} = 2.9, \text{ or 3 when rounded}$$

Thus, our group size will be 3. We now construct the frequency
table. It will contain six columns as indicated.

Class Number	Number of Students	Class Mark	Tally	Class Frequency	Relative Frequency
1	1– 3	2	|||	3	3/50
2	4– 6	5	₩	5	5/50
3	7– 9	8	||||	4	4/50
4	10–12	11			1	1/50
5	13–15	14	₩		6	6/50
6	16–18	17	₩ ||	7	7/50
7	19–21	20	₩	5	5/50
8	22–24	23	₩ ₩ ||||	14	14/50
9	25–27	26			1	1/50
10	28–30	29	||||	4	4/50

Total Frequency = 50

The relative frequency column is obtained by dividing each class
frequency by the total frequency which is 50. We draw the following
histogram, which has the frequency on the vertical line and the
number of students on the horizontal line.

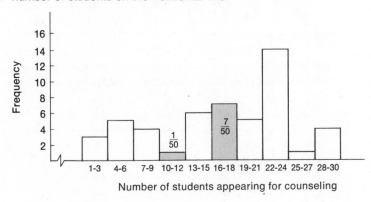

Number of students appearing for counseling

The rectangle for class 6 will contain 7/50 of the total area under the histogram. Similarly, the rectangle for class 4 contains 1/50 of that area. Finally, the rectangles for classes 1 and 2 contain

$$\frac{3}{50} + \frac{5}{50} = \frac{3+5}{50} = \frac{8}{50}$$

of the total area under the histogram. What part of the area is contained in *all* the rectangles under the histogram?

EXERCISES

For each of Exercises 1–6 construct a frequency distribution table and then draw its histogram.

1. Sam is interested in determining the average height of students in his math class. The following list gives the height, in inches, of each student in the class:

60	68	69	69	68	76
65	67	65	65	65	77
59	74	67	71	71	69
72	73	70	70	73	70
75	71	72	69	75	69
69	74	68	70	72	65

2. The number of hamburgers sold by MacDougald's, a nationwide hamburger chain, in each of the 50 states during June of last year was as follows. (*Note:* Units are in thousands.)

120	200	62	140	43	146	169	114	65	88
79	145	78	21	64	73	195	109	63	86
46	28	85	37	99	154	153	153	91	95
150	69	96	61	106	79	61	179	64	29
95	48	108	81	121	61	12	68	88	77

3. Chester Field is an habitual smoker. His wife is annoyed by his smoking and keeps a record of the number of cigarettes that he smokes each day. For the month of January Chester smoked the following number of cigarettes per day:

45	36	29	19	25	28
51	43	28	24	38	
37	39	37	18	45	
16	28	36	33	47	
29	34	39	29	49	
31	16	28	28	40	

4. In an effort to conserve energy, a department store chain offered huge discounts on its Go-Go bicycles. The following list indicates the number of bikes sold daily over the past 42 days:

5	14	6	21	1	38	8
8	40	4	5	5	16	17
16	28	3	12	9	12	24
13	31	7	15	17	15	15
12	16	9	10	16	10	6
17	21	7	8	2	9	2

5. Are drivers observing the 50 miles an hour speed limit on the state speedway? To answer this question, several legislators set up radar equipment and clocked the speed of the first forty-five cars that passed. The following speeds were recorded:

49	45	51	54	57	51	56	51	59
57	60	48	59	49	50	56	54	46
58	62	58	61	62	62	48	47	46
51	63	56	48	53	61	51	61	49
55	60	53	56	55	59	57	60	53

6. Randy is interested in buying an engagement ring for his girl friend Jennifer. Jewelers have quoted him the following prices, in dollars, for the same ring:

2050	2000	1875	2000	2098	2149
1945	1998	2249	2198	2150	2249
1989	1987	1976	1945	1898	1900
2149	2049	2004	1998	2049	2150
2200	2145	1984	2000	2200	1945

7. A student in a psychology class administered an IQ test to a group of 100 entering freshmen. The following scores were obtained:

79	108	91	118	101	83	108	109	114	117
91	107	89	117	83	77	91	129	130	129
121	111	103	111	79	94	76	139	137	131
111	114	79	114	74	88	84	141	144	115
101	97	83	107	83	91	94	97	101	103
114	110	108	121	91	92	107	102	107	111
97	121	93	124	109	108	119	103	105	87
87	132	99	99	122	104	106	78	84	104
93	127	104	87	127	105	119	118	97	83
101	116	102	111	101	113	118	93	122	96

25

    a. Construct a frequency distribution table and its histogram using 10 classes (also called intervals).

    b. Construct a frequency distribution table and its histogram using only 5 intervals.

    c. Compare the histograms of part *a* and part *b*. What information, if any, is lost by using fewer intervals?

8. Thirty students in a statistics class received the following scores on the mathematics section of the S.A.T. and on their first statistics test:

S.A.T. *Scores*					*Statistics Exam*				
586	512	478	531	631	60	89	82	78	68
483	576	493	523	428	77	28	95	57	89
612	419	601	613	388	74	76	88	43	74
510	576	526	486	449	100	91	57	71	89
633	628	612	503	573	100	74	40	84	94
463	693	572	517	486	94	90	59	87	88

    a. Construct histograms for both sets of scores.

    b. By looking at the histograms can we tell if the scores are approximately consistent?

9. The Textile Dress Company employs 40 workers, each of whom completes similar products in the company's two factories. In each factory there are 20 employees. The number of products completed by each employee in both factories is as follows:

Factory A				Factory B			
30	10	28	11	12	31	28	5
28	19	21	13	19	37	24	9
35	17	25	9	31	19	17	7
26	24	5	4	27	25	26	14
21	31	16	7	28	32	11	25

    a. Construct frequency distributions and histograms for each factory.

    b. Combine the data for both factories and construct a frequency distribution and its histogram.

10. The personnel manager of the dress company mentioned in the preceding exercise decides to change the working conditions in Factory A to determine what effect this will have on production. He installs new lighting facilities, new air conditioning, carpeting, piped music, and a new coffee machine. He now notices that the production of the employees of Factory A is

35	16	23	15
28	11	27	17
37	14	28	19
29	17	21	21
25	21	18	16

a. Construct the frequency distribution and histogram for the new data.
b. Compare the new histogram with the original histogram for Factory A and
    the histogram for the combined data. Comment.

11. The number of summones issued by two policemen, Sergeant Thursday and
    Patrolman Hannon, in the last 30 working days is as follows:

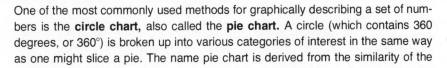

Sgt. Thursday						Ptl. Hannon					
3	10	6	3	9	10	1	8	3	4	1	6
7	0	4	0	8	7	0	6	9	7	0	3
6	1	2	10	7	9	3	4	1	6	9	2
8	8	3	7	1	5	7	2	2	5	5	7
2	5	7	6	6	4	4	5	4	8	7	9

a. Construct frequency distributions and histograms for each policeman.
b. By looking at the histograms can we determine who consistently issues
    more summones? *yes — Hannon*

12. Using the data of Table 2.1 (page 18), construct a frequency distribution and
    histogram using only
    a. 5 intervals.
    b. 15 intervals.
    c. Compare the histograms obtained in parts a and b with the histogram given
        on page 22. Comment.

13. Refer to Example 1 on page 22.
    a. What part of the area is below 10, that is, to the left of 10?
    b. What part of the area is 22 or above, that is, to the right of 22?
    c. What part of the area is between 10 and 21?
    d. What part of the area is either below 10 or above 21?
    e. What part of the area is not between 10 and 18?
    f. What part of the area is above 30?

2.2 OTHER GRAPHICAL TECHNIQUES

In the preceding section we saw how frequency distributions and histograms are
often used to picture information graphically. In this section we will discuss some
other forms of graphs which are often of great help in picturing information contained in data.

Circle Charts

One of the most commonly used methods for graphically describing a set of numbers is the **circle chart,** also called the **pie chart.** A circle (which contains 360
degrees, or 360°) is broken up into various categories of interest in the same way
as one might slice a pie. The name pie chart is derived from the similarity of the

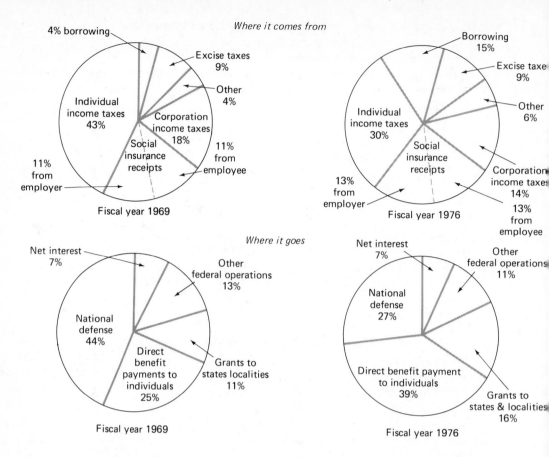

Figure 2.2
Has the U.S. government budget dollar changed? Pie charts of source and distribution of federal government revenue for fiscal year 1969 and fiscal year 1976. (Budget of the United States Government, Federal Reserve Bank of New York)

parts to pieces of a pie. Each category is assigned a certain percentage of the 360° of the total circle depending upon the data.

Figure 2.2 contains four pie charts of this type. Contrary to what many people may think, these charts show that revenue from individual income taxes dropped from 43% of the United States budget in 1969 to 30% of the budget in 1976. Similarly, direct benefit payments to individuals increased from 25% of the budget in 1969 to 39% of the budget in 1976.

The magazine clipping given on page 16 is another example of a pie chart. How do we draw such pie charts? To illustrate the procedure, consider the data of Table 2.4 which indicates the monthly living expenses in 1976 of a doctoral student at a state university.

Since the total expenditure was 360 dollars and there are 360° in a circle, we can construct the pie chart directly without any conversions. We draw a circle and partition it in such a way that each category will contain the appropriate number of degrees as shown in Figure 2.3.

TABLE 2.4

MONTHLY LIVING EXPENSES IN 1976 FOR A DOCTORAL STUDENT AT
A STATE UNIVERSITY

ITEM	AMOUNT IN DOLLARS
Food	100
Apartment	75
Car and transportation	40
Entertainment	85
Laundry	20
Miscellaneous	40
	Total = 360

It is more convenient to work with percentages than with amounts of money. Thus, we may want to convert the amount of money spent in each category into percentages. Thus, the student's money was spent as follows:

$$\frac{100}{360} = 0.2778, \text{ or } 27.78\%, \text{ for food}$$

$$\frac{75}{360} = 0.2083, \text{ or } 20.83\%, \text{ for the apartment}$$

$$\frac{40}{360} = 0.1111, \text{ or } 11.11\%, \text{ for the car}$$

$$\frac{85}{360} = 0.2361, \text{ or } 23.61\%, \text{ for entertainment}$$

$$\frac{20}{360} = 0.0556, \text{ or } 5.56\%, \text{ for laundry}$$

$$\frac{40}{360} = 0.1111, \text{ or } 11.11\%, \text{ for miscellaneous items}$$

We have indicated these percentages in the pie chart of Figure 2.3.

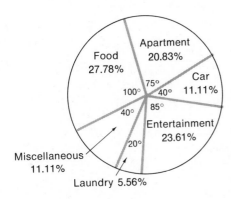

Figure 2.3

Pie chart showing the monthly living expenses in 1976 for a doctoral student at a state university.

29

Comment   We convert a fraction into a decimal by dividing the denominator, that is, bottom number, into the numerator, that is, the top number. Thus 100/360 becomes

$$
\begin{array}{r}
0.27777 \\
360\overline{)100.00000} \\
\underline{720} \\
2800 \\
\underline{2520} \\
2800 \\
\underline{2520} \\
2800
\end{array}
$$

This number is written in percentage form as 27.78%. In all problems we will usually round our numbers to two places to the right of the decimal point.

Comment   The sum of the percentages may not necessarily be 100%. This discrepancy is due to the rounding off of numbers. However, in the preceding example, the sum is 100%.

We will further illustrate the technique of drawing pie charts by working several examples.

EXAMPLE 1

During the 1976 summer season, a large retail department store chain in Chicago sold different amounts of the various brands of air-conditioners:

Brand	Number Sold
Chrysler	3100
Fedders	4800
General Electric	2000
Philco	1150
Hotpoint	850
Emerson	770
Other brands	2330
Total sold =	15,000

Draw the pie chart for this data.

SOLUTION

We first convert the numbers into percentages by dividing each by
the total 15,000. Thus, we have

Brand	Number Sold	Percentage of Total
Chrysler	3100	$\frac{3100}{15000} = 0.2067$, or 20.67%
Fedders	4800	$\frac{4800}{15000} = 0.32$, or 32%
General Electric	2000	$\frac{2000}{15000} = 0.1333$, or 13.33%
Philco	1150	$\frac{1150}{15000} = 0.0767$, or 7.67%
Hotpoint	850	$\frac{850}{15000} = 0.0567$, or 5.67%
Emerson	770	$\frac{770}{15000} = 0.0513$, or 5.13%
Other brands	2330	$\frac{2330}{15000} = 0.1553$, or 15.53%

Now we multiply each percentage by 360°, that is, the number of
degrees in a circle, to determine the number of degrees to assign to
each part. We get

$0.2067 \times 360° = 74.41°$, or 74°, for Chrysler

$0.32 \times 360° = 115.20°$, or 115°, for Fedders

$0.1333 \times 360° = 47.99°$, or 48°, for General Electric

$0.0767 \times 360° = 27.61°$, or 28°, for Philco

$0.0567 \times 360° = 20.41°$, or 20°, for Hotpoint

$0.0513 \times 360° = 18.47°$, or 18°, for Emerson

$0.1553 \times 360° = 55.91°$, or 56°, for other brands

Then, we use a protractor and compass to draw each part in order, using the appropriate number of degrees. In our case we obtain the following pie chart:

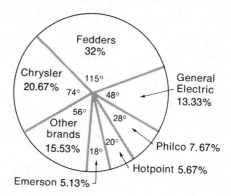

EXAMPLE 2

It has been estimated that the number of billions of barrels of oil in the western hemisphere, excluding Alaska, is given by the following pie chart:

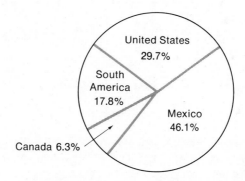

Assume there are 130 billion barrels of oil in reserve and answer the following:

a. How many barrels are in reserve in the United States?
b. How many barrels are there in Mexico?
c. How many barrels are there in Canada?

SOLUTION

a. The United States has 29.7% of the 130 billion barrels. Thus since

$$0.297 \times 130 = 38.61$$

the United States has 38.61 billion barrels of oil. (In decimal form
29.7% is written as 0.297.)

b. Mexico has 59.93 billion barrels of oil since

$$0.461 \times 130 = 59.93$$

c. Canada has 8.19 billion barrels of oil since

$$0.063 \times 130 = 8.19$$

Comment Some people believe that pie charts are more difficult to read
than other graphs; however, such charts are particularly useful when discussing
distributions of money.

Bar Graphs and Pictographs

The bar graph and a simplified version, the pictograph, are also commonly
used to graphically describe data. In such graphs **vertical bars** are used. The
height of each bar represents the number of members, that is, the frequency,
of that class. We will illustrate the use of such graphs with the following examples.

EXAMPLE 3

The number of people calling the police emergency number in
New York City for assistance during a 24 hour period on a given
day was as follows:

Time	Number of Calls Received
12 midnight– 2 A.M.	38
2 A.M.– 4 A.M.	27
4 A.M.– 6 A.M.	19
6 A.M.– 8 A.M.	20
8 A.M.–10 A.M.	22
10 A.M.–12 noon	24
12 noon– 2 P.M.	25
2 P.M.– 4 P.M.	28
4 P.M.– 6 P.M.	31
6 P.M.– 8 P.M.	39
8 P.M.–10 P.M.	41
10 P.M.–12 midnight	40

33

The bar graph for this data follows:

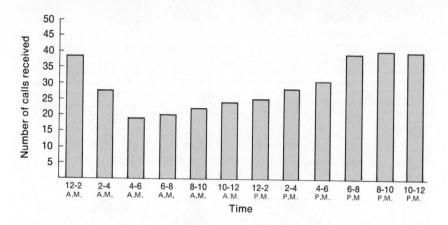

From the bar graph it is obvious that the least number of calls was received during the hours of 4 A.M.–6 A.M. The number of calls received after that period steadily increased until a maximum occurred during the 8 P.M.–10 P.M. period.

When such statistical information is presented in a bar graph, rather than in a table, it can be readily used by police officials to determine the number of policemen necessary for each time period.

EXAMPLE 4

The ABC Machine Company operates vending machines at two airports. The number of cans of soda sold during the first week of July at the airports was

	Airport 1	Airport 2
Coke	88	82
Seven-Up	79	68
Pepsi	97	79
Orange	101	83
Cola	83	71
Grape	91	84

The bar graph for this data follows:

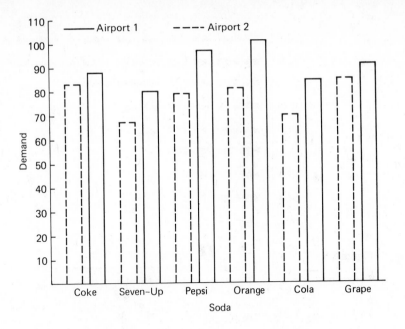

When the bar graphs for both airports are drawn on the same scale, that is, on the same graph, the company can determine which soda is in greatest demand and at which airport.

Several variations of the bar graph are commonly used. In such modifications, columns of coins, pictures, or symbols are used in place of bars. When symbols and pictures are used, the bars are sometimes drawn horizontally. We call the resulting graph a **pictograph.** Pictographs do not necessarily have to be drawn in the form of a bar graph. Several pictographs are given in Figures 2.4 and 2.5.

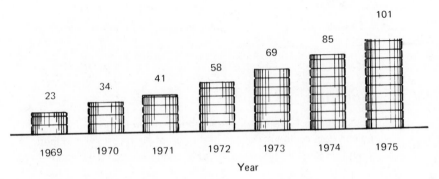

Figure 2.4

Pictograph.

35

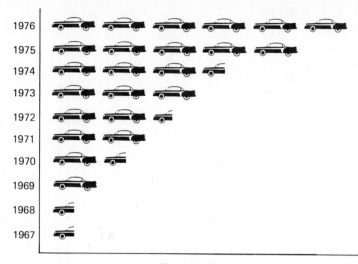

1976

1975

1974

1973

1972

1971

1970

1969

1968

1967

Figure 2.5

Pictograph.

Number of cars

Frequency Polygons

Another alternate graphical representation of the data of a frequency distribution is a **frequency polygon.** Here again the vertical line represents the frequency and the horizontal line represents the class boundaries.

DEFINITION 2.3

If the midpoints (also called class marks) of the tops of the bars in a bar graph or histogram are joined by straight lines, then the resulting figure, without the bars, is a **frequency polygon.**

EXAMPLE 5

Draw the frequency polygon for the data of Example 1 on page 22.

SOLUTION

Since the histogram has already been drawn (page 23), we place dots on the midpoints of the top of each interval and then join these dots. The result is the following frequency polygon:

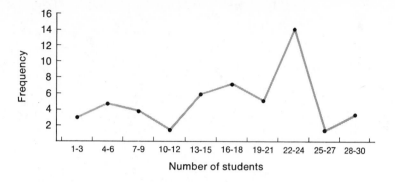

EXAMPLE 6

The Department of Agriculture compiles statistics on the amount
of rainfall in different parts of the country during the year. For a
certain community in a particular year it constructed the following
frequency polygon:

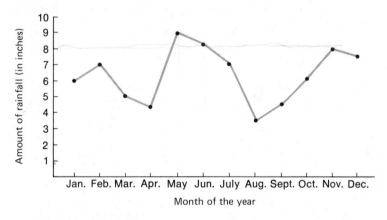

Answer the following:
a. During which months of the year is there at least 8 inches of
rain?   *MAY, JUNE*
b. During which month(s) of the year is there the most rain? *MAY*

SOLUTION

You should be able to answer this example on your own.

Since frequency polygons emphasize changes, that is, the rise and fall, in
frequency more clearly than any other graphical representation, they are used
often to display business and economic situations. However, this must be done
with great care. Figures 2.6 and 2.7 both represent the same idea, that is, the

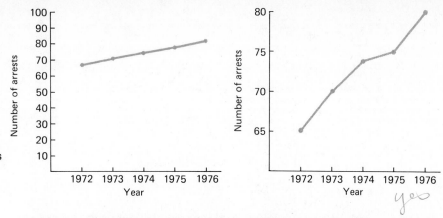

Figures 2.6 and 2.7

Two graphs of the number of students arrested for using drugs on a state university campus during the years 1972–1976.

number of students arrested for using drugs on a state university campus during the years 1972–1976. Yet Figure 2.7 seems to imply that there was a significant increase in the number of arrests over this five year period. Figure 2.6 also indicates that the number of arrests rose although not by as much. How can we have two different graphs representing the same situation? Which graph is the right one?

EXAMPLE 7

The frequency polygon showing the distribution of the heights of students at a local college is given in Figure 2.8. Analyze it carefully.

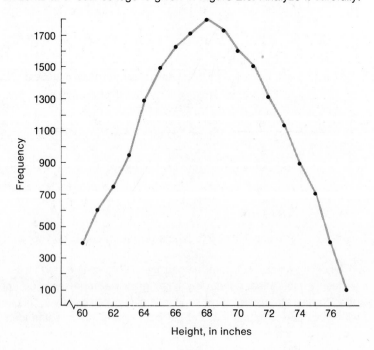

Figure 2.8

Distributions of heights of students at a local college.

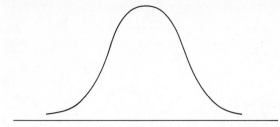

Figure 2.9

Normal curve.

There are many frequency distributions whose graphs resemble a bell shaped curve as shown in Figure 2.9. Such graphs are called **normal curves** and their frequency distributions are known as **normal distributions.** Since many things that occur in nature are normally distributed, it is no surprise that they are studied in great detail by mathematicians. We will discuss this distribution in detail in Chapter 8.

EXERCISES

1. During 1976 the student government at Tusayou University spent their $100,000 budget as follows:

Furniture and juke boxes	$ 16,000
Club activities (parties and socials)	$ 34,000
Student loans and aid	$ 8,000
Drug and sex clinic	$ 12,000
Tutoring	$ 18,000
Miscellaneous expenses	$ 12,000
	Total = $100,000

Draw a pie chart to picture the information.

2. At a certain college there are 3500 freshmen, 4480 sophomores, 3220 juniors, and 2800 seniors. Construct a pie chart which indicates the make-up of the student body at the college.

3. Three hundred and sixty students majoring in physical education were asked to indicate their favorite sport. The tallied results follow:

Basketball	72
Baseball	36
Tennis	45
Fencing	24
Football	10
Soccer	83
Swimming	90
Total =	360

Draw a pie chart to picture the information.

4. The marital status of women in a certain community is indicated by the following pie chart:

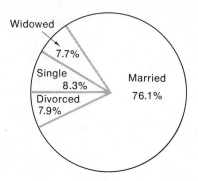

If there are 8000 women in this community, how many are married? divorced? single? widowed?

5. The Zekanim Nursing Home recently concluded a study to determine the cause of hospitalization of its senior citizens. The study was conducted so that the administrators of the home would be better able to anticipate the needs of future patients. Their results are summarized in the following pie chart:

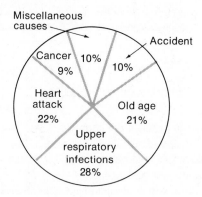

If 2000 patients were hospitalized last year in the nursing home, how many were hospitalized as a result of
a. heart attacks?
b. upper-respiratory infections or accidents?
c. miscellaneous causes?

6. During 1975 the circulation of a popular magazine was as follows:

Jan.–Feb.	9,000	July–Aug.	13,480
Mar.–Apr.	10,000	Sept.–Oct.	14,790
May–June	12,265	Nov.–Dec.	15,050

The following two frequency polygons have been prepared, one by a statistician and one by an advertising agency.

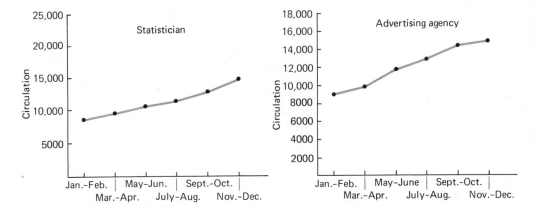

Answer the following:
a. How can two different graphs represent the same data?
b. Which frequency polygon should the management of the magazine use to determine whether the circulation has considerably increased?
c. Which is the correct frequency polygon? Explain your answer.

7. Consider the following magazine clipping and draw a bar graph picturing the amount of recycling for the different items mentioned.

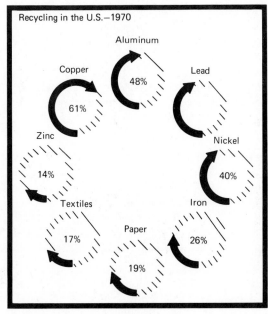

Source of Data: *Newsweek*, May 21, 1973.

From Zero Population Growth, Inc., Washington, D.C.

8. The number of hours that an average 9 year-old child watches television during a week is given in the following table:

Day	Hours
Monday	5
Tuesday	3
Wednesday	2
Thursday	4
Friday	3
Saturday	6
Sunday	5

Draw a bar graph to represent the data.

9. The distribution of IQ scores in a particular school is given in the following table:

IQ Range	Frequency
71–80	57
81–90	160
91–100	360
101–110	462
111–120	298
121–130	179
above 130	32

Draw a bar graph picturing the data.

10. Although the United States has 6% of the world population, it consumes very high percentages of the world production of certain materials.

Material	Percentage
Natural gas	57
Silver	42
Aluminum	36
Petroleum	32
Tin	32
Nickel	30
Copper	27
Steel	19

Information from *U.S. News & World Report*, December 4, 1972. Reprinted from *U.S. News & World Report*, Copyright © 1972 U.S. News & World Report, Inc.

Draw a bar graph to picture the data.

11. The following bar graph indicates the major subjects of students in a school of technology.

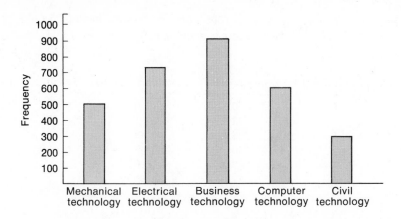

a. How many students majored in mechanical technology or electrical technology?

b. How many students majored in business?

c. How many students are in the school of technology?

12. The following pictograph indicates the population change of the United States from 1890–1970. Each symbol represents 10 million people.

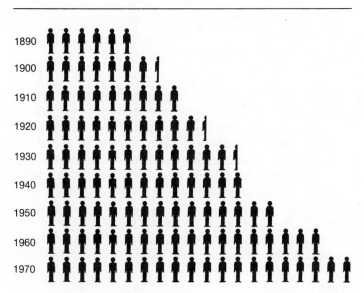

Source: United States Bureau of the Census.

a. How many people were there in 1920?

b. By approximately how many people did the population of the United States change from 1930–1970?

13. Recently a survey was taken of 1000 women who were married 25 years or more. The number of children born to each woman is indicated in the following chart:

Number of Children	Frequency	Number of Children	Frequency
0	80	6	63
1	142	7	42
2	283	8	14
3	146	9	10
4	128	10	9
5	77	Over 10	6
		Total =	1000

Construct a bar graph for the data.

14. Which of the following is (are) likely to be normally distributed?

a. Weight of individuals in your college

b. IQ scores of individuals

c. Intelligence of math teachers

d. Savings account balances for depositors in a large bank

e. Blood pressure of students in a statistics class

f. Age at which women in New York City marry for the first time

SELF-STUDY GUIDE

In this chapter we discussed the different graphical methods that one can use to picture a mass of data so that meaningful statements can be made. When given a large quantity of numbers to analyze, it is recommended that frequency tables be constructed. Some forms of graphical representation should then be used to serve as visual aids for thinking about and discussing statistical problems in a clear and easily understood manner. We discussed the different graphical techniques that one can use and also pointed out in Figures 2.6 and 2.7 on page 38 how they can be misused.

You should now be able to demonstrate your knowledge of the following ideas presented in this chapter by giving definitions, descriptions, or specific examples. Page references are given for each term so that you can easily check your answer.

Random sample (page 18) Class number (page 20)
Frequency distribution (page 18) Upper and lower limit (page 20)
Frequency (page 18) Class mark (page 20)

MASTERY TESTS

Form *A*

For questions 1–4 refer to the following histogram which gives IQ scores of second grade pupils in a local school district.

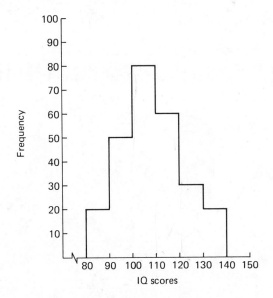

1. How many pupils have IQ scores between 101 and 120?   *e*
   a. 50 *b.* 80 *c.* 140 *d.* 130 *e.* none of these
2. How many pupils have IQ scores between 120 and 140?   *a*
   a. 50 *b.* 80 *c.* 20   *d.* 30   *e.* none of these
3. What is the relative frequency of a student whose IQ score is between 100 and 110?   *80*
4. What is the class mark for the interval 100–110? *105*
5. If we say that a score has a frequency of 28, how many students got this score?   *28*
6. The demand for electricity during a 24-hour period on a warm summer day in Fayville is given by the following frequency polygon. During which time period is electricity in greatest demand?   *4 p.m.*

ASK

2-6 p.m.

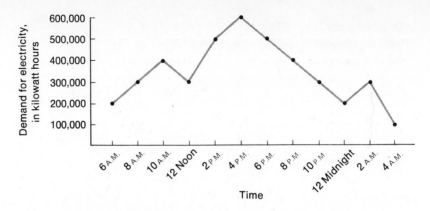

7. In the frequency polygon of question 6, what is the time period in which electricity is in least demand? *2 - 4 A.M.*

ASK ✓

8. Consider the following magazine clipping. Excluding 1975, in what year did the least number of fatalities occur? *1970*

Barring Any New Disasters . . .

1975 AIRLINE FATALITIES: HEADING FOR JET-ERA LOW

Through mid–December, U.S. airlines suffered two accidents involving fatalities, compared with nine last year. The total of 123 fatalities would be lowest since the jet era began and the lowest for any year since 1957.

BUT: In recent days, reports of "near collisions" between big jetliners have brought warnings from officials that unless steps are taken to avoid such incidents, fatalities could climb anew.

Source: National Transportation Safety Board, Federal Aviation Administration

U.S. News & World Report, December 22, 1975. Reprinted from U.S. News & World Report, Copyright © 1975 U.S. News & World Report, Inc.

9. Refer to question 8 of this test. In which year were there approximately 300 fatalities? *1961*

Form *B*

1. Consider the bar graph given in the following FBI chart. Draw a circle graph to picture the information given.

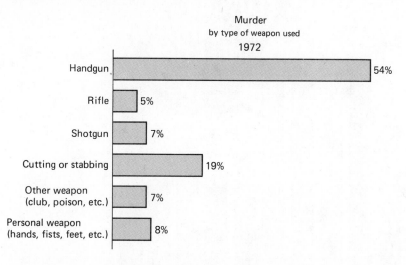

2. A coach has recorded the number of points scored by his two favorite basketball players during the past ten games:

Joe	Wilt	Joe	Wilt
32	28	35	29
44	42	40	42
26	34	38	32
30	42	22	36
27	40	34	28

We have drawn a bar graph for Joe. Draw the bar graph for Wilt on the same scale as the bar graph provided for Joe.

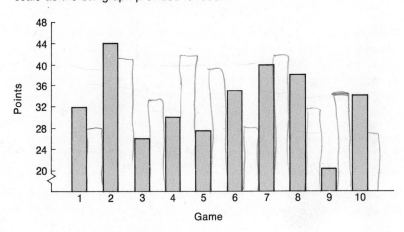

3. Consider the following magazine clipping:

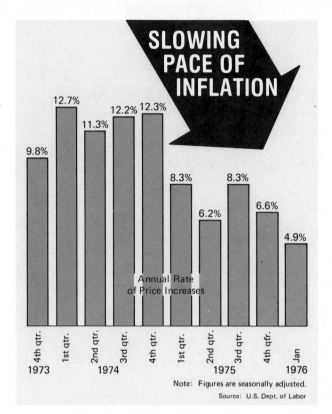

SLOWING PACE OF INFLATION

12.7%
12.2% 12.3%
11.3%
9.8%
8.3%   8.3%
6.2%   6.6%
4.9%

Annual Rate
of Price Increases

| 4th qtr. | 1st qtr. | 2nd qtr. | 3rd qtr. | 4th qtr. | 1st qtr. | 2nd qtr. | 3rd qtr. | 4th qtr. | Jan |

1973   1974   1975   1976

Note: Figures are seasonally adjusted.

Source: U.S. Dept. of Labor

U.S. News & World Report, March 8, 1976. Reprinted from *U.S. News & World Report,* Copyright © 1976 U.S. News & World Report, Inc.

a. During which two consecutive quarters did the rate of inflation increase the most? 4th '73 1st. '74

b. During which two consecutive quarters did the rate of inflation increase the least? bet. 3 & 4 '74

4. Carefully examine the FBI chart on the top of page 49. Two bar graphs are drawn side by side to indicate the number of officers killed during the years 1963–1967 and 1968–1972. Have the percentages of officers killed performing the activities listed remained the same over the years 1963–1967 and 1968–1972? Explain your answer. In most categories the rate has increased.

5. Consider the frequency distribution of family income for a large northeastern city shown on the bottom of page 49. Contrary to what you might expect, family income is not normally distributed but is skewed to the left. Can you explain why? More people are making less money.

6. The lengths of the drives of a professional golfer on a particular hole were measured, to the nearest yard, on 100 consecutive days and are as follows:

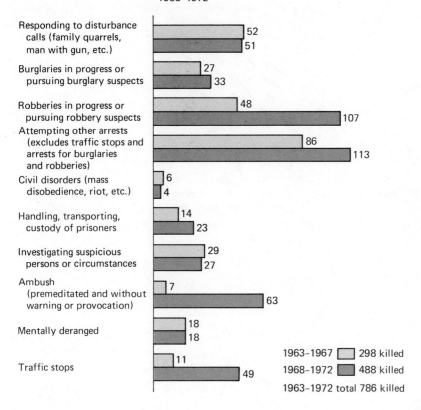

Law enforcement officers killed
by type of activity
1963–1972

Responding to disturbance calls (family quarrels, man with gun, etc.)
52
51

Burglaries in progress or pursuing burglary suspects
27
33

Robberies in progress or pursuing robbery suspects
48
107

Attempting other arrests (excludes traffic stops and arrests for burglaries and robberies)
86
113

Civil disorders (mass disobedience, riot, etc.)
6
4

Handling, transporting, custody of prisoners
14
23

Investigating suspicious persons or circumstances
29
27

Ambush (premeditated and without warning or provocation)
7
63

Mentally deranged
18
18

Traffic stops
11
49

1963–1967 ▢ 298 killed
1968–1972 ▨ 488 killed
1963–1972 total 786 killed

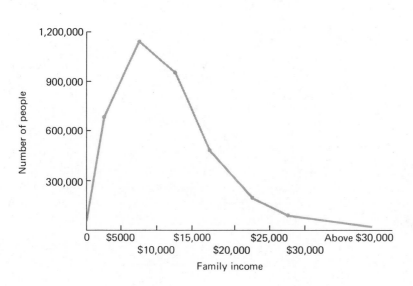

239 261 245 257 263 273 271 258 245 264
253 257 261 264 275 278 252 258 277 278
255 261 263 273 245 239 243 269 271 256
244 254 239 262 258 239 279 274 263 255
262 254 239 244 263 278 256 254 267 266
257 264 258 275 272 258 271 258 269 268
267 258 273 274 263 261 244 266 277 262
274 275 264 268 269 275 277 274 255 239
271 261 251 241 242 243 264 275 266 239
274 276 259 248 239 255 266 277 255 264

a. Using ten intervals, construct the frequency distribution.
b. Construct the histogram for the data.
c. Construct the frequency polygon for the data.
d. Is it possible to construct the frequency polygon for the data without first constructing the histogram or frequency distribution? Explain your answer.

You need the freq. dist. first in order to set up the polygon.

SUGGESTED READING

Griffin, J. I. *Statistics, Methods and Applications.* New York: Holt, Rinehart and Winston, 1962.

Huff, D. *How to Lie with Statistics.* New York: W. W. Norton & Co., 1954.

Newmark, Joseph and Lake, Frances. *Mathematics as a Second Language.* 2nd ed., Reading, Mass: Addison-Wesley Publishing Co., 1977.

O'Toole, A. L. *Elementary Practical Statistics.* New York: Macmillan Publishing Co., 1965.

Senter, R. J. *Analysis of Data: Introductory Statistics for the Behavioral Sciences.* Glenview, Illinois: Scott Foresman, 1969.

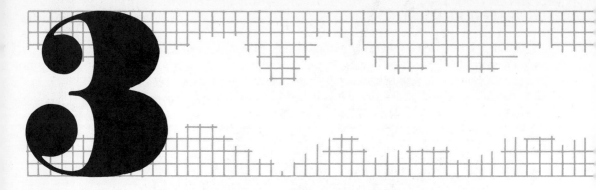

NUMERICAL METHODS FOR ANALYZING DATA

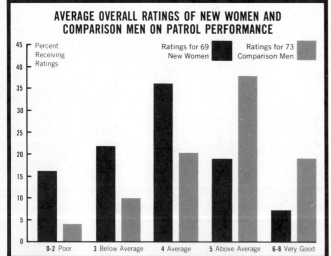

AVERAGE OVERALL RATINGS OF NEW WOMEN AND COMPARISON MEN ON PATROL PERFORMANCE

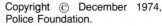

In recent years many law enforcement agencies have hired women to perform regular police duties. Many studies were undertaken to measure the overall performance of the new women police officers against that of the new men police officers. The clippings on this page indicate the results of one such study. Both the **average** and **standard deviation** are used in presenting the results of the survey.

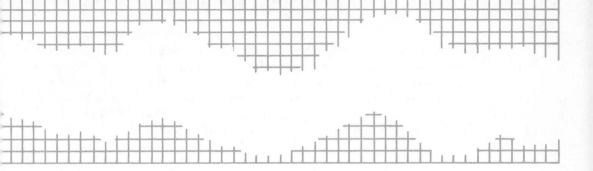

Preview

In this chapter we will discuss the following topics:

1. *Measures of Central Tendency*
We discuss the mean, median, and mode as three different ways of measuring some general trend or location of the data.

2. *Measures of Variation*
Even if we know the mean, median, or mode, we might still want to know whether the numbers are close to each other or whether they are spread out. Thus, we discuss the range, standard deviation, variance, and average deviation.

3. *Summation Notation*
Since all the measures of central tendency and variation involve adding numbers, we discuss a notation which is used for these calculations.

4. *Computational Formulas*
We discuss several shortcut formulas for calculating the mean, variance, and standard deviation.

5. *Coding*
It often pays to code the data when working with large numbers. This procedure transforms large numbers into smaller numbers.

Introduction

In Chapter 2 we learned how to summarize data and present it graphically. Although the techniques discussed there are quite useful in describing the features of a distribution, statistical inference usually requires more precise analysis of the data. In particular, we will discuss many measures that locate the *center* of a distribution of a set of data and analyze how dispersed or spread out the distribution is. Thus, we will discuss measures of central tendency and measures of variation.

Most analyses involve various arithmetic computations which must be performed on the data. In each case the operation of addition plays a key role in these calculations. Because of this we will introduce a special shorthand notation called **summation notation.** We will also discuss what is meant by **percentiles.**

After the data has been analyzed numerically, the techniques of statistical inference can be applied. This will be done in later chapters.

3.1 MEASURES OF CENTRAL TENDENCY

To help us understand what we mean by measures of central tendency, consider the Metropolis Police Department which recently purchased tires from two different manufacturers. The police department is interested in determining which tire is superior and has compiled a list on the number of miles each set of tires lasted before replacement was needed for fourteen of its identical police cars. Seven cars were fitted with Brand X tires and seven with Brand Y. The number of miles each set lasted before replacement is indicated in the following chart:

Brand X	Brand Y
14,000	10,000
12,000	8,000
12,000	14,000
14,000	10,000
14,000	8,000
11,000	40,000
14,000	8,000
Total = 91,000	98,000

It would appear that Brand Y is the better tire since the seven police cars were driven a combined total of 98,000 miles using Brand Y tires but only 91,000 miles with Brand X tires. Let us, however, analyze the data by computing the average number of miles driven with each brand of tires. We will divide each total by the number of police cars used for each brand. We have

AVERAGE

$$\text{Brand X} \qquad\qquad \text{Brand Y}$$
$$\frac{91,000}{7} = 13,000 \qquad\qquad \frac{98,000}{7} = 14,000$$

Since Brand Y tires lasted on the average 14,000 miles and Brand X on the average 13,000 miles, it again appears that Brand Y is superior. If we look at the data more carefully, however, we find that Brand X tires consistently lasted around 13,000 miles. As a matter of fact, it lasted 14,000 miles most often, 4 times. Brand Y, on the other hand, lasted 14,000 miles only once. It lasted 8,000 miles most often. Thus in terms of consistency of performance one might say that Brand X is more consistent.

Let us arrange the data for each brand in order from smallest to largest. We have

Brand X	Brand Y	
11,000	8,000	
12,000	8,000	
12,000	8,000	*median*
(14,000)	(10,000)	
14,000	10,000	
14,000	14,000	
14,000	40,000	

In this chart we have circled two numbers. These are the numbers which are in the middle for each brand. For Brand X it is 14,000 and for Brand Y it is 10,000. All the preceding ideas are summarized in the following definitions.

DEFINITION 3.1

The **mean,** also called the **average,** of a set of numbers is obtained by adding the numbers together and dividing the sum by the number of numbers added. We will denote the mean by the symbol $\bar{x}$, which is read x bar, or by the Greek letter μ, which is read mu, depending upon the situation. The use of each symbol will be explained shortly.

DEFINITION 3.2

The **mode** of a set of numbers is the number that occurs most often. If every number occurs only once, there is no mode.

DEFINITION 3.3

The **median** of a set of numbers is the number that is in the middle when the numbers are arranged in order from smallest to largest. The median is easy to calculate when we have an odd number of numbers. If we have an even number of them, the median is defined as the average of the middle two numbers when arranged in increasing order of size.

Comment A set of numbers may have more than one mode. (See Example 1 of this section.)

When the preceding definitions are applied to our example, we get

	Brand X	Brand Y
Mean	13,000	14,000
Median	14,000	10,000
Mode	14,000	8,000

Which number should the police department use to determine which tire is superior: the mean, median, or mode? You might say that the mean is not particularly helpful since one set of Brand Y tires lasted 40,000 miles and this instance had the effect of increasing the average for Brand Y considerably. In terms of consistency Brand X appears to be superior to Brand Y.

Notice that in calculating the mean we had to add numbers. Since the operation of addition plays a key role in our calculations, we now introduce **summation notation,** which we will use in this and later chapters. In this notation we use the letter x or the letter y to represent a number. When working with a distribution of many numbers, we use letters with subscripts, that is, small numbers, attached to them. Thus, we write $x_1, x_2, x_3, \ldots, x_n$, which is read as x sub one, x sub two, x sub three, $\ldots, x$ sub n. To be specific, consider the following set of numbers:

$$7, 15, 5, 3, 9, 8, 14, 21, 10$$

Here we will let x_1 denote the first number, that is, $x_1 = 7$, and we let x_2 denote the second number, that is, $x_2 = 15$. Similarly, $x_3 = 5, x_4 = 3, x_5 = 9, x_6 = 8, x_7 = 14$, $x_8 = 21$, and $x_9 = 10$. Notice that there are 9 terms so that n, which represents the total number of terms, is 9.

To indicate the operation of taking the sum of a sequence of numbers, we will use the Greek symbol Σ, which is read as sigma. Thus Σx will stand for the sum of all numbers denoted by x. In our case

$$\Sigma x = x_1 + x_2 + x_3 + x_4 + x_5 + x_6 + x_7 + x_8 + x_9$$
$$\Sigma x = 7 + 15 + 5 + 3 + 9 + 8 + 14 + 21 + 10$$
$$\Sigma x = 92$$

If we divide this sum by n, the number of terms added, the result will be the mean for these numbers.

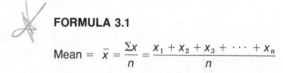

FORMULA 3.1

$$\text{Mean} = \bar{x} = \frac{\Sigma x}{n} = \frac{x_1 + x_2 + x_3 + \cdots + x_n}{n}$$

For our set of numbers the mean is

$$\bar{x} = \frac{\Sigma x}{n} = \frac{92}{9} = 10.22$$

EXAMPLE 1

Measures of
Central Tendency

The district office of a state unemployment insurance department recently hired two new employees, Rochelle and Sharon, to interview prospective aid recipients. Their supervisor is interested in determining who is the better worker. The following chart indicates the number of clients interviewed by each during one week:

Rochelle	Sharon
54	38
67	51
46	46
52	49
45	46
39	38
41	44
Total = 344	Total = 312

Calculate the mean, median, and mode for each employee.

SOLUTION

Let x represent the number of clients interviewed daily by Rochelle and let y represent the number of clients daily interviewed by Sharon. For Rochelle the mean is

$$\bar{x} = \frac{\Sigma x}{n} = \frac{x_1 + x_2 + x_3 + x_4 + x_5 + x_6 + x_7}{7}$$

$$= \frac{54 + 67 + 46 + 52 + 45 + 39 + 41}{7}$$

$$= \frac{344}{7} = 49.14$$

For Sharon the mean is

$$\bar{y} = \frac{\Sigma y}{n} = \frac{y_1 + y_2 + y_3 + y_4 + y_5 + y_6 + y_7}{7}$$

$$= \frac{38 + 51 + 46 + 49 + 46 + 38 + 44}{7}$$

$$= \frac{312}{7} = 44.57$$

Let us now arrange the numbers for each employee in order from the lowest to the highest. We get the following:

Rochelle   39, 41, 45, ⓪46, 52, 54, 67

Sharon   38, 38, 44, ⓪46, 46, 49, 51

Notice that one number has been circled for each worker. This number is in the middle and it represents the median. For both workers the median is 46.

The mode is the number that occurs most often. For Rochelle there is no mode since no number occurs more than once. For Sharon there are two modes, 38 and 46.

Which statistic should the supervisor use the mean, mode, or median in determining who is the better worker?

EXAMPLE 2

The personnel manager of the We-Getum Detective Agency has compiled the following list on the number of years former employees worked before retiring:

Worker	Number of Years of Service Before Retiring
Fay	43
Renée	38
Trudy	47
Hilda	35
Jack	42
Pedro	41
José	39
Beebabats	31

Find the mean, median, and mode.

SOLUTION

We first arrange the numbers in order from lowest to highest. The median is the number that is in the middle. Since there are an even number of terms, the median is between the two circled numbers, 39 and 41.

31,   35,   38,   ⓪39,   ⓪41,   42,   43,   47

Definition 3.3 tells us that it is the average of 39 and 41. Thus we have

$$\text{Median} = \frac{39 + 41}{2} = 40$$

Since no number occurred more than once, there is no mode. The mean is obtained by dividing the total, which is 316, by the number of terms, 8:

$$\text{Mean} = \frac{316}{8} = 39.5$$

Summarizing we have

Mean: 39.5 years
Median: 40 years
Mode: none

Which is more useful to the personnel manager? *median*

EXAMPLE 3
Calculate the mean of the following numbers:

28	19	25	17	28	19	26	17	28	25
17	28	31	22	31	17	31	28	31	14
31	14	17	28	24	31	17	14	26	24
24	24	19	24	14	28	22	31	17	22
25	12	26	19	26	12	19	26	19	28

SOLUTION

One way of calculating the mean would be to add the numbers and divide the result by 50, which is the number of terms. This takes a considerable amount of time but nevertheless we have

$$\bar{x} = \frac{\Sigma x}{n} = \frac{x_1 + x_2 + \cdots + x_{50}}{n}$$

$$= \frac{28 + 19 + 25 + \cdots + 28}{50}$$

$$= \frac{1145}{50} = 22.9$$

Thus the mean is 22.9. We can also group the data as follows:

COLUMN 1 Number (x)	COLUMN 2 Tally	COLUMN 3 Frequency (f)	COLUMN 4 x·f
12	||	2	24
14	||||	4	56
17	⊬⊬ ||	7	119
19	⊬⊬ |	6	114
22	|||	3	66
24	⊬⊬	5	120
25	|||	3	75
26	⊬⊬	5	130
28	⊬⊬ |||	8	224
31	⊬⊬ ||	7	217
		Total = 50	1145

median — 24 *mode — 28* (handwritten annotations)

Column 4 was obtained by multiplying each number by its frequency, that is, by the number of times it occurred. If we now sum column 3 and column 4 individually and divide the column 4 total by the column 3 total, our answer will be 22.9, which is the mean. Although we get the same result as we did before, it was considerably easier to obtain it by grouping the data as we did.

FORMULA 3.2

The mean of a distribution of grouped data is given by

$$\bar{x} = \frac{\Sigma xf}{\Sigma f}$$

where xf represents the product of each class mean and its frequency and Σf represents the total number of items in the distribution.

Comment   It may seem that Formula 3.2 is considerably different from Formula 3.1 for calculating the mean. In reality it is not. In ungrouped data the frequency of each observation is 1 and the total number of terms, Σf, is n. So, Formulas 3.1 and 3.2 are really the same.

To further illustrate the use of Formula 3.2, we consider the following example.

EXAMPLE 4

Tom Edison is a maintenance man for an airline. He keeps accurate records on the life of the special security light bulbs that he services. He has recorded the life of 60 light bulbs in the following chart. *Note:* A bulb that lasted exactly 50 hours is included in the 40–50 hours category. Similarly, a bulb that lasted exactly 80 hours is included in the 70–80 hours category.

$$\bar{x} = 85 + \frac{\sum df}{\sum f} \times 10$$

Life of Bulb (in hours)	Class Mark x	d	Frequency f	Product x·f	d·f
40–50	45	-4	5	225	-20
50–60	55	-3	7	385	-21
60–70	65	-2	8	520	-16
70–80	75	-1	9	675	-9
80–90	85	0	12	1020	0
90–100	95	1	9	855	9
100–110	105	2	6	630	12
110–120	115	3	4	460	12
			Total = 60	4770	

Cannot calculate median in this example accurately
OR
Calculate the mean life of a light bulb. mode

SOLUTION

In using Formula 3.2 we use the class mark for each interval. Thus we assume that any bulb that lasted between 40 and 50 hours actually lasted 45 hours. Although this may introduce a slight error, the error will be minimal when the number of bulbs is large. Thus, we use the class mark. Applying Formula 3.2 we have

$$\text{mean } \bar{x} = \frac{\Sigma xf}{\Sigma f} = \frac{225 + 385 + \cdots + 460}{5 + 7 + \cdots + 4} = \frac{4770}{60} = 79.5 \quad \text{mean}$$

Let us consider another example.

EXAMPLE 5

At a recent rock concert, ticket prices were as follows: orchestra seats, $6.00, lower mezzanine, $4.00, upper mezzanine, $3.00, balcony $2.00. Furthermore, 2000 orchestra tickets, 3000 lower mezzanine tickets, 5000 upper mezzanine tickets, and 7000 balcony tickets were sold. Find the average, also called the mean, price of a ticket.

SOLUTION

We first arrange the data in the form of a table:

COLUMN 1	COLUMN 2	COLUMN 3	COLUMN 4
Seat	*Price (x)*	*Number of Seats (f)*	*Product (x·f)*
Orchestra	$6.00	2000	$12,000
Lower mezzanine	$4.00	3000	12,000
Upper mezzanine	$3.00	5000	15,000
Balcony	$2.00	7000	14,000
		Total = 17,000	$53,000

We complete Column 4 by multiplying each entry in Column 2 with its corresponding frequency in Column 3. Then we apply Formula 3.2. We have:

$$\bar{x} = \frac{\Sigma xf}{\Sigma f} = \frac{12000 + 12000 + 15000 + 14000}{2000 + 3000 + 5000 + 7000}$$

$$= \frac{53000}{17000} = 3.12 \text{ when rounded off}$$

The average cost of a ticket was $3.12. How can the average price per ticket be $3.12 when tickets cost either $2.00, $3.00, $4.00, or $6.00?

Now let us consider the following examples in which all values are not of equal importance. To solve such examples we must use Formula 3.3.

FORMULA 3.3

If $w_1, w_2, \ldots, w_n$ are the weights assigned to the numbers $x_1, x_2, \ldots, x_n$, then the **weighted mean** denoted by the symbol $\bar{x}_w$ is given by

$$\bar{x}_w = \frac{\Sigma xw}{\Sigma w} = \frac{x_1 w_1 + x_2 w_2 + \cdots + x_n w_n}{w_1 + w_2 + \cdots + w_n}$$

This formula indicates that we multiply each number by its weight and divide the sum of these products by the sum of the weights.

EXAMPLE 6

The grades that Liz received on her exams in a statistics course and the weight assigned to each are as follows:

	Grade	Weight Assigned
Test 1	84	1
Test 2	73	2
Test 3	62	5
Test 4	91	4
Final Exam	96	3

Find Liz's average term grade.

SOLUTION

Since each test did not have the same weight, that is, count as much, Formula 3.1 or 3.2 has to be modified. This change is necessary because Formula 3.1 assumes that all numbers are of equal importance, which is not the case in this example. To calculate a weighted mean we use Formula 3.3. When this formula is applied to our example we have

$$\bar{X}_w = \frac{x_1 w_1 + x_2 w_2 + \cdots + x_5 w_5}{w_1 + w_2 + \cdots + w_5}$$

$$= \frac{(84 \cdot 1) + (73 \cdot 2) + (62 \cdot 5) + (91 \cdot 4) + (96 \cdot 3)}{1 + 2 + 5 + 4 + 3}$$

$$= \frac{1192}{15} = 79.47 \text{ when rounded off}$$

Thus the weighted mean is 79.47, not 81.2 which is obtained by adding the numbers together and dividing the sum by n, which is 5.

EXAMPLE 7

On a recent vacation trip Clark Kent kept the following record of his gasoline purchases:

	Price per Gallon (¢)	Number of Gallons Purchased	
	X	*W*	*XW*
Town 1	.61	17	10.37
Town 2	.55	21	11.55
Town 3	.58	16	9.28
Town 4	.65	11	7.15
Town 5	.57	19	10.83
	2.96	84	49.18

What is the average cost per gallon of gasoline for the entire trip?

$$\frac{49.18}{84} = 58.55¢$$

63

SOLUTION

Since Clark did not purchase an identical amount of gasoline in each town, we cannot use Formula 3.1. We use instead Formula 3.3.

$$\bar{x}_w = \frac{(61 \cdot 17) + (55 \cdot 21) + (58 \cdot 16) + (65 \cdot 11) + (57 \cdot 19)}{17 + 21 + 16 + 11 + 19}$$

$$= \frac{4918}{84} = 58.55 \text{ (rounded off)}$$

Thus the weighted average cost per gallon of gasoline for the entire trip was 58.55 cents.

Comment We mentioned earlier that both $\bar{x}$ and μ will be used to represent the mean. The symbol $\bar{x}$ is used to represent the sample mean, that is, the mean of a sample, whereas μ is used to represent the mean of the entire population. From a given problem it will usually be obvious whether we are referring to only part of the population or to the entire population. In either case the answer is obtained by the same formula. Throughout this section we used $\bar{x}$ since we were calculating sample means.

Comment The mean, median, and mode are known as **measures of central tendency,** since each measures some central or general trend, that is, location, of the data. In any particular situation one will usually be more helpful than the others.

EXERCISES

1. There are six members in the math department of a certain university. Their salaries are as follows:

Teacher	Salary (in dollars)
Alice, chairperson	18,330
Jim	13,330
Mary	14,130
Gwendolyn	15,230
Sam	14,130
John	12,150

Find the median, mean, and modal salary for the math department.

 2. The Redline Bus Corporation wishes to advertise the time it takes for its buses to travel between Norwalk and Willard. In order to do this, the company

recorded the time it took for its buses to travel between these cities for each day of February. The results are as follows:

123 157 142 152 137 168 121
-111 126 153 178 141 133 116
118 128 146 131 153 134 156
118 119 144 133 166 153 162

MEDIAN 139
MEAN 139.96
MODE 153

Calculate the mean, median, and mode for the data. Which is more useful? mode

3. Professor Smith, a prominent member of the local chapter of the American Cancer Society, believes there is too much smoking on the college campus. She petitioned the president of the college to have cigarette vending machines removed from campus. In support of her claim she obtained the following statistics on the number of packages of cigarettes sold by the machines on campus during 1976.

Jan.	89	Apr.	95	July	59	Oct.	88
Feb.	101	May	120	Aug.	48	Nov.	92
Mar.	76	June	64	Sept.	76	Dec.	117

Calculate the mean, median, and mode for the data.

4. There are six families on the thirteenth floor of a skyscraper. Their yearly incomes in dollars are as follows:

Jones	19,000	Rodriguez	17,000
Smith	17,000	Protter	22,000
Brown	16,460	Chin	17,500

Find the average salary for these families.

5. A clerk in a men's clothing store keeps a weekly list of the sizes of pants sold. The following is his list for the week of Sept. 8–14:

36	34	34	36	40	44	34	
32	36	36	38	38	38	40	
38	42	32	36	36	36	32	
34	38	42	34	44	34	34	

MEAN 36.71
median 36
mode 36

Calculate the mean, median, and mode for the data. Which is more important to the management of the store?

6. A math teacher gave the same test to his three classes. The following results were obtained:

Class 1 (38 students): Average = 76.3
Class 2 (44 students): Average = 84
Class 3 (35 students): Average = 79.1

2899.4
3696
2768.5
9363.9

Find the average grade for all three classes.

80.03

7. In a certain town an auto mechanic earns $6.00 per hour for 5 hours of work, $9.00 per hour for 3 hours of work, $10.00 per hour for 2 hours of work, and $11.00 per hour for 1 hour of work. What is his average earnings per hour?

8. Three manufacturers of stereo sets claim that the average life of their sets, under normal use, is five years. A consumer's group decides to test each manufacturer's claim. It compiles the following list on the life, in years, of stereo sets manufactured by each:

Manufacturer A: 0.5, 1.6, 2, 3.5, 4, 4.5, 6, 7, 7.9, 8, 10 *MEAN*
Manufacturer B: 4, 4, 5, 5, 5, 6, 11, 13, 14, 15, 16 *mode*
Manufacturer C: 2, 3, 4, 4, 6, 13, 14, 15 *median*

a. Which measurement of the average was each manufacturer using to support his claim?
b. From which manufacturer would you buy a stereo set? Why? *B - Consistancy*

9. A new drug treatment clinic recently opened in the eastern United States. The number of addicts treated with methadone per day during the first month was as follows:

37	15	34	26	25
38	19	22	21	28
42	18	27	32	19
17	29	28	24	35
35	20	23	36	21

Calculate the mean.

10. There are 20 people in an elevator. If their combined weight is 3000 pounds, find the average weight of a passenger.

11. Refer to Exercise 1 of this section. The union has recently negotiated a new contract, and as a result each employee will receive a $1500 raise. How are the median, mean, and mode affected by the raise?

12. An important number to hay fever sufferers is the pollen count. During the month of September the following pollen count numbers were obtained in Stevensville by a local medical center:

Pollen Count Number	Frequency
1	2
5	4
8	6
15	7
16	5
28	4
32	2

a. Find the modal pollen count.
b. Find the mean pollen count.
c. Which is the better indication of the pollen count: the mode or mean?

13. If $x_1 = 78$, $x_2 = 249$, $x_3 = 423$, $x_4 = 182$, and $x_5 = 356$, find Σx. /288
14. Tom Slender is on a diet. The number of pounds that he gained or lost weekly during the first three months of 1976 is as follows:

$$+21, \quad +10, \quad -3, \quad -5, \quad -6, \quad 0, \quad +7, \quad +9, \quad -3, \quad -8, \quad -12, \quad +14$$

Find the average weight change during the three months.

3.2 MEASURES OF VARIATION

Although the mean, median, or mode are very useful in analyzing a distribution, there are some disadvantages in using them alone. These measures only locate the center of the distribution. In certain situations location of the center may not be adequate. We need some method of analyzing variation, that is, the difference, among the terms of a distribution. In this section we will discuss some of the most commonly used methods for analyzing variation.

First let us consider Christina who is interested in determining the best way to drive to school. During one week she drove to school on the Brooks Expressway and during a second week she drove on the Kingston Expressway. The number of minutes needed to drive to school each day follows:

> *Brooks Expressway:*   15, 26, 30, 39, 45
> *Kingston Expressway:*   29, 30, 31, 32, 33

In each case the average time that it took her to drive to school was 31 minutes. Which way is better?

When she used the Brooks Expressway, the time varied from 15 to 45 minutes. We then say that for the Brooks Expressway the **range** is $45 - 15 = 30$ minutes.

On the Kingston Expressway the time varied from 29 to 33 minutes. Thus, the range is $33 - 29 = 4$ minutes.

DEFINITION 3.4

The **range** of a set of numbers is the difference between the largest number in the distribution and the smallest number.

The range is frequently used by manufacturers as a measure of dispersion, that is, spread, in specifying the variation in the quality of a product. So, although the average diameter of a drill bit may be 15/32 inches, in reality, the range in size may be enormous. The manufacturer usually specifies the range to prospective customers.

The range is also used frequently by stock brokers to describe the prices of certain stocks. One often hears such statements as "Stock X had a price range of 15 to 75 dollars, or 60 dollars, during the year."

The range is by far the simplest measure of variation to calculate since only two numbers are needed to calculate it; however, it is not commonly used because it does not tell us anything about how the other terms vary. Furthermore, if there is one extreme value in a distribution, the dispersion or the range will appear very large. If we remove the extreme term, the dispersion may become quite small. Because of this, other measures of variation such as variance, standard deviation, or average deviation are used.

To calculate the variance of a set of numbers, we first calculate the mean of the numbers. We then subtract the mean from each number and square the result. Finally we compute the average of these squares. The result is called the **variance** of the numbers. If we now take the square root of the variance, we get the **standard deviation** for the numbers.* If instead of squaring the differences from the mean we take the absolute value, that is, we neglect any negative signs, of these differences and find the average of these absolute values, the resulting number is called the **average deviation.** The symbol for absolute value is two vertical lines. Thus $|+8|$ is read as "the absolute value of $+8$."

Let us illustrate the preceding ideas by calculating the variance, standard deviation, and average deviation for the two routes that Christina uses to drive to school. Since the mean, μ, is 31 we can arrange our calculations as shown in the following chart. *Note:* Here we have used μ for mean instead of $\bar{x}$ since μ represents the population mean. See Comment on page 64.

	Time (x)	Difference from Mean $(x - \mu)$	Square of Difference $(x - \mu)^2$	Absolute Value of Difference $|x - \mu|$
	15	$15 - 31 = -16$	$(-16)^2 = 256$	16
By Way of	26	$26 - 31 = -5$	$(-5)^2 = 25$	5
Brooks	30	$30 - 31 = -1$	$(-1)^2 = 1$	1
Expressway	39	$39 - 31 = 8$	$8^2 = 64$	8
	45	$45 - 31 = 14$	$14^2 = 196$	14
		Sum $= 0$	Sum $= 542$	Sum $= 44$

Therefore, if Christina travels to school by way of the Brooks Expressway, the variance is 542/5, or 108.4, the standard deviation is $\sqrt{108.4}$, or 10.41, and the average deviation is 44/5, or 8.8.

Notice that in computing these measures of variation, we used symbols. Thus μ represents the mean, $x - \mu$ represents the difference of any number from the mean, $(x - \mu)^2$ represents the square of the difference, and $|x - \mu|$ represents the absolute value of the difference from the mean. Furthermore, the sum of the differences from the mean, $\Sigma(x - \mu)$, is 0. Can you see why?

---

*A knowledge of how to compute square roots is not needed. Such values can be obtained from Table I in the appendix on page 454.

Let us now compute the variance, standard deviation, and average deviation for traveling to school by way of the Kingston Expressway. Again we will use symbols.

	Time (x)	Difference from Mean (x − μ)	Square of Difference (x − μ)²	Absolute Value of Difference |x − μ|
	29	29 − 31 = −2	(−2)² = 4	2
By Way of	30	30 − 31 = −1	(−1)² = 1	1
Kingston	31	31 − 31 = 0	0² = 0	0
Expressway	32	32 − 31 = 1	1² = 1	1
	33	33 − 31 = 2	2² = 4	2
		Sum = 0	Sum = 10	Sum = 6

In this case the variance is 10/5, or 2, the standard deviation is $\sqrt{2}$, or 1.41, and the average deviation is 6/5, or 1.2. Here again the sum of the differences from the mean, $\Sigma(x - \mu)$, is 0. This is always the case.

We now formally define variance, standard deviation, and average deviation.

DEFINITION 3.5

The **variance** of a set of n numbers is the average of the squares of the differences of the numbers from the mean. If μ represents the mean, then

$$\text{Variance} = \frac{\Sigma(x - \mu)^2}{n}$$

69

DEFINITION 3.6

The **standard deviation** of a set of numbers is the positive square root of the variance. Thus

$$\text{Standard deviation} = \sqrt{\text{variance}}$$

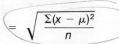

$$= \sqrt{\frac{\Sigma(x - \mu)^2}{n}}$$

We denote the **population standard deviation** by the symbol σ which is read as sigma.

$$\sigma = \sqrt{\frac{\Sigma(x - \mu)^2}{n}}$$

With this notation the variance is denoted as σ^2.

DEFINITION 3.7

The **average deviation** of a set of numbers is the average of the absolute value of the differences from the mean. Symbolically,

$$\text{Average deviation} = \frac{\Sigma|x - \mu|}{n}$$

We will illustrate the preceding ideas with another example.

EXAMPLE 1

The number of hours per day that Johnny Smash practiced playing his guitar during the past week is 5, 3, 2, 6, 4, 2, and 6. Find the variance, standard deviation, and average deviation.

SOLUTION

We arrange the data in order as shown in the following chart and perform the indicated calculations:

Number of Hours x	Difference from Mean $(x - \mu)$	Square of Difference $(x - \mu)^2$	Absolute Value of Difference $|x - \mu|$
2	$2 - 4 = -2$	$(-2)^2 = 4$	2
2	$2 - 4 = -2$	$(-2)^2 = 4$	2
3	$3 - 4 = -1$	$(-1)^2 = 1$	1
4	$4 - 4 = 0$	$0^2 = 0$	0
5	$5 - 4 = 1$	$1^2 = 1$	1
6	$6 - 4 = 2$	$2^2 = 4$	2
6	$6 - 4 = 2$	$2^2 = 4$	2
	Sum = 0	Sum = 18	Sum = 10

The mean is

$$\frac{\Sigma x}{n} = \frac{2 + 2 + 3 + 4 + 5 + 6 + 6}{7}$$

$$= \frac{28}{7} = 4$$

Our answers then are

Variance $= \dfrac{\Sigma(x - \mu)^2}{n} = \dfrac{18}{7} = 2.57$

Standard deviation $= \sqrt{\text{variance}} = \sqrt{2.57}$, or 1.60

Average deviation $= \dfrac{\Sigma|x - \mu|}{n} = \dfrac{10}{7} = 1.43$

Comment   It may seem that the standard deviation is a complicated and use-less number to calculate. At the moment let us say that it is a useful number to the statistician. When we discuss the normal distribution in later chapters, you will understand the significance and usefulness of the standard deviation.

In most statistical problems we do not have all the data for the population. Instead, we have only a small part, that is, a sample, of the population. To calculate the standard deviation in this case, the formula in Definition 3.6 has to be changed somewhat. The **sample standard deviation** is now denoted by s and is given by the following formula in which $\bar{x}$ is the mean of the sample.

FORMULA 3.4

$$s = \sqrt{\frac{\Sigma(x - \bar{x})^2}{n - 1}}$$

The difference between σ and s is whether we divide by n or by $n - 1$. For the population standard deviation we divide by n and denote our result by σ, whereas for a sample standard deviation we divide by $n - 1$ and denote our result by s. Thus s is really an estimate of σ, the population standard deviation. Very often statisticians will refer to s as *the* standard deviation, even though it is only an estimate.

3.3 COMPUTATIONAL FORMULA FOR CALCULATING THE VARIANCE

Although the formula in Definition 3.5 of the preceding section can always be used for calculating the variance, it turns out that, in practice, the calculations become quite tedious. For this reason we use a more convenient formula, which follows:

FORMULA 3.5

$$\text{Variance} = \sigma^2 = \frac{\Sigma x^2}{n} - \frac{(\Sigma x)^2}{n^2}$$

In Formula 3.5, Σx^2 means that we square each number and add the squares together.

TABLE **3.1**

SQUARES OF INTEGERS (x^2 MEANS x TIMES x)

x	x^2
1	1
2	4
3	9
4	16
5	25

Using summation notation for the data in Table 3.1, we have

$$\Sigma x = 1 + 2 + 3 + 4 + 5 = 15$$
$$\Sigma x^2 = 1^2 + 2^2 + 3^2 + 4^2 + 5^2$$
$$= 1 + 4 + 9 + 16 + 25 = 55$$

We now use Formula 3.5 to calculate the variance for the data of Table 3.1:

$$\sigma^2 = \frac{\Sigma x^2}{n} - \frac{(\Sigma x)^2}{n^2}$$

$$= \frac{55}{5} - \frac{(15)^2}{5^2}$$

$$= \frac{55}{5} - \frac{225}{25}$$

$$= 11 - 9 = 2$$

If you now calculate the variance by using the formula in Definition 3.5 of the preceding section and compare the results, your answer will be the same. It is considerably simpler, however, to get the answer by using Formula 3.5. The standard deviation is obtained by taking the square root of the variance. Thus,

$$\text{Standard deviation} = \sqrt{\text{variance}}$$
$$= \sqrt{2} \approx 1.41$$

(The symbol $\approx$ stands for approximately.)

Comment The advantage of computing the variance by Formula 3.5 is that we do not have to subtract the mean from each term of the distribution.

Beware Do not confuse the symbols Σx^2 and $(\Sigma x)^2$. The symbol Σx^2 represents the sum of the squares of each number, whereas the symbol $(\Sigma x)^2$ represents the square of the sum of the numbers. If your calculation of the variance results in a negative number, you probably have confused the two symbols.

3.4 CODING

Suppose you were asked to calculate the mean, variance, and standard deviation of the numbers 5, 8, 15, 17, and 20. We can arrange the data as shown in the following chart:

x	x^2
5	25
8	64
15	225
17	289
20	400
Sum: 65	1003

The mean is 65/5 = 13. Using Formula 3.5 of the preceding section, we find

$$\text{Variance} = \frac{1003}{5} - \frac{65^2}{5^2} = 200.6 - 169 = 31.6$$

and

$$\text{Standard deviation} = \sqrt{31.6}, \text{ or } 5.62$$

Let us now add 10 to each number of the distribution. The new numbers are 15, 18, 25, 27, and 30. What effect, if any, does this change have on the mean, variance, and standard deviation? To answer this question we will again calculate these values using y instead of x since we changed the numbers. We have

y	y^2
15	225
18	324
25	625
27	729
30	900
Sum: 115	2803

You should verify that the mean now is 23, the variance is 31.6, and the standard deviation is $\sqrt{31.6}$, or 5.62.

Notice that the mean is now 10 more than it was before. This increase represents the amount that we added to each of the original terms. The variance and standard deviation, on the other hand, remained the same. Thus only the mean is affected by adding or subtracting the same number to each term of a distribution. The variance and standard deviation are not affected; this fact provides us with a useful method of calculating the variance and standard deviation as well as the mean for these numbers. The process is called **coding.** We simply add (or subtract) the same number to (or from) each term of distribution and calculate the standard deviation and variance of the new distribution. If the number to be added (or subtracted) is chosen carefully, the resulting calculations will be simple to perform.

EXAMPLE 1

The price of a certain model used car varies from dealer to dealer. Five dealers were chosen and the prices quoted were $370, $345, $360, $350, and $375. Calculate the variance and standard deviation for the price of the used car.

SOLUTION

First we arrange the data in increasing order. Then we code the data by subtracting 360 from each number. Thus in tabular form we have:

	x	x^2
	-15	225
	-10	100
	0	0
	10	100
	15	225
Sum:	0	650

Since the variance and standard deviation remain the same for the coded data as they are for the original data, we use Formula 3.5 and find that the variance is

$$\text{Variance} = \sigma^2 = \frac{\Sigma x^2}{n} - \frac{(\Sigma x)^2}{n^2}$$

$$= \frac{650}{5} - \frac{0^2}{5^2} = 130$$

The standard deviation is $\sqrt{130}$, or 11.4. The mean is obtained by adding 360 to the mean of the new distribution which is 0. Thus, the mean of the original numbers is

$$0 + 360 = 360$$

Coding can simplify computations considerably and should be used whenever appropriate.

How are the mean, variance, and standard deviation affected if we *multiply* each term of a distribution by some number?

EXERCISES

1. A health food company employs 20 workers whose salaries (in dollars) are

14000	10000	16400	21795
17000	12000	18200	24500
16000	26000	17950	15000
18000	7000	16280	9050
9000	13500	19640	7620

Find the range of salaries for the 20 workers. $26,000 - 7000 = 19,000$

2. The following grades were obtained on a statistics test:

68	98	83	55	93
59	69	96	86	84
78	80	65	76	80
85	95	50	90	73
80	85	90	80	97

Find the range of the grades.

3. A secretary typed 10 business letters for her supervisor. The number of typing errors per letter was 4, 2, 0, 3, 8, 2, 6, 1, 9, and 0. Find the standard deviation and range for the numbers. $Stand. dev. = 9, 75$ $RANGE = 9$

4. The ages of the members of a math department are 31, 28, 44, 39, 52, 39, 58, and 29 years. Calculate the variance and standard deviation for the ages.

5. Trudy loves to speak on the telephone. Last night she made 6 calls which lasted 44, 11, 32, 27, 48, and 18 minutes. Find the mean, variance, standard deviation, and average deviation for the length of a telephone call.

6. Each year many television stations conduct fund raising drives through telethons for various charities. For the past 5 years one television station kept records on the time which passed before the first $10,000 was pledged. This information follows:

Year	Minutes Passed Before First $10,000 was Pledged
1972	19
1973	12
1974	10
1975	19
1976	5

Find the mean, variance, standard deviation, and average deviation for the numbers.

7. The number of minutes that a commuter waited for her train during one week was 10, 6, 15, 11, and 3. Calculate the mean, variance, standard deviation, and average deviation for the numbers.

8. Multiply each number in the preceding exercise by 3 and then compute the mean, variance, standard deviation, and average deviation for the new distribution. How do the results compare with those of the preceding exercise? Can you generalize?

9. A comparison shopper has compared the prices of a pound of ground beef at a number of different supermarkets. The following prices, in cents, were obtained: 89, 93, 74, 63, and 101.

    a. Calculate the mean and standard deviation by subtracting 75 cents from each price.

    b. Calculate the mean and standard deviation by subtracting 80 cents from each price.

    c. How are the mean and standard deviation affected if we subtract different numbers from each price?

10. The average cost of malpractice medical insurance last year in a certain city was $2300 for a neurosurgeon, with a standard deviation of $450. This year, rates will be increased by $600. What will the new mean and standard deviation be?

11. Last year Sam bought 5 tires which lasted 17,010, 16,080, 17,050, 16,090, and 17,000 miles. Find the variance and standard deviation for the life of the tires. (*Hint:* Code by subtracting 17,000.)

SELF-STUDY GUIDE

In this chapter we discussed various numerical methods for analyzing data. In particular, we calculated and compared three measures of central tendency: the mean, median, and mode. We pointed out that each has its advantages and disadvantages. In addition, various properties of each measure were discussed. Thus we mentioned that the mean is affected by extreme values and that the sum of the differences from the mean is zero. The mean is the most frequently used measure of central tendency. We also demonstrated how to calculate a weighted mean when the terms of a distribution are not of equal weight. In the process we introduced summation notation.

We then discussed four measures of variation which tell us how dispersed, that is, how spread out, the terms of the distribution are around the *center* of the distribution. These were the range, variance, standard deviation, and average deviation. Various shortcuts for computing the standard deviation and variance were introduced. Coding the data proved to be quite helpful. Again summation notation was used, and several of the errors commonly made using this notation were mentioned.

The following list is a summary of all formulas given in the chapter. You should be able to identify each symbol, understand the relationships among the symbols expressed in each formula, understand the significance of each formula, and use the formulas in solving problems.

1. Mean $= \dfrac{\Sigma x}{n} = \dfrac{x_1 + x_2 + \cdots + x_n}{n}$

2. Mean for grouped data $= \bar{x} = \dfrac{\Sigma xf}{\Sigma f}$

3. Weighted mean $= \bar{x}_w = \dfrac{\Sigma xw}{\Sigma w}$

4. Variance $= \dfrac{\Sigma(x - \mu)^2}{n}$

5. Standard deviation $= \sigma = \sqrt{\dfrac{\Sigma(x - \mu)^2}{n}}$

6. Average deviation $= \dfrac{\Sigma|x - \mu|}{n}$

7. Sample standard deviation $= s = \sqrt{\dfrac{\Sigma(x - \bar{x})^2}{n - 1}}$

8. Variance $= \dfrac{\Sigma x^2}{n} - \dfrac{(\Sigma x)^2}{n^2}$

You should be able to demonstrate your knowledge of the following ideas presented in this chapter by giving definitions, descriptions, or specific examples.

Mean, also called average (page 55)

Mode (page 55)

Median (page 55)

Summation notation (page 56)

Weighted mean (page 62)

Measure of central tendency
    (page 64)

Range (page 67)

Variance (page 69)

Standard deviation (page 70)

Average deviation (page 70)

Population standard deviation
    (page 70)

Sample standard deviation (page 71)

Coding (page 73)

The tests of the following section will be more helpful if you take them after you have studied the examples given in this chapter and solved the exercises at the end of each section.

MASTERY TESTS

Form *A*

1. The average grade on a math test in Mary's class is 75, and in Arlene's class it is 85. Can we say that the average grade for both classes is 80? Explain.
2. If a switchboard operator claims that the average number of calls received on a busy day is 69, to which average is she referring: the mean, median, or mode?
3. Is it possible for the range, variance, and standard deviation to be equal?
4. Is it safe for a man who is 5 feet tall and a nonswimmer to swim in a pool which has an average depth of 3 feet? Explain your answer.

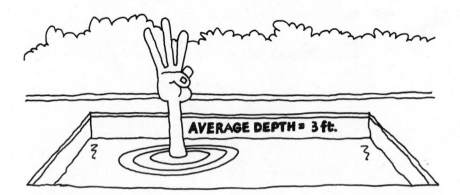

AVERAGE DEPTH = 3 ft.

5. Without actually calculating them, compare the mean and standard deviation for the following data:

$$X: \quad 7, 3, 9, 2, 6$$
$$Y: \quad 70, 30, 90, 20, 60$$

6. The sum of the deviations from the mean, that is, $\Sigma(x - \mu)$, is always
   a. 0   b. −1   c. 4   d. −3   e. none of these
7. If a distribution of 10 terms has an average of 25, then what does Σx equal?
8. In two towns four homes were polled to find out how many magazines were received by the occupants. In town A the four families received 3, 4, 5, and 6 magazines. In town B the four families received 3, 10, 17, and 24 magazines. Without actually calculating it, how does the standard deviation for town A compare with the standard deviation for town B?
9. Which of the following statements is always true?
   a. The mean has an effect on extreme scores. b. The median has an effect on extreme scores. c. Extreme scores have an effect on the mean. d. Extreme scores have an effect on the median. e. Extreme scores have an effect on the mode.

10. When a bartender claims that the average number of drinks she serves is 4, to which average is she referring?

    a. mean *b.* median *c.* mode

Form *B*

1. Bill is registered for three courses in college: economics, sociology, and philosophy. He took the midterm exam in each course and then became ill. He obtained the following information from his classmates:

        Economics: Mean grade = 78   Standard deviation = 3
        Sociology: Mean grade = 78   Standard deviation = 7
        Philosophy: Mean grade = 78   Standard deviation = 5

   In which class is his relative standing the highest? Explain your answer.

2. John is considering buying two different brands of beer. He has obtained the following data on the number of minutes that each beer will retain its flavor once opened.

        Brand A: Mean = 10   Standard deviation = 3
        Brand B: Mean = 18   Standard deviation = 2

   Which brand should he choose? Why?

3. If $x_1 = 17$, $x_2 = 9$, $x_3 = 19$, $x_4 = 5$, and $x_5 = 10$, determine the following:
   a. Σx^2 *b.* $(\Sigma x)^2$
   Compare the answers found in parts *a* and *b*. Are they the same?

4. Are Formulas 3.2 and 3.3 different or are they equivalent?

5. Show that Formula 3.5 and the formula given in Definition 3.5 are equivalent.

6. Many colleges require all entering freshmen to take the Scholastic Aptitude Test administered by the Educational Testing Service. One hundred students took this exam and their test scores follow:

Test Score	Frequency
200–below 250	1
250–below 300	2
300–below 350	6
350–below 400	10
400–below 450	17
450–below 500	19
500–below 550	19
550–below 600	14
600–below 650	9
650–below 700	3
700–below 750	1
750–below 800	0
	Sum = 100

   a. Draw a frequency distribution for the data.
   b. Calculate the mean and mode for the data. MEAN 491.25
                                     MODE cannot be calc.

7. John, who is 23, wants to date a girl who is approximately 20 years old. He is considering two computer dating services in Los Angeles, both of which claim that the average age of the girls they have on file is 20 years. John does not know which dating service to select. He believes that both are equally good. His friend persuades him to look at the individual ages of the girls and not at the average. The individual ages are as follows:

Dating Service A	15, 17, 18, 23, 27
Dating Service B	18, 19, 20, 21, 22

Is his friend right? Why?

8. The Dingling Brothers are training 100 elephants to perform a certain act in their next circus performance. The following frequency distribution indicates the number of hours required to train the elephants:

Number of Hours	Frequency
100–199	12
200–299	28
300–399	20
400–499	18
500–599	14
600–699	8
Total =	100

Calculate the mean for the data.

9. Recently, 100,000 people were given an IQ test. Thirty thousand people had an IQ score of less than 105, 40,000 had an IQ score of exactly 105, and 30,000 had an IQ score of more than 105. Find the mode and median for the distribution. Can you also find the mean?

10. Consider the following three frequency distributions:

a.

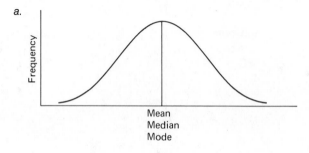

Mean
Median
Mode

b.

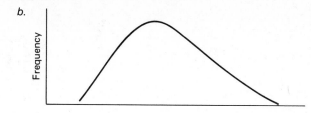

A frequency
distribution skewed
to the right.

c.

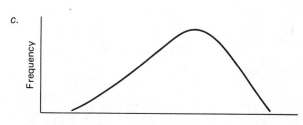

A frequency
distribution skewed
to the left.

Locate the mean, median, and mode for the frequency distributions given in parts *b* and *c*. We have already done it for the diagram of part *a*.

SUGGESTED READING

Minum, E. W. *Statistical Reasoning in Psychology and Education.* New York: Wiley, 1970.

Newmark, Joseph and Lake, Frances. *Mathematics as a Second Language* 2nd ed., Reading, Mass.: Addison-Wesley Publishing Co. 1977.

O'Toole, A. L. *Elementary Practical Statistics.* New York: Macmillan Publishing Co., 1965.

Senter, R. J. *Analysis of Data: Introductory Statistics for the Behavioral Sciences.* Glenview, Illinois: Scott Foresman, 1969.

Tanur, Judith, ed. et al. *Statistics: A Guide to the Unknown.* San Francisco: Holden-Day, Inc., 1972.

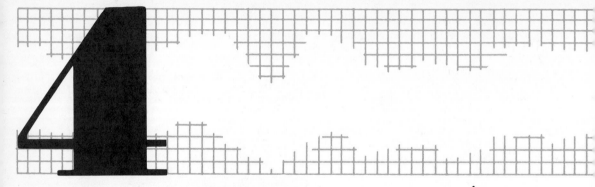

PERCENTILES
AND z-SCORES

The newspaper article and business letter indicate that percentile ranks are used not only by many colleges in determining which students will be admitted but also by many businesses. Even though all prospective job applicants were college graduates, the Statistical Corporation of America, in considering which applicant to hire, ranked all according to percentile rank.

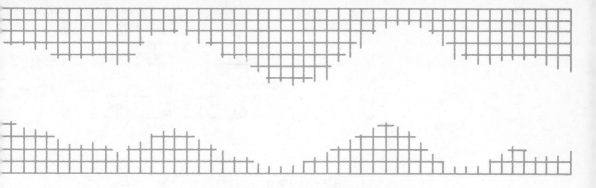

Preview

In this chapter we will discuss the following topics:

1. *Percentile*

This gives us relative standing of one score when compared to the rest. Specifically it tells us how many scores are above or below a given score.

2. *z-Score*

This gives us the relative position of a score with respect to the mean. It is expressed in terms of standard deviations.

Introduction

In previous chapters we discussed different ways of analyzing and summarizing data. Quite often one is also interested in knowing the position of a score in a distribution. Thus, if Lorraine gets a 76 on her psychology midterm exam, she would be interested in knowing how this grade compares with others in her class. For situations such as these, statisticians are interested in the **percentile rank** or **z-score** of a term in a distribution. In this chapter we will discuss these ideas and the formulas needed to compute their values.

Percent of grades below Lorraine's grade		Lorraine's grade	Percent of grades above Lorraine's grade
60%		10%	30%

Figure 4.1

Percentile rank of Lorraine

60%		30%

Figure 4.2   5% 5%

4.1 PERCENTILES AND PERCENTILE RANK

Again let us consider Lorraine who got a 76 on her midterm psychology exam. There are 150 students, including her, in the class. She knows that 60% of the class got below 76, 10% of the class got 76, and the remaining 30% got above 76.

Since 60% of the class got below her grade of 76 and 30% got above her grade, her percentile rank should be between 60 and 70. We will use 65, which is midway between 60 and 70. What we do is find the percent of scores that are below the given score and add one-half of the percent of the scores which are the same as the given score. In our case 60% of the class grades were below Lorraine's and 10% were the same as Lorraine's. Thus, the percentile rank of Lorraine's grade is

$$60 + (1/2)(10) = 65$$

Figures 4.1 and 4.2 illustrate the situation.

We now say that Lorraine's percentile rank is 65. This means that approximately 35% of the class did better than her on the exam and that she did better than 65% of the class. Essentially the percentile rank of a score tells us the percentage of the distribution which is below that score. Formally we have

DEFINITION 4.1

The **percentile rank** of a term in a distribution is found by adding the percentage of terms below it with one-half of the percentage of terms equal to the given term.

Let X be a given score, let B represent the *number* of terms below the given score X, and let E represent the *number* of terms equal to the given score X. If there are n terms altogether, then the percentile rank of X is given by Formula 4.1.

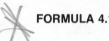

FORMULA 4.1

$$\text{Percentile rank of } X = \frac{B + \frac{1}{2}E}{n} \cdot 100$$

We will now illustrate the use of Formula 4.1.

EXAMPLE 1

Bill and Jill are twins, but both are in different classes. Recently they both got 80 on a math test. The grades of the other students in their class were:

Jill's class: 64, 67, 73, 73, 73, 74, 77, 77, 78, 78, 79, 80, 80, 82, 91, 94, 100

Bill's class: 43, 65, 68, 73, 75, 76, 76, 77, 79, 80, 80, 80, 80, 85, 86, 87, 88, 90, 92, 96

Find the percentile rank of each student.

SOLUTION

We will use Formula 4.1. Jill's grade is 80. There were two 80s (including Jill's) in the class, so $E = 2$. There were 11 grades below 80, so $B = 11$. Since there are 17 students in the class altogether, $n = 17$. Thus,

$$\text{Jill's percentile rank} = \frac{11 + \frac{1}{2}(2)}{17} \cdot 100$$

$$= \frac{11 + 1}{17} \cdot 100 = \frac{12}{17} \cdot 100$$

$$= \frac{1200}{17} = 70.59$$

Jill's percentile rank is 70.59. Using a similar procedure for Bill's class, we find that $B = 9$, $E = 4$, and $n = 20$. Thus,

$$P = \frac{B + E}{n} \cdot 100$$

$$\text{Bill's percentile rank} = \frac{9 + \frac{1}{2}(4)}{20} \cdot 100$$

$$= \frac{9 + 2}{20} \cdot 100 = \frac{11}{20} \cdot 100$$

$$= \frac{1100}{20} = 55$$

Bill's percentile rank is 55.

Comment The percentile rank of an individual score is often more helpful than the particular score value. Although both Bill and Jill had grades of 80, Jill's percentile rank is considerably higher. If we assume that the levels of competition are equivalent in both classes, this may indicate that Jill's performance is superior to Bill's performance when compared to the rest of their respective classes.

We often use the word **percentile** to refer directly to a score in a distribution. So, instead of saying that the percentile rank of Lorraine's grade is 65, we would say that her grade is in the 65th percentile. Similarly, if a term has a percentile rank of 40, we would say that it is in the 40th percentile.

Percentiles are used quite frequently to describe the results of achievement tests and the subsequent ranking of people taking those tests. This is especially true when applying for many civil service jobs. If there are more applicants than available jobs, candidates are often ranked according to percentiles. Many colleges use only percentile ranks, rather than the numerical high school average, to determine which candidates to admit. The reason is that percentile ranks of a student's high school average reflect how they did with respect to their classmates, whereas numerical averages only indicate an individual student's performance. The newspaper article and the business letter presented at the beginning of this chapter clearly illustrate this use of percentiles.

Since percentiles are numbers that divide the set of data into 100 equal parts, we can easily compare percentiles. Thus, in Example 1 we were able to find the percentile rank of Jill and Bill, even though they both were in different classes.

During World War II the United States Army administered the Army General Classification Tests (AGCT) to thousands of enlisted men. The results showed

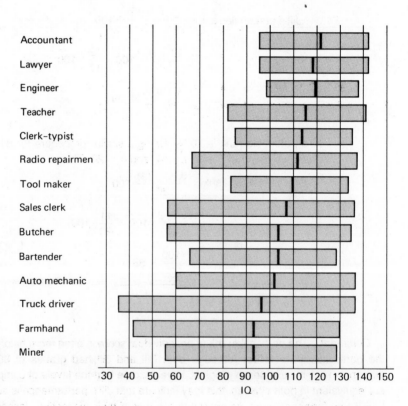

Figure 4.3

Each bar shows the IQ range between the 10th and 90th percentiles for men in that occupation. The vertical bars represent the 50th percentiles. Note that although the average IQ score of accountants was 121 and that of miners 93, some miners had higher IQ scores than accountants.

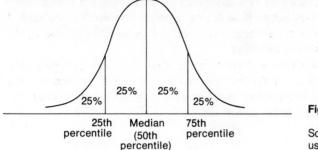

Figure 4.4

Some of the frequently used percentiles.

important differences in the average IQ of men in various jobs, ranging from 93 for miners and farmhands to around 120 for accountants, lawyers, and engineers. Figure 4.3 shows the IQ range between the 10th and 90th percentiles for workers in various occupations. Furthermore, the vertical bars represent the 50th percentile or median scores. Very often, when such tests are administered to large groups of people, the results are given in terms of percentile bands as shown in Figure 4.3. Since percentiles are used quite often, special names are given to the 25th and 75th percentiles of a distribution. Of course, the *50th percentile is the median of the distribution.* See Figure 4.4.

DEFINITION 4.2

The 25th percentile of a distribution is called the **lower quartile.** It is denoted by P_{25}. Thus, 25% of the terms are below the lower quartile and 75% of the terms are above it.

DEFINITION 4.3

The 75th percentile of a distribution is called the **upper quartile.** It is denoted by P_{75}. Thus, 75% of the terms are below the upper quartile and 25% of the terms are above it.

Notation The various percentiles are denoted by the letter *P* with the appropriate subscript. Hence, P_{37} denotes the thirty-seventh percentile, P_{99} denotes the 99th percentile, etc.

4.2 z-Scores

As we saw in the preceding section, one way of measuring the performance of an individual score in a population is by determining its percentile rank. Using percentile rank alone, however, can sometimes be misleading. For example, two students in different classes may have the same percentile rank. Yet one student may be far superior to his or her competitors, whereas the second may only slightly surpass the others in his or her class.

Statisticians have another very important way of measuring the performance of an individual score in a population. This measure is called the **z-score.** The z-score measures how many standard deviations an individual score is away from the mean. We define it formally as follows:

DEFINITION 4.4 — FORMULA 4.2

The z-score of any number X in a distribution whose mean is μ and whose standard deviation is σ is given by

$$z = \frac{X - \mu}{\sigma}$$

where X = value of number in original units

μ = population mean

σ = population standard deviation

Comment   The z-score of a number in a population is sometimes called the **z-value** or **measurement in standard units.**

Comment   Since σ is always a positive number, z will be a negative number whenever X is less than μ, since $X - \mu$ is then a negative number. A z-score of 0 implies that the term has the same value as the mean.

We now illustrate how to calculate z-scores with several examples.

EXAMPLE 1

A certain brand of flashlight battery has a mean life, μ, of 40 hours and a standard deviation of 5 hours. Find the z-score of a battery which lasts

a. 50 hours   *b.* 35 hours   *c.* 40 hours.

Since $\mu = 40$ and $\sigma = 5$, we will use Formula 4.2.

a. The z-score of 50 is

$$\frac{50 - 40}{5} = \frac{10}{5} = 2$$

b. The z-score of 35 is

$$\frac{35 - 40}{5} = \frac{-5}{5} = -1$$

c. The z-score of 40 is

$$\frac{40 - 40}{5} = 0$$

EXAMPLE 2

Two consumer's groups, one in New York and one in California, recently tested, at numerous local colleges, a number of different brands of canned tuna fish for taste appeal. Each consumer group used a different rating system. The following results were obtained:

New York	Rating	California	Rating
A	1	M	25
B	10	N	35
C	15	P	45
D	21	Q	50
E	28	R	70

Which brand has the greatest taste appeal?

SOLUTION

At first glance it would appear that California brand R is superior since its rating was 70. However, it is obvious from the rating and from the given information that the two consumer's groups awarded their points differently, so that the point value alone is not enough of a basis for deciding among the different brands. We will therefore convert each of the ratings into standard scores. These calculations are shown in Tables 4.1 and 4.2.

TABLE 4.1

RATING OF NEW YORK BRAND TUNA FISH ($\mu = 15$, $\sigma = 9.22$)

BRAND	RATING X	MEAN μ	DIFFERENCE FROM MEAN $(X - \mu)$	z-SCORE $z = \dfrac{X - \mu}{\sigma}$
A	1	15	$1 - 15 = -14$	$\dfrac{-14}{9.22} = -1.52$
B	10	15	$10 - 15 = -5$	$\dfrac{-5}{9.22} = -0.54$
C	15	15	$15 - 15 = 0$	$\dfrac{0}{9.22} = 0$
D	21	15	$21 - 15 = 6$	$\dfrac{6}{9.22} = 0.65$
E	28	15	$28 - 15 = 13$	$\dfrac{13}{9.22} = 1.41$

TABLE 4.2

RATING OF CALIFORNIA BRAND TUNA FISH ($\mu = 45$, $\sigma = 15.17$)

BRAND	RATING X	MEAN μ	DIFFERENCE FROM MEAN $(X - \mu)$	z-SCORE $z = \dfrac{X - \mu}{\sigma}$
M	25	45	$25 - 45 = -20$	$\dfrac{-20}{15.17} = -1.32$
N	35	45	$35 - 45 = -10$	$\dfrac{-10}{15.17} = -0.66$
P	45	45	$45 - 45 = 0$	$\dfrac{0}{15.17} = 0$
Q	50	45	$50 - 45 = 5$	$\dfrac{5}{15.17} = 0.33$
R	70	45	$70 - 45 = 25$	$\dfrac{25}{15.17} = 1.65$

We can now use the z-scores as a basis for comparison of the different brands. Clearly, California brand R for which $z = 1.65$ is superior to New York brand E for which $z = 1.41$.

Notice that the sum of the z-scores for the New York brands is 0:

$$(-1.52) + (-0.54) + (0) + (0.65) + (1.41) = 0$$

This means that the average of the z-scores is 0, since 0 divided by 5, the number of z-scores, is 0. Also the sum of the z-scores for the California brands is 0:

$$(-1.32) + (-0.66) + (0) + (0.33) + (1.65) = 0$$

Therefore the mean is 0. If you now compute the standard deviations of the z-scores in Tables 4.1 and 4.2, you will find that the standard deviation in each case is 1. We summarize these facts in the following rule.

Rule In any distribution the mean of the z-scores is 0 and the standard deviation of the z-scores is 1.

Formula 4.2 can be changed so that if we are given a particular z-score, we can calculate the corresponding original score. The changed formula is as follows:

FORMULA 4.3 $X = \mu + z\sigma$

EXAMPLE 3

In a recent swimming contest the mean score was 40 and the standard deviation was 4. If Carlos had a z-score of -1.2, how many points did he score?

SOLUTION

Since $\mu = 40$, $\sigma = 4$, and $z = -1.2$, we can use Formula 4.3. Thus we have

$$X = 40 + (-1.2)(4)$$
$$= 40 - 4.8$$
$$= 35.2$$

Carlos' score was 35.2.

EXERCISES

1. Twenty students in an ecology class have been collecting old newspapers to be used for recycling. The number of pounds of paper collected by each student in one week follows:

91

Pedro–423	Frances–384	Chris–289	Jimmy–384
Bill–277	Doris–352	Jack–301	Bob–352
George–325	Tessie–378	Tina–299	Mabel–398
Bruce–426	Ann–396	Lois–378	Maurice–407
Marie–293	Gwendolyn–401	Clark–304	Richard–288

Find the percentile rank of Doris and the percentile rank of Jimmy.

2. In Sheldon's physical education class of 50 students, 36 students can do more push-ups than he can and 2 students, including Sheldon, can do as many push-ups as he can. Find his percentile rank.

3. One hundred prisoners of the state correctional facility at Newport volunteered to be subjects for an endurance test performed by a panel of psychologists. The following scores were obtained:

Score	Frequency
10–19	6
20–29	9
30–39	12
40–49	19
50–59	24
60–69	18
70–79	9
80–89	3
Total = 100	

$$\frac{21}{100} \cdot 100 = 21$$

$$\frac{79}{100} \cdot 100 = 79$$

a. Calculate the percentile rank of Hal who scored 34 on the test. 21
b. Calculate the percentile rank of Sal who scored 64 on the test. 79

4. The following statistics on the family income for the Old Falls community are available:

$$P_{25} = \$ 7,500$$
$$P_{50} = \$11,300$$
$$I_{75} = \$12,800$$
$$P_{85} = \$21,030$$

What percent of the families earn
a. less than $7500? 25%
b. more than $21,030? 15%
c. less than $12,800? 75%
d. more than $11,300? 50%
e. between $7500 and $21,030? 60%

5. In a recent weight-lifting contest the average weight lifted was 250 pounds with a standard deviation of 20 pounds. Find the z-score of $z = \frac{X - \mu}{\sigma}$
a. José who lifted 290 pounds. 2
b. Bob who lifted 220 pounds. -1.5
c. Mark who lifted 310 pounds. 3
d. Mike who lifted 250 pounds. 0

6. The following z-scores were obtained by 5 faculty members of the math department on their teacher evaluations at Ticky University. In the university there are 400 faculty members.

    Jean 2.63, Mabel −1.36, Sherry −1.02, Sydney 1.69, Isaac 0.97

    a. Rank these people from highest to lowest.
    b. Which of these teachers was above the mean?
    c. Which of these teachers was below the mean?

7. A skill test was given to 100 secretaries. The mean score was 35 and the standard deviation was 2.6. $X = \mu + z\sigma$

    a. If Steve had a z-score of −0.84, what was his actual score? $32.816 = 32.82$
    b. If Alice had a z-score of 2.48, what was her actual score? $41.448 = 41.45$
    c. If Bob had a z-score of −3.03, what was his actual score? $27.122 = 27.12$

8. The following information concerning grades are posted on the bulletin board:

Test Grade	z-Score	Percentile Rank
55	−2	2
65	−1	16
75	0	30
85	1	50
95	2	85

    a. What was the average test grade?
    b. What percent of the class got between 65 and 85?
    c. What test score has a z-value of −2.76?

9. Vera and Judy have both applied for the same job. Vera scored 80 on a state aptitude test where the mean was 70 and the standard deviation was 4.2. Judy scored 510 on the company exam where the mean was 490 and the standard deviation was 10.3. Assuming that the company uses these test results as the only criterion for hiring new employees and that both tests are considered as equal by company officials, who will get the job? Explain your answer.

10. The average weight of a student in a statistics course is 150 pounds with a standard deviation of 4.2 pounds. Find the weights of Lou, Drew, and Sue if $X = \mu + z\sigma$ their z-scores are 2.5, −0.96, and 1.48 respectively.

11. Mary Ruth just entered college and has not decided on a major. She decides to take an aptitude test in order to help her select a major. Her results, as well as the results of the other candidates, are as follows:

Talent	Mean	Standard Deviation	Mary Ruth's Score
Writing	58	3.7	46
Acting	84	7.9	85
Medicine	37	2.8	44
Law	49	4.7	41

a. Transform each of Mary Ruth's scores into a z-score.
b. In which field does she have the most talent?
c. In which field does she have the least talent?

SELF-STUDY GUIDE

In this chapter we saw that an individual score is sometimes meaningless unless it is accompanied by a percentile rank or z-score. When scores are converted into percentile ranks or z-values, we can then make meaningful statements about them and compare them with other scores.

We discussed and demonstrated how to calculate percentile ranks as well as z-scores. The latter play an important role in the normal distribution to be discussed in a later chapter.

At this point you have learned some of the common terms used in statistical analysis and some of the graphic techniques. The formulas are important too. Identify each symbol in the following formulas paying special attention to the relationships among the symbols expressed in the formulas and to the significance of each formula. Become more familiar with these formulas by using them to solve problems.

$$\text{Percentile rank of } X = \frac{B + \frac{1}{2} E}{n} \cdot 100$$

$$z\text{-score} \quad z = \frac{X - \mu}{\sigma}$$

$$\text{Original score } X = \mu + z\sigma$$

You should now be able to demonstrate your knowledge of the following ideas presented in this chapter by giving definitions, descriptions, or specific examples.

Percentile rank (page 84)   z-score (page 88)

Percentile (page 86)   z-value (page 88)

Lower quartile (page 87)   Standard units (page 88)

Upper quartile (page 87)

The tests of the following section will be more useful if you take them after you have studied the examples and solved the exercises given in this chapter.

Form *A*

1. All negative z-scores fall below the ___.
2. A z-score of 0 coincides with the ___.
3. When the difference from the mean is positive in sign, the z-score is
   a. positive *b.* negative *c.* zero
4. In a normal distribution what is the z-score equivalent of the 50th percentile?
5. True or false: Negative terms can never have positive z-scores.
6. If we know the z-score of a term and we wish to calculate the raw score, we multiply the z-score with the standard deviation and add the
   a. mean *b.* variance *c.* median *d.* standard score *e.* none of these
7. In a certain distribution, $\mu = 25$ and $\sigma = 0$. What is the z-score of any term in this distribution?
8. A z-score indicates how many standard deviations a term is above or below the
   a. mean *b.* median *c.* mode *d.* none of these
9. A student took the Educational Testing Service exam and scored 700. If the mean grade is 500 with a standard deviation of 100, the student's z-score is
   a. +1 *b.* +2 *c.* −2 *d.* −4 *e.* none of these
10. What percentage of scores in a distribution are between P_{45} and P_{65}?

Form *B*

1. Joe has a high school average of 86. The college that he wishes to attend will not accept any applicant with a percentile rank below 80. Is Joe sure he will be accepted by this college or is it possible he will be denied admission? Explain your answer.

2. Can a percentile rank of 55 have a negative z-score? Explain your answer. *yes*

3. Verify that the standard deviation of the z-score of Tables 4.1 and 4.2 (page 90)
is 1. *as the mean of the is 0.*

*4. According to police records, the following distribution represents the number of
accidents per 100,000 people in a certain city:

$$P = \dfrac{B + \frac{1}{2}E}{n} \times 100$$

Number of Accidents		Frequency
10–14.9	12.45	4
15–19.9	17.45	9
20–24.9	22.45	7
25–29.9	27.45	2
		22

Find P_{55} for the data.

5. Is it true that the median (the 50th percentile) is the average of the lower quartile
(the 25th percentile) and upper quartile (the 75th percentile)? In other words, is
it true that

no

$$\frac{P_{25} + P_{15}}{2} = P_{50}$$

Explain your answer. *no*

6. The following distribution indicates the range of grades that can be expected on
many intelligence tests. Notice that the scores are normally distributed.

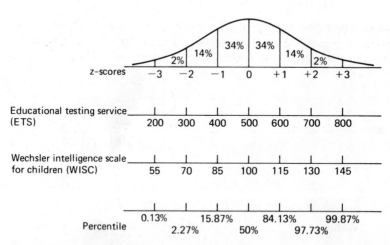

a. If someone scores 400 on the ETS exam, what is the corresponding
percentile rank? *15.87%*

b. What percent of the children taking the WISC exam will score higher than
115? *16%*

c. Find the percentile rank of a score which has a z-value of +2. *97.73*

*7. The following relative frequencies were obtained by entering college students from a certain school on the Scholastic Aptitude Test administered by the Educational Testing Service.

Test Score	Relative Frequency
200–249	0.01
250–299	0.02
300–349	0.06
350–399	0.10
400–449	0.17
450–499	0.19
500–549	0.19
550–599	0.14
600–649	0.08
650–699	0.03
700–749	0.01
750–799	0

Approximately what score must a student have in order to be in the

a. 75th percentile?   554

b. 90th percentile?   612

SUGGESTED READING

Harrell, T. W. and Harrell, M. S. "Army General Classification Test Scores for Civilian Occupations" in *Educational and Psychological Measurement.* 1945, **5,** pp. 225–239.

Mendenhall, W. *Introduction to Probability and Statistics.* 4th ed., Belmont, California: Duxbury Press, 1975.

Mode, E. B. *Elements of Probability and Statistics.* Englewood Cliffs, New Jersey: Prentice-Hall, 1966.

Wonnacott, Thomas and Wonnacott, Ronald J. *Introductory Statistics.* New York: Wiley, 1969.

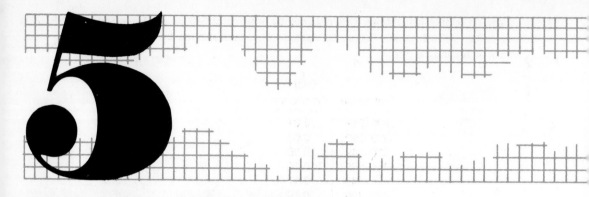

PROBABILITY

U.S. News & World Report,
December 15, 1969. Reprinted
from *U.S. News & World Report,*
Copyright © 1969 U.S. News &
World Report, Inc.

1970 Draft Lottery

Birth-day	January Draft-priority number	Birth-day	February Draft-priority number	Birth-day	March Draft-priority number	Birth-day	April Draft-priority number	Birth-day	May Draft-priority number	Birth-day	June Draft-priority number
1	•305	1	86•	1	108•	1	32•	1	•330	1	•249
2	159•	2	144•	2	29•	2	•271	2	•298	2	•228
3	•251	3	•297	3	•267	3	83•	3	40•	3	•301
4	•215	4	•210	4	•275	4	81•	4	•276	4	20•
5	101•	5	•214	5	•293	5	•269	5	•364	5	28•
6	•224	6	•347	6	139•	6	•253	6	155•	6	110•
7	•306	7	91•	7	122•	7	147•	7	35•	7	85•
8	•199	8	181•	8	•213	8	•312	8	•321	8	•366
9	•194	9	•338	9	•317	9	•219	9	•197	9	•335
10	•325	10	•216	10	•323	10	•218	10	65•	10	•206
11	•329	11	150•	11	136•	11	14•	11	37•	11	134•
12	•221	12	68•	12	•300	12	•346	12	133•	12	•272
13	•318	13	152•	13	•259	13	124•	13	•295	13	69•
14	•238	14	4•	14	•354	14	•231	14	178•	14	•356
15	17•	15	89•	15	169•	15	•273	15	130•	15	180•
16	121•	16	•212	16	166•	16	148•	16	55•	16	•274
17	•235	17	•189	17	33•	17	•260	17	112•	17	73•
18	140•	18	•292	18	•332	18	90•	18	•278	18	•341
19	58•	19	25•	19	•200	19	•336	19	75•	19	104•
20	•280	20	•302	20	•239	20	•345	20	183•	20	•360
21	•186	21	•363	21	•334	21	62•	21	•250	21	60•
22	•337	22	•290	22	•265	22	•316	22	•326	22	•247
23	118•	23	57•	23	•256	23	•252	23	•319	23	109•
24	59•	24	•236	24	•258	24	2•	24	31•	24	•358
25	52•	25	179•	25	•343	25	•351	25	•361	25	137•
26	92•	26	•365	26	170•	26	•340	26	•357	26	22•
27	•355	27	•205	27	•268	27	74•	27	•296	27	64•
28	77•	28	•299	28	•223	28	•262	28	•308	28	•222
29	•349	29	•285	29	•362	29	•191	29	•226	29	•353
30	164•			30	•217	30	•208	30	103•	30	•209
31	•211			31	30•			31	•313		

In December 1969 the United States Selective Service established a priority system for determining which young men would be drafted into the army. Capsules representing each birth date were placed in a drum and selected at random. Those men whose numbers were selected first were almost certain that they would be drafted. The different birthdays were given draft priority numbers as indicated in the clipping on the draft lottery. Supposedly, each birthday had an equally likely

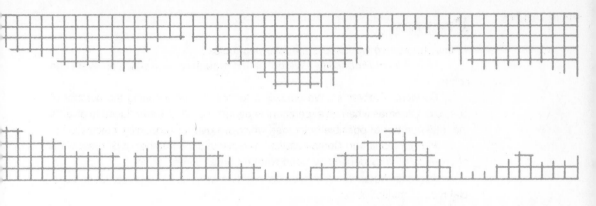

1970 Draft Lottery

Birth-day	July Draft-priority number	Birth-day	August Draft-priority number	Birth-day	September Draft-priority number	Birth-day	October Draft-priority number	Birth-day	November Draft-priority number	Birth-day	December Draft-priority number
1	93•	1	111•	1	•225	1	•359	1	19•	1	129•
2	•350	2	45•	2	161•	2	125•	2	34•	2	•328
3	115•	3	•261	3	49•	3	•244	3	•348	3	157•
4	•279	4	145•	4	•322	4	•202	4	•266	4	165•
5	•188	5	54•	5	82•	5	24•	5	•310	5	56•
6	•327	6	114•	6	6•	6	87•	6	76•	6	10•
7	50•	7	168•	7	8•	7	•234	7	51•	7	12•
8	13•	8	48•	8	•184	8	•283	8	97•	8	105•
9	•277	9	106•	9	•263	9	•342	9	80•	9	43•
10	•284	10	21•	10	71•	10	•220	10	•282	10	41•
11	•248	11	•324	11	158•	11	•237	11	46•	11	39•
12	15•	12	142•	12	•242	12	72•	12	66•	12	•314
13	42•	13	•307	13	175•	13	138•	13	126•	13	163•
14	•331	14	•198	14	1•	14	•294	14	127•	14	26•
15	•322	15	102•	15	113•	15	171•	15	131•	15	•320
16	120•	16	44•	16	•207	16	•254	16	107•	16	96•
17	98•	17	154•	17	•255	17	•288	17	143•	17	•304
18	•190	18	141•	18	•246	18	5•	18	146•	18	128•
19	•227	19	•311	19	177•	19	•241	19	•203	19	•240
20	•187	20	•344	20	63•	20	•192	20	•185	20	135•
21	27•	21	•291	21	•204	21	•243	21	156•	21	70•
22	153•	22	•339	22	160•	22	117•	22	9•	22	53•
23	172•	23	116•	23	119•	23	•201	23	182•	23	162•
24	23•	24	36•	24	•195	24	•196	24	•230	24	95•
25	67•	25	•286	25	149•	25	176•	25	132•	25	84•
26	•303	26	•245	26	18•	26	7•	26	•309	26	173•
27	•289	27	•352	27	•233	27	•264	27	47•	27	78•
28	88•	28	167•	28	•257	28	94•	28	•281	28	123•
29	•270	29	61•	29	151•	29	•229	29	99•	29	16•
30	•287	30	•333	30	•315	30	38•	30	174•	30	3•
31	•193	31	11•			31	79•			31	100•

probability of being selected. However, by analyzing the numbers carefully, we find that the majority of those birthdays that occurred later in the year had a higher priority number than those that occurred earlier in the year. Did each birthday have an equal probability of being selected?

The other newspaper clipping gives a list of the winning lottery numbers. If you buy a lottery ticket, what is the probability that you will win the lottery?

Preview

In this chapter we will discuss the following topics:

1. *Definition of Probability*   We define the probability with which an event can occur.

2. *Counting Problems*   We discuss a formula for determining the number of possible outcomes when an experiment is performed. This is introduced to give us the total number of possible outcomes which we use for probability calculations.

3. *Permutations and Combinations*   We discuss the number of different ways of arranging things depending upon whether order counts or not.

4. *Factorial Notation*   A convenient notation which is used to represent a special type of multiplication.

5. *Pascal's Triangle*   A computational device for calculating the number of possible combinations.

6. *Odds and Mathematical Expectation* These words, often used by gamblers, represent the payoff for a situation and the likelihood of obtaining it.

Introduction

Although the word *probability* may sound strange to you, it is not as unfamiliar as you may think. In everyday life situations we frequently make decisions and take action as a result of the probability of certain events. Thus, if the weatherman forecasts rain with a probability of 80%, we undoubtedly would prepare ourselves accordingly.

Let us, however, analyze the weatherman's forecast. What he really means is that based upon past records, 80% of the time when the weather conditions have been as they are today, rain has followed. Thus, his probability calculations and resultant forecasts are based upon past records. They are based upon the assumption that since in the *past* rain has occurred a certain percentage of the time, it will occur the same percentage of times in the *future*. This is but one usage of probability. It is based on **relative frequency.** We will explain this idea in greater detail shortly.

Probability is also used in statements which express a personal judgment or conviction. This can be best illustrated by the following statements: "If the United States had not dropped the atomic bomb on Japan, World War II would *probably* have lasted several more years" or "If all the New York Mets players had been healthy the entire season, they *probably* would have won the pennant last year."

Probability can also be used in other situations. For example, if a fair coin* is tossed, we would all agree that the probability is 1/2 that heads comes up. This is because there are only two possible outcomes when we flip a coin, heads or tails.

Now consider the following conversation overheard in a student cafeteria.

---

*A fair coin is a coin which has the same chance of landing on heads as on tails. Throughout this book we will always assume that we have fair coins unless told otherwise.

Bill: I am going to cut math today.

Eric: Why?

Bill: I didn't do my homework.

Eric: So what? I didn't either.

Bill: Since the teacher calls on at least half of the class each day for answers, he will *probably* call on me today and find out that I am not prepared.

Eric: The teacher called on me yesterday, so he *probably* won't get me today. I am going to class.

In the preceding situation each student is making a decision based upon probability.

Since probability has so many possible meanings and uses, we will first analyze the nature of probability and how to calculate it. This will be done in this chapter. In the next chapter we will discuss various rules which allow us to calculate probabilities for many different situations.

Historically, probability had its origin in the gambling rooms. As mentioned in Chapter 1, the Chevalier de Méré started the mathematical theory of probability in 1654 through his correspondence with Pascal and Fermat. More than three centuries ago the great Italian scientist Galileo was asked to explain why a throw of three dice turns up a sum of nine less often than a sum of ten. Over the course of many years probability theory has left the gambling rooms and has grown to be an important and ever expanding branch of mathematics.

5.1 DEFINITION OF PROBABILITY

Probability theory can be thought of as that branch of mathematics that is concerned with calculating the probability of outcomes of experiments.

Since many ideas of probability were derived from gambling situations, let us consider the following experiment. An honest die (the plural is dice) was rolled many times and the number of 1s that came up were recorded. The results are

Number of 1s that came up:	1	11	18	99	1001	10,001
Number of rolls of the die:	6	60	120	600	6000	60,000

Notice that in each case the number of 1s that appeared is approximately 1/6 of the total number of tosses of the die. It would then be reasonable to conclude that the probability of a 1 appearing is 1/6.

Although when a die is rolled there are six equally likely possible outcomes if it is an honest die (see Figure 5.1), we are concerned with the number of 1s appearing. Each time a 1 appears, we call it a **favorable outcome.** There are six possible outcomes of which only one is favorable. The probability is thus the number of favorable outcomes divided by the total number of possible outcomes which is 1 divided by 6, or 1/6. The preceding chart indicates that our guess, that the probability is approximately 1/6, is correct.

Figure 5.1

Possible outcomes when die is rolled once.

PROBABILITY

Similarly if a coin is tossed once, we would say that the probability of getting a head is 1/2 since there are two possible outcomes, heads and tails, and only one is favorable. These are the only two possible outcomes in this case. All the possible outcomes of an experiment are referred to as the **sample space** of the experiment. We will usually be interested in only some of the outcomes of the experiment. The outcomes which are of interest to us will be referred to as an **event.** Thus, in flipping a coin once, the sample space is Heads or Tails, abbreviated as H,T. The event of interest is H.

In the rolling of a die the sample space consists of six possible outcomes: a 1, a 2, a 3, a 4, a 5, and a 6. We may be interested in the event, "getting a 1."

If we toss a coin twice, the sample space is HH, HT, TH, and TT. There are four possibilities. In this abbreviated notation, HT means heads on the first toss and tails on the second toss, whereas TH means tails on the first toss and heads on the second toss. The event "getting a head on both tosses" is denoted by HH. The event "no head" is TT.

To further illustrate the idea of sample space and event, consider the following examples.

EXAMPLE 1

Two dice are rolled at the same time. Find the sample space.

Figure 5.2

Thirty-six possible outcomes when two dice are rolled at the same time.

SOLUTION

There are 36 possible outcomes as pictured in Figure 5.2.

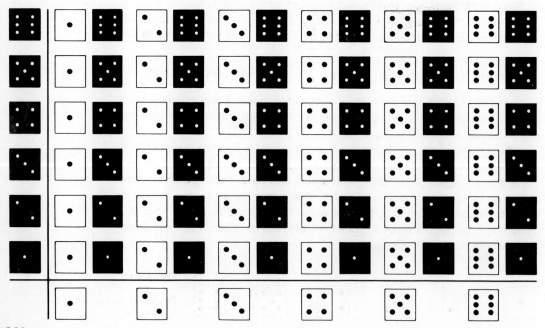

The possible outcomes are summarized as follows:

Die 1	Die 2	Die 1	Die 2	Die 1	Die 2	Die 1	Die 2	Die 1	Die 2	Die 1	Die 2
1	1	2	1	3	1	4	1	5	1	6	1
1	2	2	2	3	2	4	2	5	2	6	2
1	3	2	3	3	3	4	3	5	3	6	3
1	4	2	4	3	4	4	4	5	4	6	4
1	5	2	5	3	5	4	5	5	5	6	5
1	6	2	6	3	6	4	6	5	6	6	6

The event "sum of 7 on both dice together" can happen in six ways. These are

Die 1	Die 2	Die 1	Die 2
1	6	3	4
6	1	5	2
4	3	2	5

Similarly the event "sum of 9 on both dice together" can happen in four ways. The event "sum of 13 together" can happen in zero ways.

EXAMPLE 2

Two contestants, Jack and Jill, are on a quiz show. Each is in a soundproof booth and cannot hear what the other is saying. They are each asked to select a number from 1 to 3. They each win $10,000 if they select the same number. In how many different ways can they win the prize?

SOLUTION

They will win the prize if they both say 1, 2 or 3. The sample space for this experiment is

Jill guesses	Jack guesses	Jill guesses	Jack guesses	Jill guesses	Jack guesses
1	1	2	1	3	1
1	2	2	2	3	2
1	3	2	3	3	3

Thus the event "winning the prize" can occur in three possible ways.

We now define formally what we mean by probability.

DEFINITION 5.1

If an event can occur in n equally likely ways and if f of these ways are considered favorable, then the **probability** of getting a favorable outcome is

$$\frac{\text{Number of favorable outcomes}}{\text{Total number of outcomes}} = \frac{f}{n}$$

Thus, the probability of any event equals the number of favorable outcomes divided by the total number of possible outcomes.

We use the symbol $p(A)$ to stand for "the probability of event A."

Comment   Definition 5.1 is but one of the several definitions of probability that exist. It is based upon the assumption that if an event has probability f/n, then in the long run there will be f favorable outcomes of the event out of n trials. For this reason probability is referred to as the **relative frequency** of the event since it represents the percentage of times that the event will happen in repeated experiments.

When we say that the probability of getting heads is 1/2 when tossing a coin, we mean that in the long run the number of heads appearing divided by the total number of tosses will be approximately 1/2. Similarly, if the probability of getting a 1 when an honest die is rolled is 1/6, then in the long run the number of times that a 1 will appear divided by the total number of rolls will be approximately 1/6. Thus, the probability of an outcome can be thought of as the fraction of times that the outcome will occur in a long series of repetitions of the experiment. This is why this idea is sometimes referred to as the relative frequency concept of probability. Probability can also be defined from a strictly mathematical, that is, axiomatic, point of view; however, this is beyond the scope of this text.

Let us now illustrate the concept of probability with several examples.

EXAMPLE 3

A family has three children. What is the probability that all three children are girls?

SOLUTION

We will use Definition 5.1. We first find the total number of ways of having three children, that is, the sample space. There are eight possibilities as shown in the following table:

Child 1	Child 2	Child 3
Boy	Boy	Boy
Boy	Girl	Girl
Boy	Girl	Boy
Boy	Boy	Girl
Girl	Boy	Boy
Girl	Girl	Boy
Girl	Boy	Girl
Girl	Girl	Girl

Of these only one is favorable, namely the outcome Girl, Girl, Girl. Thus,

$$p(3 \text{ girls}) = 1/8$$

EXAMPLE 4

A card is selected from an ordinary deck of 52 cards. What is the probability of getting

a. a queen?
b. a diamond?
c. a black card?
d. a picture card?
e. the king of clubs?

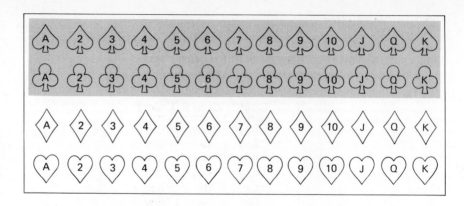

Figure 5.3

The sample space when a card is drawn from an ordinary deck of fifty-two cards.

SOLUTION

Since there are 52 cards in the deck, the total number of outcomes is 52. This is shown in Figure 5.3.

a. As shown in Figure 5.3, there are 4 queens in the deck, so there are 4 favorable outcomes. Definition 5.1 tells us that

$$p(\text{queen}) = \frac{4}{52} = \frac{1}{13}$$

b. There are 13 diamonds in the deck, so there are 13 favorable outcomes. Therefore,

$$p(\text{diamonds}) = \frac{13}{52} = \frac{1}{4}$$

c. Since a black card can be either a spade or a club, there are 26 black cards in the deck as shown in Figure 5.3. Therefore,

$$p(\text{black card}) = \frac{26}{52} = \frac{1}{2}$$

d. There are 12 picture cards (4 jacks, 4 queens, and 4 kings), so there are 12 favorable outcomes. Therefore,

$$p(\text{picture card}) = \frac{12}{52} = \frac{3}{13}$$

e. There is only one king of clubs in a deck of 52 cards. Thus,

$$p(\text{king of clubs}) = \frac{1}{52}$$

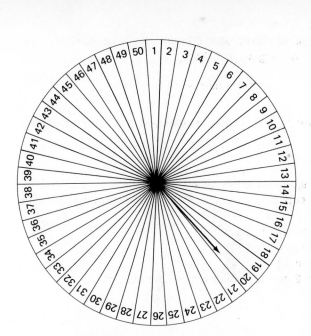

Figure 5.4

Roulette wheel.

EXAMPLE 5

A roulette wheel has the numbers 1 through 50 painted on it as
shown in Figure 5.4. Tickets numbered 1 through 50 have been
sold. The wheel will be rotated and the number on which the pointer
lands will be the winning number. If the pointer stops on the line
between two numbers, the wheel must be turned again. A prize of a
new car will be awarded to the person whose ticket number
matches the winning number. What is the probability that

a. ticket number 42 wins?

b. a ticket between the numbers 26 and 39 wins?

c. ticket number 51 wins?

SOLUTION

Since 50 tickets were sold, the total number of possible outcomes,
that is, the sample space, is 50.

a. Only 1 ticket numbered 42 was sold. There is only one favorable
outcome. Thus,

$$p(\text{ticket number 42 wins}) = \frac{1}{50}$$

b. There are 12 ticket numbers between 26 and 39, not including
26 and 39. Thus,

$$p(\text{a ticket between the numbers 26 and 39 wins}) = \frac{12}{50} = \frac{6}{25}$$

107

c. Since tickets numbered up to 50 were sold, ticket number 51 can never be a winning number. Thus,

$$p(\text{ticket number 51 wins}) = 0$$

An event that can never happen is called a **null event** and its probability is 0.

EXAMPLE 6

Two fair dice are rolled at the same time and the number of dots appearing on both dice is counted. Find the probability that this sum
a. is 7.
b. is an odd number larger than 6.
c. is less than 2.
d. is more than 12.
e. is between 2 and 12, including these two numbers.

SOLUTION

When two dice are rolled, there are 36 possible outcomes, that is, the sample space has 36 possibilities. These were listed in Example 1 on page 103.
a. The sum of 7 on both dice together can happen in 6 ways so that

$$p(\text{sum of 7}) = \frac{6}{36}, \text{ or } \frac{1}{6}$$

b. The statement "a sum which is an odd number larger than 6" means a sum of 7, a sum of 9, or a sum of 11. A sum of 7 on both dice together can happen in 6 ways. Similarly, a sum of 9 on both dice together can happen in 4 ways, and a sum of 11 can happen in 2 ways. There are then 12 favorable outcomes out of 36 possibilities. Thus,

$$p \text{ (a sum which is an odd number larger than 6)} = \frac{12}{36} = \frac{1}{3}$$

c. When two dice are rolled, the minimum sum is 2, and we cannot obtain a sum which is less than 2. There are *no* favorable events. This is the null event. Hence,

$$p(\text{a sum which is less than 2}) = 0$$

d. When two dice are rolled, the maximum sum is 12. We cannot obtain a sum which is more than 12. This is the null event. Thus,

$$p(\text{a sum which is more than 12}) = 0$$

e. When two dice are rolled we *must* obtain a sum which is between 2 and 12, including the numbers 2 and 12. There are 36 possible outcomes and *all* these are favorable. Thus,

$$p(\text{a sum between 2 and 12, including 2 and 12}) = \frac{36}{36} = 1$$

Therefore, a favorable outcome *must* occur in this case.

An event that is certain to occur is called the **certain event** or **definite event** and its probability is 1.

Comment Any event, call it A, may or may not occur. If it is sure to occur, we have the certain event and its probability is 1. If it will never occur, we have the null event and its probability is 0. Thus, if we are given any event A, then we know that its probability *must* be between 0 and 1 and possibly equal to 0 or 1. This is because the event may or may not occur. Probability can *never* be a negative number.

Comment We mentioned earlier that probability can be thought of as the fraction of times that an outcome will occur in a long series of repetitions of an experiment. However, there may be certain experiments which cannot be repeated. For example, if Gary's kidney has to be removed surgically, we cannot think of this as an experiment that can be repeated over and over again, at least as far as Gary is concerned. How do we assign probabilities in this case? This is not an easy task. Calculating the probability in such situations requires the judgment and experience of a doctor familiar with *many* experiments of a similar type. Thus, if doctors tell you that you have an 80% chance of surviving the operation, they mean that based upon their previous experiences with such situations, 80% of the patients with similar operations have survived. Usually an experienced surgeon can assign a fairly reasonable probability to the success of a nonrepeatable operation.

EXERCISES

1. A die is rolled once.
   a. List all possible outcomes.
   b. Find the probability that an even number comes up.
   c. Find the probability that the number that shows is larger than 4.
   d. Find the probability of getting a 6.
2. Two dice are rolled. What is the probability that
   a. the same number appears on both dice? 6/36
   b. two fives appear? 1/36
   c. the sum will be an odd number? 18/36
   d. the sum will be a number larger than 8? 10/36
   e. the number appearing on each die is even? 9/36

3. Two dice are altered by painting an additional dot on each face that originally had only 1 dot. If these changed dice are now rolled once,
   a. list all possible sums that can be obtained with these dice. 4 – 12 incl.
   b. find the probability of rolling a sum of 2. 0
   c. find the probability of rolling a sum between 4 and 12, including 4 and 12. 36/36
   d. find the probability that the sum is an even number. 20/36

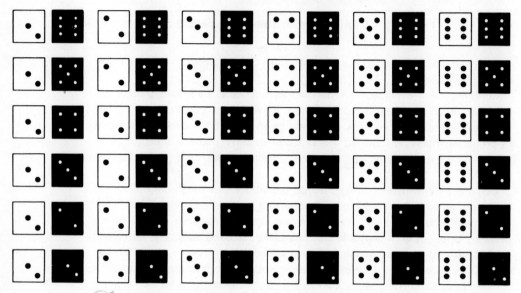

4. A card is drawn from a deck of 52 cards. What is the probability that it is
   a. the jack of hearts? 1/52
   b. a queen? 4/52
   c. a club? 13/52
   d. a card numbered between 5 and 10, including 5 and 10? 24/52

5. A nurse has collected 200 pints of blood from student volunteers and stores them in 200 identical pint bottles. The bottles are not labeled. A bottle is selected at random. If it is known that 3 of the bottles contain contaminated blood, find the probability that the selected bottle is *not* contaminated. 197/200

6. A woman has 7 nickels, 4 pennies, 3 dimes, and 6 quarters in her purse. Her young child selects a coin at random from her purse. Find the probability that the coin selected is
   a. a penny 4/20 b. a nickel 7/20 c. a dime 3/20 d. a half-dollar. 0

7. A man and a woman get into an elevator and can get off at any floor from 1 to 6. What is the probability that they both get off on the same floor? 6/36

8. A secretary types 3 letters and addresses 3 envelopes. Before inserting each letter into the appropriate envelope, she drops them all on the floor. If she picks up all the letters and envelopes and inserts a letter at random (without looking) into an envelope, what is the probability that each letter will be inserted in its correct envelope? *Hint:* Find the sample space.

1/6

see notes

9. The following table indicates the principal cause of accidental deaths in the United States during 1973:

Cause	Number
Motor vehicle	55,800
Falls	16,900
Burns	6,400
Drowning	8,700
Firearms	2,700
Poison gases	1,500
Other poisons	3,700
Total =	95,700

Source: Division of Vital Statistics, National Center for Health Statistics.

What is the probability that an accident victim died as a result of
a. falls b. burns c. a machinery accident

10. Robert received the names of 5 girls for possible blind dates. Since he does not know any of the girls, he decides to use a spinner to determine which girl to date. Let's assume that he will date only these girls. Find the probability that he

a. dates Laura.
b. dates Mary.
c. dates Laura or Mary.
d. does not date Laura or Mary.
e. dates Betty.

111

11. Which of the following numbers cannot be the probability of some event?
    a. −1/2
    b. 1.32
    c. 6/7
    d. 0
    e. −1
    f. 0.93

12. The following is a breakdown of the number of people in the various ranks in the mathematics department at Binomial University:

Rank	Number of Males	Females	
Instructor	3	5	8
Assistant Professor	12	7	19
Associate Professor	6	9	15
Professor	5	8	13 / 55

What is the probability that an employee selected at random from the math department will be
    a. a male associate professor? 6/55
    b. a female professor? 8/55
    c. a professor? 13/55
    d. a male? 26/55
    e. a lecturer? 0

13. The Smiths have 4 children. What is the probability that they have 2 boys and 2 girls? (*Hint:* Find the sample space.) 6/16

14. Four jets (a TWA jet, a United airlines jet, an Eastern airlines jet and a Pan American jet) are circling Kennedy Airport waiting to land. What is the probability that they land in the following order?

> first plane to land– Eastern
> second plane to land– TWA
> third plane to land– Pan American
> fourth plane to land– United

15. Bob is taking a true–false test and has no idea of the answers to the last three questions. He decides to guess at the answers.
    a. If c stands for a correct answer and w stands for a wrong answer, list all the possible outcomes. see notebook
    b. Find the probability that he guesses all three answers correctly. 1/8
    c. Find the probability that at least one of his answers is correct. 7/8
    d. Find the probability that all three answers are wrong. 1/8

16. A slot machine has 3 wheels and each wheel has a 0, 1, and 2 painted on it.
When a button is pushed, each wheel will display a number. Each wheel
operates independently. The complete number shown is the amount of money
won. Thus if the wheels together show 120, $120 is won.

   a. List all the possible outcomes for this machine.
   b. Find the probability of winning $111.
   c. Find the probability of winning nothing.
   d. Find the probability of winning at least one dollar.

17. Five musicians have been nominated for awards for their contributions to the
field of music. Their names are Roy Rodriguez, Terry Jones, Mary Simmons,
Johnny Gash, and Pat Baboon. If only 3 of these musicians will actually
receive an award, what is the probability that the winners are Mary, Johnny
and Terry? R, T, M, J, B

RTM
RmJ
RJB
RTJ
RTB
TmJ
TJB
mJB

see notes

113

5.2 COUNTING PROBLEMS

In determining the probability of an event, we must first know the total number of possible outcomes. In many situations it is a rather simple task to list all the possible outcomes and then to determine how many of these are favorable. In other situations there may be so many possible outcomes that it would be too time consuming to list all of them. Thus when two dice are rolled, there are 36 possible outcomes. These are listed on page 103. Exercise 16 of Section 5.1 has 27 possible outcomes. When there are too many possibilities to list, we can use rules, which will be given shortly, to determine the actual number.

One technique that is sometimes used to determine the number of possible outcomes is to construct a **tree diagram.** The following examples will illustrate how this is done.

EXAMPLE 1

By means of a tree diagram we can determine the number of possible outcomes when a coin is repeatedly tossed. If one coin is tossed, there are two possible outcomes, heads or tails, as shown in the tree diagram in Figure 5.5. If two coins are tossed, there are now four possible outcomes. These are HH, HT, TH, and TT, since

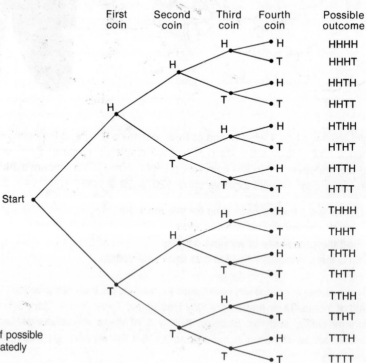

Figure 5.5

Tree diagram of the number of possible outcomes when a coin is repeatedly tossed.

each of the possible outcomes on the first toss can occur with each of the two possibilities on the second toss. So, if heads appeared on the first toss, we may get heads or tails on the second toss. The same is true if tails appeared on the first toss. The tree diagram shows that there are four possibilities. When three coins are tossed, the diagram shows that there are eight possible outcomes. Also, there are sixteen possible outcomes when four coins are tossed.

EXAMPLE 2

Hedda Lettuce is about to order dinner in the student cafeteria. She can choose any one of three main courses and any one of four desserts:

Main Course	Dessert
Hamburgers	Hot-fudge Sundae
Steak	Jello
Southern-fried Chicken	Cake
	Fruit

Using a tree diagram, find all the possible dinners that Hedda can order.

SOLUTION

Hedda may order any one of the three dishes as a main course. With each main course she may order any one of four desserts. These possibilities are pictured in Figure 5.6. The diagram shows that there are twelve possible meals that Hedda can order.

Comment Although counting the number of possible outcomes by using a tree diagram is not difficult when there are only several possibilities, it becomes very impractical to construct a tree when there are many possibilities. For example, if 10 coins are tossed, there are 1024 different outcomes. Similarly, if a die is rolled 4 times, there are 1296 different outcomes. For situations such as these we need a rule to help us determine the number of possible outcomes.

Before stating the rule, however, let us analyze the following situations. When one die is rolled, there are six possible outcomes, 1, 2, 3, 4, 5, 6 When a second die is rolled, there are 6 × 6, or 36, possible outcomes. These are listed on page 103.

If again we analyze Exercise 16 of Section 5.1, we find that the first wheel can show 0, 1, or 2. This gives us three possibilities. Similarly, the second wheel can also show 0, 1, or 2. The same is true for the third wheel. This gives us 3 possibilities for wheel 1, 3 possibilities for wheel 2, and 3 possibilities for wheel 3. We then have a total of 27 possible outcomes since

$$3 \times 3 \times 3 = 27$$

This leads us to the following rule.

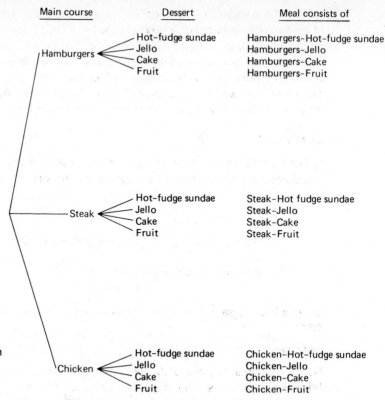

Main course   Dessert   Meal consists of

Hamburgers — Hot-fudge sundae — Hamburgers-Hot-fudge sundae
Jello — Hamburgers-Jello
Cake — Hamburgers-Cake
Fruit — Hamburgers-Fruit

Steak — Hot-fudge sundae — Steak-Hot fudge sundae
Jello — Steak-Jello
Cake — Steak-Cake
Fruit — Steak-Fruit

Chicken — Hot-fudge sundae — Chicken-Hot-fudge sundae
Jello — Chicken-Jello
Cake — Chicken-Cake
Fruit — Chicken-Fruit

Figure 5.6

Tree diagram of the possible dinners that can be ordered from a choice of any one of three main courses and any one of four desserts.

RULE

If one thing can be done in *m* ways, and if after this is done, something else can be done in *n* ways, then both things can be done in a total of *m · n* different ways in the stated order.

EXAMPLE 3

A geology teacher plans to travel from New York to Florida and then on to Mexico to collect rock specimens for his class. From New York to Florida he can travel by train, airplane, boat, or car. However, from Florida to Mexico he cannot travel by train. In how many different ways can he make the trip?

SOLUTION

Since he can travel from New York to Florida in 4 different ways and from Florida to Mexico in 3 different ways, he can make the trip in 4 × 3, or 12, different ways.

EXAMPLE 4

In a certain state, license plates have 3 letters followed by 2 digits. If the first digit cannot be 0, how many different license plates can be made if
a. repetitions of letters or numbers are allowed?
b. repetitions of letters are not allowed?

SOLUTION

There are 26 letters and 10 possible digits (0, 1, 2, . . . , 9).
a. If repetitions are allowed, the same letter can be used again. Since 0 cannot be used as the first digit, the total number of different license plates is

$$26 \times 26 \times 26 \times \textcircled{9} \times 10 = 1,581,840$$

Thus, 1,581,840 different license plates are possible. Note that the circled position has only 9 possibilities. Why?
b. If repetition of letters is not allowed, there are 26 possibilities for the first letter, but only 25 possibilities for the second letter since once a letter is used it may not be used again. For the third letter there are only 24 possibilities. There are then a total of 1,404,000 different license plates since

$$26 \times 25 \times 24 \times 9 \times 10 = 1,404,000$$

EXAMPLE 5

In Example 1 (see page 114) there are 2 possible outcomes for the first toss, 2 possibilities for the second toss, and 2 possibilities for the third toss. Thus there are a total of 8 possible outcomes when three coins are tossed or when one coin is tossed three times since

$$2 \times 2 \times 2 = 8$$

This is the same result we obtained using tree diagrams. It is considerably easier to do it this way.

EXERCISES

1. A certain model car comes with one of three possible engine sizes and with or without air conditioning. Furthermore, it is equipped with automatic or standard transmission. In how many different ways can a buyer select a car?

$$3 \times 2 \times 2 = 12$$

2. In the student cafeteria at Nutrition University the following items can be purchased from the vending machines:

Juice	Sandwich	Beverage
Orange	Tuna fish	Pepsi
Apple	Cheese	Coke
Tomato	Ham	Orange Soda
Pineapple		Tea
		Coffee

$4 \times 3 \times 5 =$

60

If a student selects one item from each category, how many different meals can be obtained?

3. In a certain college all liberal arts students must successfully pass one course chosen from each of the following categories in order to graduate:

Category A	Category B	Category C
Math 1	Introduction to Ecology	Art of Self-defense
Physics 10	Family Planning	Home Economics
Chemistry 100	Conservation	Child Care
Astronomy 12		Basket Weaving
Geology 17		Needlepoint
Computer 5		Belly Dancing

$6 \times 3 \times 6 = 108$

In how many different ways can a student satisfy the college requirement?

4. In how many different ways can the letters of the word DRUGS be arranged if
   a. repetition is allowed? 3125
   b. repetition is not allowed? 120

   Note: Each arrangement does not have to form a word.

5. Construct a tree diagram showing all the possibilities, male or female, for a family that has five children.

6. In the baseball playoffs the first team that wins three out of five games is the league winner. Make a tree diagram showing the possible ways in which the playoffs can end.

7. Five people are standing in line waiting to file their 1976 Federal Income Tax return. In how many different ways can they stand in line? $5 \times 4 \times 3 \times 2 \times 1 = 120$

8. How many different numbers greater than 3000 can be formed with the digits 2, 3, 5, and 9 if no repetitions are allowed? 18

9. In a certain state, license plates consist of five numbers followed by a letter. How many different license plates can be formed if
   a. the first number must be a 3 (repetitions are allowed)? $260,000$
   b. no digits can be repeated? $786,240$
   c. there are no restrictions at all? $2,600,000$

10. Melba is getting dressed. She can select any one of four pairs of slacks and any one of three blouses depending upon their color as shown below. Make a tree diagram showing the possible outfits that she may select.

Slacks	Blouse
Black	White print
White	Red print
Blue	Black
Red	

11. In an effort to protect the consumer, the agriculture department of a certain state requires that all canned foods manufactured with in the state be stamped with numbers or letters as follows:

a. The company number as assigned by the department; there are 9 companies within the state. $9 \times 8 = 72$

b. The plant location within the state; the state is divided into 8 geographical areas labeled A, B, C, D, E, F, G, or H. $72 \times 12 =$

c. The date of manufacture: month and last 2 digits of year.   12

For example, a can with the code 1A0277 on it means that it was manufactured by company 1 located in area A during the month of February 1977. Using the above scheme, how many different codes are possible for cans manufactured during the years 1976–1978?   3   2592

12. How many different 9-digit social security numbers are there? If a tenth digit is added, how many different social security numbers will there be? Assume that there are no restrictions at all.

13. Two cards are to be selected from a deck of 52. The first card is not replaced before the second card is drawn. In how many different ways can the two cards be selected if

a. both must be diamonds? $13 \times 12 = 156$

b. the first card must be a club and the second must be a face card? $13 \times 11 = 153$

c. neither one of the cards can be an ace? 48×47 2256 ways

club no   club
   face   face

| 10 | 12 | = 120 |
| 3 | 11 | = 33 |

120
33
153

5.3 PERMUTATIONS

Consider the following situation. A gas station is running short of gas and has only enough gas to fill up a car, a bus, and a truck. In how many different ways can the three vehicles line up for gas? There are six ways. These are as follows:

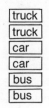

truck	car	bus
truck	bus	car
car	truck	bus
car	bus	truck
bus	truck	car
bus	car	truck

C B T
B C T
T C B
B T C
T B C
C T B

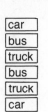

119

Notice that in each arrangement order is important. Thus the arrangement truck car bus means that the truck is filled first, then the car, and finally the bus. This leads us to a very useful idea in mathematics known as **permutations.** It is defined in Definition 5.2.

DEFINITION 5.2

A **permutation** is any arrangement of objects in a particular order.

EXAMPLE 1

How many different permutations can be formed from the letters of the word SEX?

SOLUTION

There are six permutations. These are

SEX, SXE, XSE, XES, EXS, ESX

EXAMPLE 2

How many different three member committees can be formed among Mabel, Helen, Maurice, and Ben if in each committee the first person selected is to be the chairperson and the second person selected is to be the secretary?

SOLUTION

There are 24 such committees, and we list them as follows:

(1) Mabel, Helen, Ben	(2) Helen, Ben, Mabel
(3) Mabel, Ben, Helen	(4) Helen, Mabel, Ben
(5) Ben, Mabel, Helen	(6) Ben, Helen, Mabel
(7) Maurice, Mabel, Helen	(8) Maurice, Helen, Mabel
(9) Helen, Maurice, Mabel	(10) Helen, Mabel, Maurice
(11) Mabel, Maurice, Helen	(12) Mabel, Helen, Maurice
(13) Ben, Maurice, Helen	(14) Ben, Helen, Maurice
(15) Helen, Ben, Maurice	(16) Helen, Maurice, Ben
(17) Maurice, Ben, Helen	(18) Maurice, Helen, Ben
(19) Ben, Mabel, Maurice	(20) Ben, Maurice, Mabel
(21) Mabel, Ben, Maurice	(22) Mabel, Maurice, Ben
(23) Maurice, Ben, Mabel	(24) Maurice, Mabel, Ben

In each of these committees, that is, permutations, order is important since the first person selected is to be chairperson and the second is to be secretary. Thus, the committee Ben, Maurice, Helen for example is different from the committee Ben, Helen, Maurice.

Perhaps you are wondering how we knew that there would be 24 permutations. Are there any formulas that can be used to determine the number of possible permutations? The answer is yes, but before discussing such formulas we introduce some special symbolism. This is the **factorial notation.** In this notation we use symbols such as $n!$ and read this as n factorial. We use this notation when we wish to multiply a series of numbers together starting from a given number and multiplying this by the number which is 1 less than that and so on until we get to the number one, which is where we stop. Thus, 3!, read as 3 factorial, means $3 \times 2 \times 1$.

Therefore,
$$3! = 3 \times 2 \times 1 = 6$$

Also
$$5! \text{ means } 5 \times 4 \times 3 \times 2 \times 1, \text{ or } 120$$
$$4! \text{ means } 4 \times 3 \times 2 \times 1, \text{ or } 24$$
$$6! \text{ means } 6 \times 5 \times 4 \times 3 \times 2 \times 1, \text{ or } 720$$

In this notation, 0! is defined as 1.

Let us now return to Example 2 of this section. In this example we were interested in the number of three member committees that can be formed among four people. This actually represents the number of permutations of 4 things taken 3 at a time or the number of possible permutations of 3 things that can be formed out of 4 possible things. For situations such as this one and similar ones we introduce the notation $_4P_3$ which will stand for the number of permutations of 3 things out of 4 possible things. Similarly, $_6P_4$ will represent the number of possible permutations of 6 things taken 4 at a time. More generally, $_nP_r$ means the number of possible permutations of n things taken r at a time.

To determine how many possible permutations there actually are for n things taken r at a time, we use the following convenient formula.

FORMULA 5.1 $\qquad _nP_r = \dfrac{n!}{(n-r)!}$

EXAMPLE 3

a. Find $_6P_4$. b. Find $_7P_5$. c. Find $_5P_5$.

SOLUTION

a. $_6P_4$ means the number of possible permutations of 6 things taken 4 at a time. Using Formula 5.1 we have $n = 6$, $r = 4$, and $n - r = 6 - 4 = 2$. Thus,

$$_6P_4 = \frac{6!}{(6-4)!} = \frac{6!}{2!} = \frac{6 \cdot 5 \cdot 4 \cdot 3 \cdot 2 \cdot 1}{2 \cdot 1} = \frac{6 \cdot 5 \cdot 4 \cdot 3 \cdot \cancel{2} \cdot \cancel{1}}{\cancel{2} \cdot \cancel{1}}$$
$$= 6 \cdot 5 \cdot 4 \cdot 3 = 360$$

Thus, $_6P_4 = 360$.

b. $_7P_5$ means the number of permutations of 7 things taken 5 at a time, so that $n = 7$ and $r = 5$. Using Formula 5.1 we have

$$_7P_5 = \frac{7!}{(7-5)!} = \frac{7!}{2!} = \frac{7 \cdot 6 \cdot 5 \cdot 4 \cdot 3 \cdot 2 \cdot 1}{2 \cdot 1}$$

$$= \frac{7 \cdot 6 \cdot 5 \cdot 4 \cdot 3 \cdot 2 \cdot 1}{2 \cdot 1}$$

$$= 7 \cdot 6 \cdot 5 \cdot 4 \cdot 3 = 2520$$

Thus, $_7P_5 = 2520$.

c. $_5P_5$ means the number of permutations of 5 things taken 5 at a time, so that $n = 5$ and $r = 5$. Using formula 5.1 we have

$$_5P_5 = \frac{5!}{(5-5)!} = \frac{5!}{0!}$$

$$= \frac{5!}{1} \quad \text{(Since 0! is 1)}$$

$$= 5! = 5 \cdot 4 \cdot 3 \cdot 2 \cdot 1$$

$$= 120$$

Thus, $_5P_5 = 5! = 120$.

Comment The number of permutations of 5 things taken 5 at a time is represented by $_5P_5$. This really means the number of different ways of arranging 5 things. Our answer turned out to be 5!. This leads us to Formula 5.2.

FORMULA 5.2

The number of possible permutations of n things taken n at a time, denoted as $_nP_n$, is $n!$.

EXAMPLE 4

In Example 1 on page 120 we were asked to calculate the number of different permutations that can be formed from the three letter word SEX. This actually represents the number of possible permutations of 3 letters taken 3 at a time. Formula 5.2 tells us that there are 3!, or 6, possible permutations since

$$3! = 3 \cdot 2 \cdot 1 = 6$$

These were listed on page 120.

Suppose a librarian has 2 identical algebra books and 3 identical geometry books to be shelved. In how many different ways can this be done? Since there are 5 books altogether and all have to be shelved, we are tempted to use Formula 5.2 or Formula 5.1. Unfortunately since the 2 algebra books are identical, we cannot tell them apart. There are actually 10 possible permutations. These are:

algebra, algebra, geometry, geometry, geometry
algebra, geometry, geometry, geometry, algebra
algebra, geometry, algebra, geometry, geometry
algebra, geometry, geometry, algebra, geometry
geometry, geometry, geometry, algebra, algebra
geometry, geometry, algebra, algebra, geometry
geometry, geometry, algebra, geometry, algebra
geometry, algebra, algebra, geometry, geometry,
geometry, algebra, geometry, algebra, geometry
geometry, algebra, geometry, geometry, algebra

Of course, if we label the algebra books as copy 1 and copy 2, which is sometimes done by some libraries, we can use Formula 5.1.
Thus,

| Algebra copy 1 | Algebra copy 2 | Geometry copy 1 | Geometry copy 2 | Geometry copy 3 |

would be a different permutation than

| Algebra copy 2 | Algebra copy 1 | Geometry copy 1 | Geometry copy 2 | Geometry copy 3 |

However, this is usually not done. Thus these two permutations have to be counted as the same or as only one permutation. Formulas 5.1 and 5.2 have to be revised somewhat to allow for the possibility of repetitions. This leads us to the following formula:

FORMULA 5.3

The number of different permutations of n things of which p are alike, q are alike, or r are alike, etc. is

$$\frac{n!}{p!q!r!\ldots}$$

It is understood that $p + q + r + \cdots = n$

123

In our example we have 5 books to be shelved, so $n = 5$. Since the 2 algebra books are identical and the 3 geometry books are also identical, $p = 2$ and $q = 3$. Formula 5.3 tells us that the number of permutations is

$$\frac{5!}{2! \cdot 3!}$$

$$= \frac{5 \cdot 4 \cdot 3 \cdot 2 \cdot 1}{2 \cdot 1 \cdot 3 \cdot 2 \cdot 1}$$

$$= \frac{5 \cdot \overset{2}{\cancel{4}} \cdot \cancel{3} \cdot \cancel{2} \cdot \cancel{1}}{\cancel{2} \cdot \cancel{1} \cdot \cancel{3} \cdot \cancel{2} \cdot \cancel{1}}$$

$$= 5 \cdot 2 = 10$$

Thus, there are 10 permutations, which were listed previously.

EXAMPLE 5

How many different permutations are there of the letters in the word
a. IDIOT? b. STATISTICS?

SOLUTION

a. The word IDIOT has 5 letters, so $n = 5$. There are 2 i's, so $p = 2$. Formula 5.3 tells us that the number of permutations is

$$\frac{5!}{2!}$$

$$= \frac{5 \cdot 4 \cdot 3 \cdot \cancel{2} \cdot \cancel{1}}{\cancel{2} \cdot \cancel{1}}$$

$$= 5 \cdot 4 \cdot 3 = 60$$

There are 60 permutations.

b. Since the word STATISTICS has 10 letters, $n = 10$. The letter S is repeated 3 times, T is repeated 3 times, and I is repeated twice, so $p = 3$, $q = 3$, and $r = 2$. Formula 5.3 tells us that the number of permutations is

$$\frac{10!}{3!3!2!}$$

$$= \frac{10 \cdot 9 \cdot 8 \cdot 7 \cdot \cancel{6} \cdot 5 \cdot \overset{2}{\cancel{4}} \cdot \cancel{3} \cdot \cancel{2} \cdot \cancel{1}}{\cancel{3} \cdot \cancel{2} \cdot \cancel{1} \cdot \cancel{3} \cdot \cancel{2} \cdot \cancel{1} \cdot \cancel{2} \cdot \cancel{1}}$$

$$= 10 \cdot 9 \cdot 8 \cdot 7 \cdot 5 \cdot 2 = 50,400$$

There are 50,400 permutations.

EXERCISES

1. Evaluate each of the following symbols:

   a. $7!$

   b. $6!$

   c. $2!$

   d. $\dfrac{5!}{5!}$

   e. $\dfrac{0!}{3}$

   f. $\dfrac{4!}{2!2!}$

   g. $\dfrac{7!}{3!4!}$

   h. $\dfrac{6!}{3!3!}$

   i. $_7P_6$

   j. $_5P_4$

   k. $_6P_2$

   l. $_3P_2$

   m. $_7P_7$

   n. $_2P_2$

   o. $_0P_0$

   p. $_4P_0$

2. In how many different ways can seven students line up to register for a math course?

3. Five people are seated at a bar. In how many different ways can Pete, the bartender, select three patrons to whom he will serve either a scotch, a rye, or a bourbon whiskey? A patron cannot order more than one drink.

4. Mrs. Gibbs goes to the playground where seven of her children are playing. She needs three volunteers: one to go to the grocery store, one to do the dishes, and one to do the family laundry. In how many different ways can she get the three volunteers?

5. Mary is interviewing seven people for a new job. In how many different ways can she interview the job applicants?

6. On a recent skiing vacation Bill broke his foot. Eight of his friends have bought a get well card and a gift. In how many different ways can his friends sign the card?

7. Ralph Raider wishes to road test the following seven cars: a Chevelle, a Plymouth Duster, a Dodge Valiant, a Skylark, a Volkswagen, a Toyota, and a Gremlin. In how many orders can he road test four of these cars?

8. Dr. Miller, a leading veterinarian, has treated five cats today. The cats are Felix, Morris, Pussy, Kitty, and Fluffy. The doctor is scheduling appointments for next week's treatments for each.

   a. In how many different ways can these appointments be arranged?

   ⋆b. What is the probability that Felix is scheduled for the first appointment?

9. The student governing board at Diploma Mill University consists of eight members. A new president, vice president, secretary, and treasurer are to be elected for the fall semester.

   a. How many permutations of the list of these officers is possible from the 8 governing members only?

   ⋆b. If Manya is a member of the governing board, what is the probability that she will be elected president?

   ⋆c. What is the probability that she will be elected to any office?

⋆ This symbol indicates that the question requires particularly careful thought.

10. How many different permutations are there of the letters in the word
    a. MARIJUANA? b. GONORRHEA? c. GOOFBALL? d. MISSISSIPPI?

11. Mr. Crystal has taken his seventeen children to a photographer for a family picture. In how many different ways can they line up for the picture?

12. In how many different ways can the letters of the word SMOKING be arranged if a vowel must be the first letter of each permutation? (Repetitions are not allowed)

13. How many different five digit telephone numbers can be formed if
    a. the first number must be 8 (repetitions are allowed)?
    b. no digit can be repeated?
    c. there are no restrictions at all?

14. Recently, a wealthy man donated nine famous paintings to a museum. In how many different ways can three of these paintings be exhibited on the first floor of the museum if one is to be placed by the main entrance, one by the elevators, and one by the information desk?

15. Four math books, three ecology books, two music books, and three economics books are to be arranged on a bookshelf. None of the books are identical.
    a. How many different permutations of these books are there?
    b. How many different permutations of these books are there if books on the same subject are to be grouped together?

16. In how many different ways can the letters of the word HELP be arranged?

5.4 COMBINATIONS

Imagine that we have a 6-person rescue party that is climbing a mountain in search of survivors of an airplane crash. They suddenly spot the plane on a ledge but the passageway to the ledge is very narrow and only 4 people can proceed. The remaining 2 people will have to return. How many different 4-person rescue groups can be formed to reach the ledge?

In this situation we are obviously interested in selecting 4 out of 6 people. However, since the order in which the selection is to be made is not important, Formula 5.1 of Section 5.3 (see page 121) has to be changed.

Any selection of things in which the order is not important is called a **combination.** Let us determine how many possible combinations there actually are. If the names of the people of the 6-person rescue party are Alice, Betty, Calvin, Drew, Ellen, and Frank, denoted A, B, C, D, E, and F, then Formula 5.1 of Section 5.3 tells that there are $_6P_4$ possible ways of selecting the 4 people out of a possible 6. This yields

$$_6P_4 = \frac{6!}{(6-4)!}$$

$$= \frac{6!}{2!}$$

$$= 360 \text{ possibilities}$$

However, we know that permutations take order into account. Since we are not interested in order, there cannot be 360 possible combinations. Thus, if the 4-person rescue party consists of A, B, C, and D, there would be the following 24 different permutations.

A B C D	B A C D	C A B D	D A B C
A B D C	B A D C	C A D B	D A C B
A D C B	B D A C	C B A D	D B A C
A D B C	B D C A	C B D A	D B C A
A C B D	B C A D	C D A B	D C B A
A C D B	B C D A	C D B A	D C A B

Since these permutations consist of the same 4 people, A, B, C, and D, we consider them as only 1 combination of these people.

Similarly for any other combination of 4 people there are 24 permutations of these people. Thus, it would seem reasonable to divide the 360 by 24 getting 15 and concluding that there are only 15 different combinations.

Notice that 24 = 4!. Therefore, the number of combinations of 6 things taken 4 at a time is

$$\frac{_6P_4}{4!} = \frac{6!}{4!2!}$$

$$= 15$$

In general, consider the problem of selecting r objects from a possible n objects.

We have the following definitions and formula.

DEFINITION 5.3

A **combination** is a selection from a collection of distinct objects where order is not important.

DEFINITION 5.4

The number of different ways of selecting r objects from a possible n objects, where the order is not important, is called the **number of combinations of n things taken r at a time** and is denoted as $_nC_r$. Some books use the symbol $\binom{n}{r}$ instead of $_nC_r$.

FORMULA 5.4

The number of combinations of n things taken r at a time is

$$_nC_r = \frac{n!}{r!(n-r)!}$$

Formula 5.4 is especially useful when calculating the probability of certain events. We illustrate the use of this formula with several examples.

EXAMPLE 1

Consider any 5 people whom we shall name A, B, C, D, and E.
a. In how many ways can a committee of 3 be selected from among them?
b. What is the probability of selecting the 3-person committee consisting of A, B, and C?
c. In how many ways can a committee of 5 be selected from among them?

SOLUTION

a. We must select any 3 people from a possible 5 and order does not matter. Since this is the number of combinations of 5 things taken 3 at a time, we want $_5C_3$. Using Formula 5.4 with $n = 5$ and $r = 3$, we have

$$_5C_3 = \frac{5!}{3!(5-3)!} = \frac{5!}{3!2!} = \frac{5 \cdot \overset{2}{\cancel{4}} \cdot \cancel{3} \cdot \cancel{2} \cdot \cancel{1}}{\cancel{3} \cdot \cancel{2} \cdot \cancel{1} \cdot \cancel{2} \cdot \cancel{1}} = 10$$

There are 10 possible 3-person committees that can be formed. We can verify this answer by listing them:

A B C   A B E   A C E   B C D   B D E
A B D   A C D   A D E   B C E   C D E

b. There are 10 possible 3-person committees that can be formed as listed in part a. Of these only one consists of A, B, and C. Thus,

$$p(\text{committee consists of A, B, and C}) = 1/10$$

c. We are interested in selecting any 5 people from a possible 5 and order does not matter. This is $_5C_5$. Thus,

$$_5C_5 = \frac{5!}{5!(5-5)!}$$

$$= \frac{5!}{5!0!} \quad (\text{Remember } 0! = 1)$$

$$= \frac{5 \cdot 4 \cdot 3 \cdot 2 \cdot 1}{5 \cdot 4 \cdot 3 \cdot 2 \cdot 1 \cdot 1}$$

$$= 1$$

So there is only 1 combination containing all 5 people.

EXAMPLE 2

How many different 11-member football teams can be formed from a possible 20 players if any player can play any position?

SOLUTION

We are interested in the number of combinations of 20 players taken 11 at a time. So $n = 20$ and $r = 11$. Thus,

$$_{20}C_{11} = \frac{20!}{11!(20-11)!}$$

$$= \frac{20!}{11!9!}$$

$$= 167,960$$

There are 167,960 possible 11-member football teams.

EXAMPLE 3

a. How many different poker hands consisting of 5 cards can be dealt from a deck of 52 cards?
b. What is the probability of being dealt a royal flush in 5-card poker? A royal flush consists of the ten, jack, queen, king, and ace of the same suit.

SOLUTION

a. Since order is not important, we are interested in the number of combinations of 5 things out of a possible 52. So $n = 52$ and $r = 5$. Thus,

$$_{52}C_5 = \frac{52!}{5!(52 - 5)!}$$

$$= \frac{52!}{5!47!}$$

$$= \frac{52 \cdot 51 \cdot 50 \cdot 49 \cdot 48 \cdot \cancel{47!}}{5 \cdot 4 \cdot 3 \cdot 2 \cdot 1 \cdot \cancel{47!}}$$

$$= 2{,}598{,}960.$$

There are 2,598,960 possible poker hands.
b. Out of the possible 2,598,960 different poker hands only 4 are favorable. These are ten, jack, queen, king, and ace of hearts; ten, jack, queen, king, and ace of clubs; ten, jack, queen, king, and ace of diamonds; and ten, jack, queen, king, and ace of spades.
Thus,

$$p(\text{royal flush}) = \frac{4}{2{,}598{,}960} = \frac{1}{649{,}740}$$

EXAMPLE 4

John has 10 single dollar bills of which 3 are counterfeit. If he selects 4 of them at random, what is the probability of getting 2 good bills and 2 counterfeit bills?

SOLUTION

We are interested in selecting 4 bills from a possible 10. Thus the number of possible outcomes is $_{10}C_4$. The 2 good bills to be drawn must be drawn from the 7 good ones. This can happen in $_7C_2$

ways. Also, the 2 counterfeit bills to be drawn must be drawn from the 3 counterfeit ones. This can happen in $_3C_2$ ways. Thus,

$$p(2 \text{ good bills and 2 counterfeit bills}) = \frac{_7C_2 \cdot _3C_2}{_{10}C_4}$$

Now

$$_7C_2 = \frac{7!}{2!(7-2)!}$$

$$= \frac{7!}{2!5!} = 21$$

and

$$_3C_2 = \frac{3!}{2!(3-2)!}$$

$$= \frac{3!}{2!1!} = 3$$

Also

$$_{10}C_4 = \frac{10!}{4!(10-4)!}$$

$$= \frac{10!}{4!6!} = 210$$

Therefore,

$$p(2 \text{ good bills and 2 counterfeit bills}) = \frac{21 \cdot 3}{210}$$

$$= \frac{3}{10}$$

EXAMPLE 5

Find the probability of selecting all good bills in the preceding example.

SOLUTION

Since we are interested in selecting only good bills, we will select 4 good bills from a possible 7 and no counterfeit bills. Thus

$$p(\text{all good bills}) = \frac{_7C_4 \cdot _3C_0}{_{10}C_4}$$

$$= \frac{35 \cdot 1}{210} = 1/6$$

The probability of selecting all good bills is 1/6.

Figure 5.7

Pascal's triangle.

There is an alternate method for computing the number of possible combinations of n things taken r at a time, $_nC_r$, which completely avoids the factorial notation. This can be accomplished by using **Pascal's triangle.** Such a triangle is shown in Figure 5.7. How do we construct such a triangle?

Each row has a 1 on either end. All the in-between entries are obtained by adding the numbers immediately above and directly to the right and left of them as shown by the arrows in the diagram of Figure 5.8. For example, to obtain the entries for the 8th row we first put a 1 on each end (moving over slightly). Then we add the 1 and 7 from row 7 getting 8 as shown. We then add 7 and 21 getting 28, then, we add 21 to 35 getting 56, and so on. Remember to place the 1s on each end. The numbers must be lined up as shown in the diagram. Such a triangle of numbers is known as **Pascal's triangle** in honor of the French mathematician who

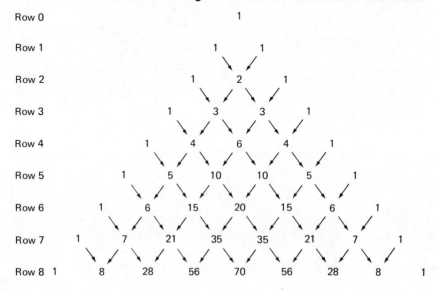

Figure 5.8

Construction of Pascal's triangle

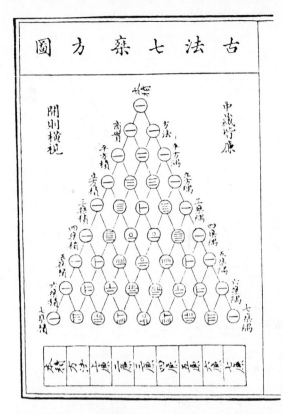

Figure 5.9

The Pascal triangle as used in 1303. (Redrawn from Joseph Needham, *Science and Civilization in China*, III, Cambridge University Press, 135.)

found many applications for it. Although the Chinese knew of this triangle several centuries earlier (see Figure 5.9), it is named for Pascal.

Let us now apply Pascal's triangle to solve problems in combinations.

EXAMPLE 6

Professor Mathew Matics tells 4 jokes a semester. His policy is never to repeat any combination of 4 jokes once they are used. How many semesters will 8 different jokes last the professor?

SOLUTION

Since order is not important, we are interested in the number of combinations of 8 things taken 4 at a time. So $n = 8$ and $r = 4$. Therefore, we must evaluate $_8C_4$. We will find $_8C_4$ first using Formula 5.4 and then using Pascal's triangle. Using Formula 5.4 we have

$$_8C_4 = \frac{8!}{4!(8-4)!} = \frac{8!}{4!4!} = 70$$

Thus, the 8 jokes will last him for 70 semesters.

Row 0   1

Row 1   1   1

Row 2   1   2   1

Row 3   1   3   3   1

Row 4   1   4   6   4   1

Row 5   1   5   10   10   5   1

Row 6   1   6   15   20   15   6   1

Row 7   1   7   21   35   35   21   7   1

Row 8   1   8   28   56   70   56   28   8   1

Figure 5.10

First eight rows of Pascal's triangle.

Let us now evaluate $_8C_4$ using Pascal's triangle. Since $n = 8$, we write down the first 8 rows of Pascal's triangle as shown in Figure 5.10.

Now we look at row 8. Since we are interested in selecting 4 things from a possible 8 things, we skip the 1 and move to the fourth entry appearing after 1. It is 70. This number represents the value of $_8C_4$. Thus $_8C_4 = 70$. Similarly, to find $_8C_6$ we move to the sixth entry after the end 1 on row 8. It is 28. Thus $_8C_6 = 28$. Also the first entry after the end 1 is 8, so that $_8C_1 = 8$. If we wanted $_8C_8$, we would move to the eighth entry after the end 1. It is 1, so that $_8C_8 = 1$. To find $_8C_0$ we must move to the zeroth entry after the end 1. This means we do not go anywhere; we just stay at the 1. Thus, $_8C_0 = 1$.

Comment When you use Pascal's triangle, remember that the rows are numbered from 0 rather than from 1, so that the rows are labeled row 0, row 1, row 2, etc.

EXAMPLE 7

Using Pascal's triangle find
a. $_5C_2$, b. $_5C_3$, c. $_6C_6$, d. $_7C_0$.

SOLUTION

We will use the Pascal triangle shown in Figure 5.10.
a. To find $_5C_2$, go to row 5 and move across to the second number from the left after the end 1. The entry is 10. Thus $_5C_2 = 10$.

b. To find $_5C_3$, go to row 5 and move across to the third number from the left after the end 1. The entry is 10. Thus $_5C_3 = 10$.

c. To find $_6C_6$, go to row 6 and move across to the sixth number from the left after the end 1. The entry is 1. Thus $_6C_6 = 1$.

d. To find $_7C_0$, go to row 7 and do not move anywhere after the end 1. Remain there. Thus $_7C_0 = 1$.

EXERCISES

1. Evaluate each of the following:

a. $_7C_6$ b. $_6C_4$

c. $_8C_3$ d. $_{10}C_0$

e. $_5C_1$ f. $_9C_6$

g. $_7C_2$ h. $_8C_4$

i. $_7C_8$ j. $_7C_7$

2. I.M. Slender notices 10 cakes displayed in the showcase window of the neighborhood bakery. In how many ways can 4 of these cakes be purchased by I.M.?

3. After seizing the prison warden and 6 other guards, several rebellious prisoners are demanding a getaway car. In how many different ways can they form a committee consisting of 2 hostages to relay their demands to the police?

4. Each year a state tax department audits the books of 5 of the 15 largest companies doing business within the state. In how many different ways can the tax department select the 5 companies to be audited?

5. To investigate charges of discrimination against minorities, a task force consisting of 3 women and 4 men is to be chosen from a total of 7 women and 8 men from minority groups. In how many ways can this task force be selected?

6. Ten people have entered a beauty contest from which 3 winners will be selected. In how many different ways can the winners be selected to share the $10,000 prize equally?

7. John Daring drives his pickup truck, which has 6 tires, over some glass.
   a. In how many different ways can he get two flat tires?
   b. In how many different ways can 4 of the 6 tires not have flats?
   c. Compare your answers in parts a and b. What do you notice?

8. A disc jockey has been given 12 records to play on the air. However, he has time to play only 8. In how many different ways can he select the 8 records to be played?

9. Professor Bergen, a psychology teacher, needs 5 student volunteers to be subjects for her latest personality tests. If there are 15 students in her class who have volunteered, in how many different ways can she select the 5 subjects for the experiment?

10. Smilin' Sam owns 10 horses which he wishes to enter in this Saturday's races at the racetrack. However, the regulations of the racetrack do not allow any one owner to enter more than 3 horses on a given day.

    a. In how many different ways can he select the 3 horses that are to be entered?

    b. In how many different ways can he select the 7 horses that are not to be entered?

11. Medical researchers are testing a new drug for its effect on decreasing high blood pressure. From a control group of 12 people, 3 will be given the new drug and 9 will be given a placebo. In how many different ways can the 3 subjects be selected?

12. How many different faculty-student disciplinary committees can be formed from a total of 8 students and 7 faculty members if each committee is to consist of 4 students and 3 faculty members?

13. Bill's Music Shack employs 8 workers in its morning shift and 6 workers in its afternoon shift. As an economy move, the management decides to fire 2 workers. What is the probability that

    a. both fired employees will be from the morning shift?

    b. both fired employees will be from the afternoon shift?

    c. one of the fired employees will be from the morning shift and one will be from the afternoon shift?

14. In Example 7 on page 134, we calculated $_5C_2$ and $_5C_3$. Our answer was the same for both. Can you explain why?

5.5 ODDS AND MATHEMATICAL EXPECTATION

Gamblers are frequently interested in determining which games are profitable to them. What they would really like to know is how much money can be earned in the long run from a particular game. If, in the long run nothing can be won, that is, if as much money that is won will be lost, then why play at all? Similarly, if in the long run no money can be won but money can be lost, then the gambler will not play such a game. He will play any game only if money can be won. We refer to the amount of money to be won as the **mathematical expectation** of the game and define it as follows.

DEFINITION 5.5

The **mathematical expectation** of an event is the amount of money to be won or lost in the long run.

DEFINITION 5.6

Consider an event which has probability p of occurring and which has a payoff, that is, the amount won, m_1. Consider also a second event with probability p_2 and payoff m_2, a third event with probability p_3 and payoff m_3, etc. The mathematical expectation of the event is

$$m_1p_1 + m_2p_2 + m_3p_3 + \cdots$$

The numbers attached to the letters are called subscripts. They have no special significance. We use them only to avoid using too many different letters.

We illustrate these definitions with several examples.

EXAMPLE 1

A large company is considering opening 2 new factories in different towns. If it opens in town A, it can expect to make $63,000 profit per year with a probability of 4/7. However, if it opens in town B, it can expect to make a profit of $77,000 with a probability of only 3/7. What is the company's mathematical expectation?

SOLUTION

We use Definition 5.6. We have

$$(63{,}000)\left(\frac{4}{7}\right) + (77{,}000)\left(\frac{3}{7}\right)$$

$$= 36{,}000 + 33{,}000 = 69{,}000$$

Thus, the company's mathematical expectation is $69,000.

EXAMPLE 2

A contractor is bidding on a road construction job which promises a profit of $200,000 with a probability of 7/10 and a loss, due to strikes, weather conditions, late arrival of building materials, etc., of $40,000 with a probability of 3/10. What is his mathematical expectation?

SOLUTION

A + sign will denote a gain and a − sign will denote a loss. Using Definition 5.6, we find the contractor's mathematical expectation is

$$(+\$200{,}000)\left(\frac{7}{10}\right) + (-\$40{,}000)\left(\frac{3}{10}\right)$$

$$= \$140{,}000 - \$12{,}000 = \$128{,}000$$

The contractor's mathematical expectation is $128,000.

Another interesting application of probability is concerned with betting. Gamblers frequently speak of the **odds** of a game. To best understand this idea, consider George who believes that whenever he washes his car, it usually rains the following day. If George has just washed his car and if the probability of it raining tomorrow is 3/10, gamblers would say that the odds in favor of it raining are 3 to 7 and the odds against it raining are 7 to 3. The 7 represents the 7 chances of it not raining. Formally we have the following definitions.

DEFINITION 5.7

The **odds in favor** of an event occurring are *p* to *q* where *p* is *the number of favorable outcomes* and *q* is *the number of unfavorable outcomes.*

DEFINITION 5.8

If *p* and *q* are the same as in Definition 5.7, the **odds against** an event happening are *q* to *p*.

We now illustrate these definitions with several examples.

EXAMPLE 3

What are the odds in favor of the New York Mets winning the world series if the probability of them winning is 4/7 and the probability of them losing is 3/7?

SOLUTION

Since the probability of them winning is 4/7, this means that out of 7 possibilities 4 are favorable and 3 are nonfavorable. Thus, the odds in favor of the New York Mets winning the series are 4 to 3.

EXAMPLE 4

Leon is in a restaurant. He decides to give the waiter a tip consisting of only 1 coin selected randomly from among the 6 that he has in his pocket. What are the odds against him giving the waiter a penny tip, if he has a penny, a nickel, a dime, a quarter, a half-dollar, or a dollar piece?

SOLUTION

There are 6 coins; 5 are favorable and 1 is unfavorable. Thus the odds against giving the waiter a penny tip are 5 to 1.

EXAMPLE 5

What are the odds in favor of getting a face card when selecting a
card at random from a deck of 52 cards?

SOLUTION

There are 52 possibilities; 12 are favorable and 40 are unfavorable.
Thus, the odds in favor of getting a face card are 12 to 40.

EXERCISES

1. A box contains 8 green marbles, 7 yellow marbles, and 5 black marbles. One
   marble is to be selected from the box. You get $10 if the marble selected is
   black, but you lose either $3 if the marble selected is green or $5 if it is yellow.
   What is your mathematical expectation?
2. A card is selected at random from a deck of cards. If the card selected is a face
   card, Trudy wins $10. Otherwise, she loses $3. What is her mathematical
   expectation?
3. A publisher has invested $50,000 to print a math book. If the book sells more
   than 100,000 copies over a 3-year period, the publisher will make a $300,000
   profit. Otherwise, the publisher will lose the $50,000. If the probability that the
   book sells more than 100,000 copies is 3/5, what is the mathematical expectation?
4. If graduation ceremonies at a certain college are conducted outside, $5800
   can be raised for alumni activities. If it rains, graduation ceremonies have to be
   conducted in the school auditorium and only $2000 can be raised. If the
   probability of rain is 3/5 and the probability of nice weather is 2/5, how much
   money can the college expect to raise?
5. A coin is tossed 3 times. If heads appears on all 3 tosses, Mary will win $2. If
   heads appears on at least 2 of the tosses, she will win $16. Otherwise she
   loses $8. What is her mathematical expectation?
6. The probability that Tim is speaking on the telephone is 3/8. What are the odds
   in favor of his mother being able to get through to him when she calls him?
7. During the 1974 fiscal year the Internal Revenue Service audited 2.5% of all
   individual tax returns filed in California. John Williams lives in California. What
   are the odds against his tax return being audited?
8. Joe took pictures of his daughter's birthday cake with a defective camera. If
   the probability that the camera functioned properly is 3/10, what are the odds
   in favor of the pictures coming out?
9. What are the odds in favor of getting a black card when selecting 1 card at
   random from a deck of 52 cards?

10. On a grocer's shelf there are 3 stale loaves of bread and 15 fresh loaves. Stephen selects a loaf at random. What are the odds in favor of his getting a fresh loaf?

11. Along with 9 other guests, Isaac places his hat on the table. All the hats look alike. The hostess puts all the hats in the closet. After the party Isaac returns and selects a hat at random. What are the odds against his getting his own hat?

12. Based upon past experience we know the probability that Elaine's teacher will be late to class is 4/17. What are the odds in favor of her teacher being late?

SELF-STUDY GUIDE

In this chapter we discussed various aspects of probability. We noticed that probability is concerned with outcomes of experiments. Among the different ways of defining probability, we defined probability in terms of relative frequency. Thus, we had

$$p = \frac{\text{number of outcomes favoring an event}}{\text{total number of possible outcomes}}$$

We noticed that the probability of an event was between 0 and 1, the null event and the definite event, respectively.

To enable us to determine the total number of possible outcomes, we analyzed various counting techniques. Tree diagrams, permutations, and combinations were introduced and discussed in detail. Permutations represent arrangements of objects where order *is* important, whereas combinations represent selections of objects where order is *not* important. Applications of permutations and combinations to many different situations were given, in addition to the usual gambling problems.

Finally, probability was applied to determine the amount of money to be won in the long run in various situations. This was called the mathematical expectation of the event. We also discussed what is meant by statements such as odds in favor of an event and odds against an event. Definitions were given which allow us to calculate these odds.

You should be able to identify each symbol in the following formulas, understand the relationships among the symbols expressed in each formula, understand the significance of each formula, and use the formulas in solving problems.

1. $p = \dfrac{\text{number of favorable outcomes}}{\text{total number of possible outcomes}} = \dfrac{f}{n}$

2. $_nP_r = \dfrac{n!}{(n-r)!}$ The number of permutations of n things taken r at a time.

3. $_nP_n = n!$ The number of permutations of n things taken n at a time.

4. $\dfrac{n!}{p!q!r!\cdots}$ etc. The number of permutations of n things where p are alike, q are alike,

5. $_nC_r = \dfrac{n!}{r!(n-r)!}$ The number of combinations of n things taken r at a time.

6. $m_1p_1 + m_2p_2 + m_3p_3 + \cdots$ Mathematical expectation of an event.

7. Odds in favor of an event are p to q. $\begin{cases} p = \text{the number of favorable outcomes} \\ q = \text{the number of unfavorable outcomes} \end{cases}$
   Odds against an event are q to p.

   You should now be able to demonstrate your knowledge of the following ideas presented in this chapter by giving definitions, descriptions, or specific examples. Page references are given for each term so that you can check your answer.

Relative frequency (page 100)
Favorable outcome (page 101)
Sample space (page 102)
Event (page 102)
Probability (page 104)
Null event (page 108)
Certain event or definite event
   (page 109)

Tree diagrams (page 114)
Permutation (page 120)
Factorial notation (page 121)
Combination (page 127)
Pascal's triangle (page 132)
Mathematical expectation (page 136)
Odds in favor of an event (page 138)
Odds against an event (page 138)

   The tests of the following section will be more useful if you take them after you have studied the examples and solved the exercises given in this chapter.

MASTERY TESTS

Form A

1. True or false: 0! and 1! have the same value.
2. What is the probability that your math teacher's birthday is April Fool's day?
3. Mr. Smith was hit by a car. He believes that the probability that he lives is ½ since there are 2 possible outcomes, namely that he will live or die. Do you agree with this reasoning? Explain your answer.
4. If the probability of an event is one in a million, is the event unlikely or impossible to happen? Explain.

5. The number of possible variations in the genetic make-up of any one child is 2^{48}. If the Jones' have 2 children, what is the probability that the 2 children will be exactly alike?

6. Consider the following spinner. It is spun twice. List the sample space.

7. Find the probability that the sum of the outcomes on both spins in test question 6 is more than 3.

8. In a statistics class of 50 students the following information was obtained:

	Sex	
	Male	Female
Color of Eyes Blue	14	16
Green	8	12

Find the probability that a student has blue eyes.

9. Find the probability that a student in the class described in test question 8 is a male.

10. In a certain state, license plates consist of 6 letters with repetition allowed. What is the probability that your license plates will be SHNOOK?

Form *B*

1. In how many different ways can a sports fan enter the stadium by one entrance and leave by another if the stadium has twelve entrances?

2. In how many different ways can 8 people be seated on a bench if there is room for only 2 people?

3. A prison is considering assigning identification tags to each prisoner using one of two schemes: *a.* 3 letters followed by 2 numbers or *b.* 2 letters followed by 3 numbers. Which of these schemes will result in more identification tags?

4. Construct a tree diagram to determine the number of possible ways that a coin can be flipped 4 times in succession so that throughout the flips there are always at least as many heads as tails.

5. How many different committees of 6 people can be formed from 6 Democrats and 8 Republicans if each committee is to consist of at least 3 Republicans?

6. How many three-letter words can be formed if the middle letter must be a vowel (A, E, I, O, U) and the other two letters must be consonants with no repetition allowed?

7. A chest contains 2 red pennants, 4 white pennants, and 6 blue pennants. Three pennants are selected and arranged on a pole. In how many different ways can this be done if order counts? Pennants of the same color are not identical and are distinguishable.

8. Consider the information given in the following chart. Is it true that the probability that a newborn child is a boy is 0.5?

Births in the United States

Year	Number of Males	Number of Females	Total Number
1955	2,073,719	1,973,576	4,047,295
1960	2,179,708	2,078,142	4,257,850
1965	1,927,054	1,833,304	3,760,358
1970	1,915,378	1,816,008	3,731,386
1971	1,822,910	1,733,060	3,555,970
1972	1,669,927	1,588,484	3,258,411
1973	1,608,326	1,528,639	3,136,395

Source: Division of Vital Statistics, National Center for Health Statistics, Public Health Service.

9. Each of the 13,475 students at a state university goes to school by bicycle. A 3-digit number is carved on each bike to discourage bicycle theft. Will a 3-digit number be sufficient for all the bikes? Explain your answer.

10. Suppose that school officials decide to carve a 3-letter code on each bike described in test question 9. Will 3 letters provide a sufficient code for each bike?

11. Mr. and Mrs. Johnson own 4 cars which they park in front of their house each evening. If they own a Chevrolet, Plymouth, Cadillac, and a Toyota, what is the probability that the cars are parked in the following order:

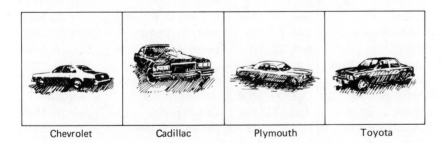

| Chevrolet | Cadillac | Plymouth | Toyota |

12. Many people believe that an unlucky event occurs when the thirteenth of the month is on a Friday. Moreover, they believe that this occurs very rarely. To

check this, consider the following chart which indicates the day of the week on which the thirteenth occurred during the last 400 years:

| | How often the 13th of the |
Day of the week	month occurred on this day
Sunday	687
Monday	685
Tuesday	685
Wednesday	687
Thursday	684
Friday	688
Saturday	684

By computing the probability that the thirteenth of the month will occur on a Friday, determine if this probability is greater than, less than, or equal to the probability of it falling on any other day of the week.

13. The following is a list of the 38 presidents of the United States and their birthdays:

President	Birthday	President	Birthday
George Washington	Feb. 22, 1732	James Garfield	Nov. 19, 1831
John Adams	Oct. 30, 1735	Chester Arthur	Oct. 5, 1830
Thomas Jefferson	Apr. 13, 1743	Grover Cleveland	Mar. 18, 1837
James Madison	Mar. 16, 1751	Benjamin Harrison	Aug. 20, 1833
James Monroe	Apr. 28, 1758	William McKinley	Jan. 29, 1843
John Quincy Adams	July 11, 1767	Theodore Roosevelt	Oct. 27, 1858
Andrew Jackson	Mar. 15, 1767	William Taft	Sept. 15, 1857
Martin Van Buren	Dec. 5, 1782	Woodrow Wilson	Dec. 28, 1857
William H. Harrison	Feb. 9, 1773	Warren Harding	Nov. 2, 1865
John Tyler	Mar. 29, 1790	Calvin Coolidge	July 4, 1872
James Knox Polk	Nov. 2, 1795	Herbert Hoover	Aug. 10, 1874
Zachary Taylor	Nov. 24, 1784	Franklin D. Roosevelt	Jan. 30, 1882
Millard Fillmore	Jan. 7, 1800	Harry S. Truman	May 8, 1884
Franklin Pierce	Nov. 23, 1804	Dwight D. Eisenhower	Oct. 14, 1890
James Buchanan	Apr. 23, 1791	John F. Kennedy	May 29, 1917
Abraham Lincoln	Feb. 12, 1809	Lyndon B. Johnson	Aug. 27, 1908
Andrew Johnson	Dec. 29, 1808	Richard M. Nixon	Jan. 9, 1913
Ulysses S. Grant	Apr. 27, 1822	Gerald Ford	July 14, 1913
Rutherford Hayes	Oct. 4, 1822	James E. Carter	Oct. 1, 1924

You will notice that no president was born during the month of June. Assuming that a president has an equally likely chance of being born in any of the 12 months, what is the probability that a president was born in June?

SUGGESTED READING

Adler, I. *Probability and Statistics for Everyman.* New York: New American Library, 1966.

Bell, E. T. *Men of Mathematics.* New York: Simon & Schuster, 1961. Chapter 5 contains a bibliography of Pascal.

Bergamini D. and eds. *Life. Mathematics* (Life-Science Library). New York: Time Life Books, 1970. Pages 126 to 147 discuss figuring the odds in an uncertain world.

Epstein, R. A. *Theory of Gambling and Statistical Logic.* New York: Academic Press, 1967. Contains an interesting discussion on the fairness of coins.

Havermann, E. "Wonderful Wizard of Odds" in *Life* vol. 51, no. 14, Oct. 6, 1961, Page 30 and those following contain a discussion of odds.

Huff, Darrell. *How to Take a Chance.* New York: Norton, 1959.

Kasner, E. and Newman, J. *Mathematics and the Imagination.* New York: Simon & Schuster, 1940. See the chapter on chance and probability.

Mathematics in the Modern World (Readings from *Scientific American*) San Francisco: W. H. Freeman, 1968. Article 22 discusses chance, Articles 23 and 24 discuss probability.

Newmark, Joseph and Lake, Frances. *Mathematics as a Second Language.* 2nd ed., Reading, Mass.: Addison-Wesley, 1977. Chapter 8 discusses probability and its applications.

Ore, Øystern. *Cardano, the Gambling Scholar.* New York: Dover, 1965.

Polya, G. *Mathematics and Plausible Reasoning.* Princeton, N.J.: Princeton University Press, 1957.

Weaver, W. *Lady Luck.* New York: Anchor Books, also Doubleday, 1963.

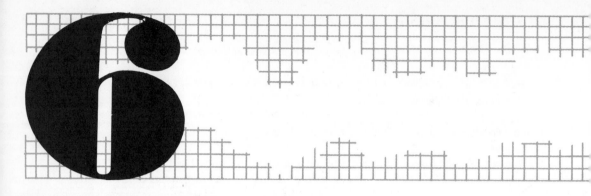

RULES OF PROBABILITY

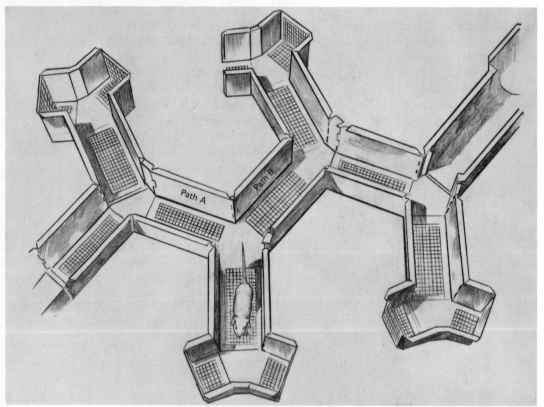

Path A

Path B

A leading psychologist is experimenting with rats to determine how quickly they learn maze patterns. Mazes used to study human learning are similar, in principle, to those used with animals. The rat indicated in the picture has already arrived at the food box. What is the probability that the rat came there from path A, not from path B?

Preview

In this chapter we will discuss the following topics:

    1. *Addition Rules*
These rules allow us to determine the probability of either event *A*, event *B*, or both happening.

    2. *Mutually Exclusive Events*
We discuss whether an event is or is not mutually exclusive and how this determines which addition rule to use.

    3. *Conditional Probability*
We discuss how one event is affected by the occurrence or non-occurrence of another.

    4. *Independent Events*
When one event is not affected at all by the occurrence of another, we have independent events and a simplified multiplication rule.

    5. *Bayes' Rule*
If we know the outcome of some experiment, we might be interested in determining the probability that it occurred because of some specific event.

Introduction

In Chapter 5 we discussed the nature of probability and how to calculate its value. However, in many situations it turns out that in order to determine the probability of an event, we must first calculate the probability of other related events and then combine them somehow. In this chapter we will discuss in detail the exact nature of several important rules for combining probabilities; these are rules for addition, rules for multiplication, conditional probability rules, and Bayes' rule. Depending upon the situation these rules enable us to combine probabilities so that we may determine the probability of some event of interest.

6.1 ADDITION RULES

Let us look in on Charlie who is playing cards. He is about to select 1 card from an ordinary deck of 52 playing cards. His opponent Dick will pay him $50 if the card selected is a face card (that is, a jack, queen, or king) *or* an ace. What is the probability that Charlie wins the $50?

To answer the question we first notice that a card selected cannot be a face card and an ace at the same time. Mathematically we say that the events of "drawing a face card" and of "drawing an ace" are **mutually exclusive.**

Since there are 12 face cards in a deck (4 jacks, 4 queens, and 4 kings), the probability of getting a face card is 12/52, or 3/13. Similarly, the probability of getting an ace is 4/52, or 1/13, since there are 4 aces in the deck. There are then 12 face cards and 4 aces. Thus there are 16 favorable outcomes out of a possible 52 cards in the deck. Applying the definition of probability we get

$$p(\text{face card or ace}) = \frac{16}{52} = \frac{4}{13}$$

Let us now add the probability of getting a face card with the probability of getting an ace. We have

$$p(\text{face card}) + p(\text{ace}) = \frac{12}{52} + \frac{4}{52}$$

$$= \frac{12 + 4}{52} \quad \text{(since the denominators are the same)}$$

$$= \frac{16}{52} = \frac{4}{13}$$

This indicates that

$$p(\text{face card or ace}) = p(\text{face card}) + p(\text{ace})$$

The same reasoning can be applied for any mutually exclusive events. First we have the following definition and formula.

DEFINITION 6.1

Consider any two events A and B. If both events cannot occur at the same time, we say that the events A and B are **mutually exclusive.**

FORMULA 6.1

If A and B are mutually exclusive, then

$$p(A \text{ or } B) = p(A) + p(B)$$

We illustrate the use of Formula 6.1 with several examples.

EXAMPLE 1

Louis has been shopping around for a calculator and decides to buy the scientific model produced by either the Texas Instruments Company or the Hewlett Packard Company. The probability that he will buy the Hewlett Packard model is 1/9 and the probability that he will buy the Texas Instrument model is 4/9. What is the probability that he will buy either of these two models?

SOLUTION

Since Louis will buy only one calculator, the events "buys the Texas Instruments model" and "buys the Hewlett Packard model" are mutually exclusive. Thus Formula 6.1 can be used. We have

p(buys either model) = p(buys Texas Instrument model) + p(buys Hewlett Packard model)

$$= \frac{4}{9} + \frac{1}{9}$$

$$= \frac{4+1}{9} = \frac{5}{9}$$

Therefore, p(Louis buys either model) = 5/9.

EXAMPLE 2

Two dice are rolled. What is the probability that the sum of the dots appearing on both dice together is 9 or 11?

SOLUTION

Since the events "getting a sum of 9" and "getting a sum of 11" are mutually exclusive, Formula 6.1 can be used. When 2 dice are rolled there are 36 possible outcomes. These were listed on page 103. There are 4 possible ways of getting a sum of 9. Thus

$$p(\text{sum of 9}) = \frac{4}{36}$$

Also, there are only 2 possible ways of getting a sum of 11. Thus

$$p(\text{sum of 11}) = \frac{2}{36}$$

Therefore,

$$p(\text{sum of 9 or 11}) = p(\text{sum of 9}) + p(\text{sum of 11})$$

$$= \frac{4}{36} + \frac{2}{36}$$

$$= \frac{6}{36} = \frac{1}{6}$$

Hence the probability that the sum is 9 or 11 is 1/6.

EXAMPLE 3

Doris and her friends plan to travel to Florida during the winter intersession period. The probability that they go by car is 2/3, and the probability that they go by plane is 1/5. What is the probability that they travel to Florida by car or plane?

SOLUTION

Since they plan to travel to Florida either by car or by plane, not by both, we are dealing with mutually exclusive events. Formula 6.1 can be used. Therefore,

$$p(\text{go by car or plane}) = p(\text{go by car}) + p(\text{go by plane})$$

$$= \frac{2}{3} + \frac{1}{5}$$

$$= \frac{10}{15} + \frac{3}{15}$$

$$= \frac{13}{15}$$

Note: We cannot add the fractions 2/3 and 1/5 together as they are since they do not have the same denominators. We thus change each fraction to a fraction with the same denominator. Thus 2/3 becomes 10/15 and 1/5 becomes 3/15. Hence, the probability that they go to Florida by car or plane is 13/15.

EXAMPLE 4

The probability that Mohammed will win the boxing match this coming Monday is 7/8. What is the probability that he will not win the fight?

Since the events "Mohammed wins the fight" and "Mohammed does not win the fight" are mutually exclusive, we can use Formula 6.1. One of these events must occur so that the event "Mohammed wins the fight or does not win the fight" is the definite event. We know that the definite event has probability 1 (see page 109). Thus,

p(Mohammed wins fight or does not win fight) = p(wins fight) + p(does not win fight)

$$1 = \frac{7}{8} + p(\text{does not win fight})$$

$$1 - \frac{7}{8} = p(\text{does not win fight})$$

$$\frac{8}{8} - \frac{7}{8} = p(\text{does not win fight})$$

$$\frac{1}{8} = p(\text{does not win fight})$$

Therefore, the probability that Mohammed does not win the fight is 1/8.

More generally, consider any event A. Let $p(A)$ be the probability that A happens and let $p(A')$, read as the probability of A prime, be the probability that A does not happen. Since either A happens or does not happen, we can use Formula 6.1. Thus,

$p(A$ happens or does not happen) = $p(A$ happens) + $p(A$ does not happen)
$$1 = p(A) + p(A')$$
$$1 - p(A) = p(A') \quad \text{(We subtract } p(A) \text{ from both sides.)}$$

Therefore, the probability of A not happening is $1 - p(A)$. *Note:* Some books refer to the event A' as the **complement** of event A.

Now consider the following problem. One card is drawn from a deck of cards. What is the probability of getting a king or a red card? At first thought we might say that since there are 4 kings and 26 red cards, then

$$p(\text{king or red card}) = p(\text{king}) + p(\text{red card})$$

$$= \frac{4}{52} + \frac{26}{52}$$

$$= \frac{30}{52} = \frac{15}{26}$$

151

Thus, we would say that the probability of getting a king or a red card is 15/26. Notice, however, that in arriving at this answer we have counted some cards twice. The 2 red kings have been counted as both kings and red cards. Obviously we must count them only once in probability calculations. The events "getting a king" and "getting a red card" are not mutually exclusive. We therefore have to revise our original estimate of the total number of favorable outcomes by deducting the number of cards that have been counted twice. We will subtract 2. When this is done, we get

$$p(\text{king or red card}) = p(\text{king}) + p(\text{red card}) - p(\text{king also a red card})$$

$$= \frac{4}{52} + \frac{26}{52} - \frac{2}{52}$$

$$= \frac{4 + 26 - 2}{52}$$

$$= \frac{28}{52} = \frac{7}{13}$$

Thus, the probability of getting a king or a red card is 7/13. This leads us to a more general formula.

ADDITION RULE

If A and B are events, the probability of obtaining either of them is equal to the probability of A plus the probability of B minus the probability of both occurring at the same time.

Symbolically, the addition rule is as follows:

FORMULA 6.2

If A and B are *any* events, then

$$p(A \text{ or } B) = p(A) + p(B) - p(A \text{ and } B)$$

We now apply Formula 6.2 in several examples.

EXAMPLE 5

The probability that Sylvester plays a guitar is 1/4, and the probability that he plays a clarinet is 5/8. If the probability that he plays both these instruments is 5/24, what is the probability that he plays the guitar or that he plays the clarinet?

Since it is possible that Sylvester plays both these instruments, these events are not mutually exclusive. Thus we must use Formula 6.2. We have

p(plays guitar or clarinet) $= p$(plays guitar) $+ p$(plays clarinet) $- p$(plays guitar and clarinet)

$$= \frac{1}{4} + \frac{5}{8} - \frac{5}{24}$$

$$= \frac{6}{24} + \frac{15}{24} - \frac{5}{24}$$

$$= \frac{6 + 15 - 5}{24}$$

$$= \frac{16}{24} = \frac{2}{3}$$

Thus, the probability that he plays either instrument is 2/3.

EXAMPLE 6

The student government at Zeesa University is conducting an election. The probability that Jeff votes in the election is 3/11 and the probability that his girlfriend Fern votes is 4/7. If the probability that both vote is 5/77, find the probability that either votes.

SOLUTION

Since both may vote in the election, the events are not mutually exclusive. When we use Formula 6.2, we have

p(Jeff or Fern votes) $= p$(Jeff votes) $+ p$(Fern votes) $- p$(both vote)

$$= \frac{3}{11} + \frac{4}{7} - \frac{5}{77}$$

$$= \frac{21}{77} + \frac{44}{77} - \frac{5}{77}$$

$$= \frac{60}{77}$$

Thus, the probability that either votes is 60/77.

EXAMPLE 7

Environmentalists have accused a large company in the eastern United States of dumping nuclear waste material in a local river. The probability that either the fish in the river or the animals that drink from the river will die is 11/21. The probability that only the fish will die is 1/3 and the probability that only the animals that drink from the river will die is 2/7. What is the probability that both the fish *and* the animals that drink from the river will die?

SOLUTION

Since both the fish and animals may die, the events are not mutually exclusive. We then use Formula 6.2. We have

p(fish or animals die) = p(fish die) + p(animals die) $-$ p(both fish and animals die)

$$\frac{11}{21} = \frac{1}{3} + \frac{2}{7} - p(\text{both die})$$

$$\frac{11}{21} = \frac{7}{21} + \frac{6}{21} - p(\text{both die})$$

$$\frac{11}{21} = \frac{13}{21} - p(\text{both die})$$

$$\frac{11}{21} + p(\text{both die}) = \frac{13}{21}$$

$$p(\text{both die}) = \frac{13}{21} - \frac{11}{21} = \frac{2}{21}$$

Thus, the probability that both the fish and the animals that drink from the river will die is 2/21.

Comment Although you may think that Formulas 6.1 and 6.2 are different, this is not the case. Formula 6.1 is just a special case of Formula 6.2. Formula 6.2 can always be used, since if the events A and B are mutually exclusive, the probability of them happening together is 0. In this case Formula 6.2 becomes

$$p(A \text{ or } B) = p(A) + p(B) - 0$$

which is exactly Formula 6.1.

EXERCISES

1. Determine which of the following events are mutually exclusive and which are not:
   a. going on vacation and quitting a job;
   b. becoming pregnant and getting a nervous breakdown;
   c. having high blood pressure and having a heart condition;
   d. deciding to move to California and deciding to move to New York;
   e. breaking a leg and breaking an arm on a skiing vacation;
   f. taking a shower and listening to the radio;
   g. having long hair and having a beard;
   h. studying for a math exam and getting 100 on the exam;
   i. being a male and being a female;
   j. winning first prize and winning the second prize in the state lottery.

2. The probability that Audrey has a cat in her house is 5/14. The probability that she has a dog in her house is 3/7 and the probability that she has both a dog and a cat is 1/7. What is the probability that Audrey has either a cat or a dog in her house?

3. The probability that Jane forgot to take her sunglasses on her vacation is 7/10. The probability that she forgot to take her suntan lotion is 3/5. The probability that she forgot to take both items is 4/10. What is the probability that she forgot to take either item?

4. Willie has arranged to meet his girl friend Melissa either in the library or in the student lounge. The probability that he meets her in the lounge is 1/3 and the probability that he meets her in the library is 2/9. What is the probability that he meets her in the library or the lounge?

5. A customer enters a restaurant. The probability that the customer orders a steak or a hamburger is 8/11. The probability that the customer orders a steak is 2/11 and the probability that the customer orders a hamburger is 7/11. What is the probability that the customer orders both a steak and a hamburger?

6. In a certain school district the probability that a father will come to school on open school day is 3/11. The probability that a mother will come to school is 4/9. If the probability that either parent comes to school is 19/33, what is the probability that both parents come to school?

7. Connie Sumer has taken the day off from work to await the delivery of her new washing machine and new color television. The probability that the washing machine arrives in the morning is 0.64. The probability that the new color television arrives in the morning is 0.8. If the probability that either the washing machine or the television arrives in the morning is 0.93, what is the probability that they both arrive in the morning?

8. Dr. C. Nile is planning a summer vacation. The probability of his visiting France is 0.2 and the probability of his visiting Germany is 0.3. What is the probability of his visiting either country if the probability of visiting both is 0.1?

9. Josiah S. Carberry is traveling on an airplane. He asks the steward to bring him a copy of *Time* magazine or *U.S. News and World Report.* The probability

that *Time* magazine is being used by another traveler is 0.6. The probability that *U.S. News and World Report* magazine is being used by another traveler is 0.4. If the probability that either magazine is being used by another traveler is 0.9, what is the probability that both magazines are being used?

10. Ed is planning to buy his wife either a bouquet of flowers or a bottle of perfume, but not both, for her birthday. The probability that he buys her the flowers is 0.32. If the probability that he buys her either the flowers or the perfume is 0.79, what is the probability that he buys her the perfume?

For any three events *A, B,* and *C* the probability of *A* or *B* or *C* is given by the formula

$$p(A \text{ or } B \text{ or } C) = p(A) + p(B) + p(C) - p(A \text{ and } B) - p(A \text{ and } C)$$
$$- p(B \text{ and } C) + p(A \text{ and } B \text{ and } C)$$

Use this formula to solve the following problems.

11. Sherry is majoring in foreign languages. The probability that she speaks several different languages is as follows:

Language(s)	Probability
Russian	18/39
Swahili	22/39
Chinese	2/3
Chinese and Swahili	11/39
Chinese and Russian	1/3
Russian and Swahili	3/13
Russian, Chinese, and Swahili	2/13

Find the probability that she speaks Russian, Swahili, or Chinese.

12. Pat smokes 3 different brands of cigarettes with the following probabilities:

Brand(s)	Probability
Winston	2/3
Raleigh	37/60
Kent	29/60
Winston and Raleigh	19/60
Raleigh and Kent	3/10
Winston and Kent	1/3
Winston, Kent, or Raleigh	1

Find the probability that she smokes all 3 brands.

6.2 CONDITIONAL PROBABILITY

Although the addition rule given in Section 6.1 applies to many different situations, there are still other problems that cannot be solved by that formula. It is for this reason that we introduce conditional probability. Let us first consider the following problem.

EXAMPLE 1

The administration of Podunk University is considering the establishment of a drug clinic to help students "break the habit." However, not all the students at the school favor the proposal. As a result, a survey of the one thousand students at the school is conducted to determine student opinion about the proposal. The following table summarizes the results of the survey:

	Against Drug Clinic	For Drug Clinic	No Opinion	Total
Freshmen	23	122	18	163
Sophomores	39	165	27	231
Juniors	58	238	46	342
Seniors	71	127	66	264
Total	191	652	157	1000

Answer the following questions:
a. What is the probability that a student selected at random voted against the establishment of the drug clinic is 191/1000.
b. If a student is a freshman, what is the probability that the student voted for the drug clinic?
c. If a senior is selected at random, what is the probability that the senior has no opinion about the clinic?

SOLUTION

a. Since there was a total of 191 students who voted against the establishment of the drug clinic out of a possible 1000 students, we apply the definition of probability and get

$$p(\text{student voted against the drug clinic}) = \frac{191}{1000}$$

Thus, the probability that a student selected at random voted against the establishment of the drug clinic is 191/1000.
b. There are 163 freshmen in the school. One hundred twenty-two of them voted for the drug clinic. Since we are concerned with

freshmen only, the number of possible outcomes of interest to us is 163, not 1000. Out of these, 122 are favorable. Thus the probability that a student voted for the drug clinic given that the student is a freshman is 122/163.

c. In this case the information given narrows the sample space to 264 seniors, 66 of which had no opinion. Thus the probability that a student has no opinion given that the student is a senior is 66/264, or 33/132.

The situation of part *b* or that of part *c* in Example 1 is called a **conditional probability** because we are interested in the probability of a student voting in favor of the establishment of the drug clinic given that, or conditional upon the fact that, the student is a freshman. We express this condition mathematically by using a vertical line "|" to stand for the words "given that" or "if we know that." We then write

$$p(\text{student voted in favor of drug clinic} \mid \text{student is a freshman}) = \frac{122}{163}$$

Similarly for part *c* we write

$$p(\text{student had no opinion} \mid \text{student is a senior}) = \frac{66}{264} = \frac{33}{132}$$

EXAMPLE 2

Sherman is repairing his car. He has removed the 6 spark plugs. Four are good and 2 are defective. He now selects 1 plug and then, without replacing it, selects a second plug. What is the probability that both spark plugs selected are good?

SOLUTION

We will list the possible outcomes and then count all the favorable ones. To do this we label the good spark plugs as $g_1, g_2, g_3,$ and $g_4,$ and the defective ones as d_1 and d_2. The possible outcomes are

g_1, g_2	g_2, g_1	g_3, g_1	g_4, g_1	d_1, g_1	d_2, g_1
g_1, g_3	g_2, g_3	g_3, g_2	g_4, g_2	d_1, g_2	d_2, g_2
g_1, g_4	g_2, g_4	g_3, g_4	g_4, g_3	d_1, g_3	d_2, g_3
g_1, d_1	g_2, d_1	g_3, d_1	g_4, d_1	d_1, g_4	d_2, g_4
g_1, d_2	g_2, d_2	g_3, d_2	g_4, d_2	d_1, d_2	d_2, d_1

There are 30 possible outcomes. Twelve of these are favorable. These are the circled ones. They represent the outcome that both spark plugs are good. Thus

$$p(\text{both spark plugs selected are good}) = \frac{12}{30} = \frac{2}{5}$$

EXAMPLE 3

In Example 2 what is the probability that both spark plugs selected
are good if we know that the first plug selected is good?

SOLUTION

Again we list all the possible outcomes and count the number of
favorable ones.

$$\boxed{g_1, g_2}\quad \boxed{g_2, g_1}\quad \boxed{g_3, g_1}\quad \boxed{g_4, g_1}$$
$$\boxed{g_1, g_3}\quad \boxed{g_2, g_3}\quad \boxed{g_3, g_2}\quad \boxed{g_4, g_2}$$
$$\boxed{g_1, g_4}\quad \boxed{g_2, g_4}\quad \boxed{g_3, g_4}\quad \boxed{g_4, g_3}$$
$$g_1, d_1\quad g_2, d_1\quad g_3, d_1\quad g_4, d_1$$
$$g_1, d_2\quad g_2, d_2\quad g_3, d_2\quad g_4, d_2$$

Since we know that the first plug selected is good, there are only 20
possible outcomes. Of these, 12 are favorable. These are the
circled ones. Thus the probability that both spark plugs are good if
we know that the first plug is good is

$$\frac{12}{20} = \frac{3}{5}$$

Using the conditional probability notation, this result can be written
as

$$p(\text{both spark plugs are good} \,|\, \text{first spark plug is good}) = \frac{3}{5}$$

Comment Example 3 differs from Example 2 since in Example 3 we are interested in determining the probability of getting 2 good spark plugs once we know
that the first one selected is good. On the other hand, in Example 2 we were
interested in determining the probability of getting 2 good plugs without knowing
whether the first plug is defective or not.

Let us analyze the problem discussed at the beginning of this section in detail.
There are a total of 163 freshmen out of a possible 1000 students in the school.
Thus

$$p(\text{freshman}) = \frac{163}{1000}$$

Also, there were 122 freshmen who voted in favor of the clinic. Thus,

$$p(\text{freshman and voted in favor of clinic}) = \frac{122}{1000}$$

Summarizing these results we have

$$p(\text{freshman}) = \frac{163}{1000}$$

$$p(\text{freshman and voted in favor of clinic}) = \frac{122}{1000}$$

Let us now divide p(freshman *and* voted in favor of clinic) by p(freshman). We get

$$\frac{p(\text{freshman and voted in favor of clinic})}{p(\text{freshman})}$$

$$= \frac{122/1000}{163/1000}$$

$$= \frac{122}{1000} \div \frac{163}{1000}$$

$$= \frac{122}{1000} \cdot \frac{1000}{163}$$

$$= \frac{122}{163}$$

This is the same result as p(student voted in favor of drug clinic | student is a freshman). In both cases the answer is 122/163.

If we let A stand for "student voted in favor of drug clinic" and B stand for "student is a freshman," then the previous result suggests that

$$p(A|B) = \frac{p(A \text{ and } B)}{p(B)}$$

We can apply the same analysis for part c of the problem. We have

$$p(\text{senior}) = \frac{264}{1000} \quad \text{and} \quad p(\text{senior and no opinion}) = \frac{66}{1000}$$

If we divide p(senior and no opinion) by p(senior) we get

$$\frac{p(\text{senior and no opinion})}{p(\text{senior})} = \frac{66/1000}{264/1000}$$

$$= \frac{66}{1000} \div \frac{264}{1000}$$

$$= \frac{66}{1000} \cdot \frac{1000}{264} = \frac{66}{264} = \frac{33}{132}$$

Thus,

$$p(\text{student had no opinion}\,|\,\text{student is a senior}) = \frac{p(\text{senior and no opinion})}{p(\text{senior})}$$

We can generalize our discussion by using a formula which is called the **conditional probability formula.**

FORMULA 6.3

If A and B are any events, then

$$p(A\,|\,B) = \frac{p(A \text{ and } B)}{p(B)}$$

We illustrate the use of Formula 6.3 with several examples.

EXAMPLE 3

In Ashville the probability that a married man smokes is 0.90. If the probability that a married man *and* his wife smoke is 0.45, what is the probability that his wife smokes, given that he smokes?

SOLUTION

We will use Formula 6.3. We are told that

$$p(\text{man smokes}) = 0.90$$

and

$$p(\text{man and wife smoke}) = 0.45$$

Thus

$$p(\text{wife smokes}\,|\,\text{man smokes}) = \frac{p(\text{man and wife smoke})}{p(\text{man smokes})}$$

$$= \frac{0.45}{0.90}$$

$$= \frac{45}{90} \quad \text{(We multiply numerator and denominator by 100)}$$

$$= \frac{1}{2}$$

Thus, $p(\text{wife smokes}\,|\,\text{man smokes}) = 1/2$.

EXAMPLE 4

Joe often speeds while driving to school in order to arrive on time. The probability that he will speed to school is 0.75. If the probability that he speeds and gets stopped is 0.25, find the probability that he is stopped, given that he is speeding.

SOLUTION

We use Formula 6.3. We are told that p(Joe speeds) is 0.75 and p(speeds and is stopped) is 0.25. Thus

$$p(\text{he is stopped} \mid \text{he speeds}) = \frac{p(\text{speeds and is stopped})}{p(\text{speeds})}$$

$$= \frac{0.25}{0.75}$$

$$= \frac{25}{75}$$

$$= \frac{1}{3}$$

Thus p(he is stopped $\mid$ he speeds) = 1/3.

EXAMPLE 5

Janet likes to study. The probability that she studies *and* passes her math test is 0.40. If the probability that she studies is 0.83, what is the probability that she passes the math test, given that she has studied?

SOLUTION

We use Formula 6.3. We have

$$p(\text{passes math test} \mid \text{she studied}) = \frac{p(\text{studies and passes math test})}{p(\text{she studies})}$$

$$= \frac{0.40}{0.83}$$

$$= \frac{40}{83}$$

Thus, p(she passes math test $\mid$ she has studied) = 40/83.

EXERCISES

1. The probability that Harry will visit the dentist *and* the doctor today is 0.21. If the probability that he visits the doctor is 0.63, what is the probability that he visits the dentist, given that he has just visited the doctor?

2. In a certain community the probability that a man smokes cigars *and* pipes is 1/5. The probability that a man smokes only cigars is 3/5. If a man is seen smoking a cigar, what is the probability that the man also smokes a pipe?

3. The probability that Susan will be talking on the telephone to her boyfriend Steve on a Tuesday evening is 0.44. The probability that she will be talking on the telephone, but not to Steve, is 0.66. If Susan is observed speaking on the telephone Tuesday evening, what is the probability that she is speaking to Steve?

4. Leon is doing the family laundry. The probability that he forgets to use bleach *and* fabric softener is 1/3. The probability that he forgets to use only bleach is 4/9. If we know that Leon forgot to use bleach, what is the probability that he also forgot to use fabric softener?

5. Miguel's car will not start. The probability that he has a dead battery *and* that the car is out of gas is 0.23. The probability that he has a dead battery is 0.53. If he knows that he has a dead battery, what is the probability that the car is also out of gas?

6. Alice is standing in front of a phone booth. There is a piece of paper on the floor which says "Phone Out of Order." Is this an old sign or not? Should she deposit a dime? The probability that the phone is out of order is 8/9. The probability that the phone is out of order *and* that she loses her dime when she deposits it is 7/18. If the phone actually is out of order, what is the probability that she loses her dime?

7. Bruce has been invited to perform his act as part of this year's fund raising campaign. However, Bruce has been very ill recently. The probability that Bruce appears and performs his act is 0.39. The probability that he appears only but does not perform is 0.78. If Bruce is observed entering the studio where the fund raising campaign is being held, what is the probability that he will perform his act?

8. Marilyn is returning from her Florida vacation. The probability that the plane arrives late and that her baggage is lost is 0.07. The probability that the plane is late is 0.11. If the plane arrives late, what is the probability that her baggage is lost?

9. Twelve percent of all senior students at a certain college have tried drugs at one time or another. Twenty-two percent of the students at the college are seniors. If a student is selected at random, what is the probability that the student is a senior who has tried drugs?

10. The probability that a coed at State University wears contact lenses is 0.08. Sixty-one percent of all students at State University are coed. What is the probability that a randomly selected student is a coed who wears contact lenses?

11. The probability that Gerald goes water skiing and skin diving over the weekend is 0.28. The probability that he only goes skin diving is 0.86. If Gerald is seen skin diving, what is the probability he will also go water skiing?

6.3 INDEPENDENT EVENTS

In Section 6.2 we discussed the conditional probability formula and how it is used. In this section we will discuss a variation of the conditional probability formula known as the multiplication rule.

Consider a large electric company in northeastern United States. In recent years it has been unable to meet the demand for electricity. To prevent any cable damage and blackouts as a result of overload, that is, too much electrical demand, it has installed two special switching devices to automatically shut off the flow of electricity and thus prevent cable damage when an overload occurs. The probability that the first switch will not work properly is 0.4, and the probability that the second switch will not work properly, given that the first switch fails is 0.3. What is the probability that both switches will fail?

Let us look at Formula 6.3 in Section 6.2. It says that for any events A and B

$$p(A|B) = \frac{p(A \text{ and } B)}{p(B)}$$

If we multiply both sides of this equation by $p(B)$ we get

$$p(A|B) \cdot p(B) = p(A \text{ and } B)$$

This equation is called the **multiplication rule.** We state this formally as follows:

FORMULA 6.4

Multiplication Rule: If A and B are any events, then

$$p(A \text{ and } B) = p(A|B) \cdot p(B)$$

If we now apply Formula 6.4 to our example we get

$p(\text{both switches fail}) = p(\text{switch 2 fails}|\text{switch 1 has failed}) \cdot p(\text{switch 1 fails})$
$= (0.3)(0.4)$
$= 0.12$

Thus, the probability that both switches fail is 0.12.

EXAMPLE 1

In a certain community the probability that a man over 40 years old is overweight is 0.42. The probability that his blood pressure is high given that he is overweight is 0.67. If a man over 40 years of age is selected at random, what is the probability that he is overweight and that he has high blood pressure?

SOLUTION

We use Formula 6.4. We have

$$
\begin{aligned}
p(\text{overweight and high blood pressure}) &= p(\text{high blood pressure}|\text{overweight}) \cdot p(\text{overweight}) \\
&= (0.67)(0.42) \\
&= 0.2814
\end{aligned}
$$

Thus, the probability that a man over 40 is overweight and has high blood pressure is approximately 0.28.

EXAMPLE 2

A new cleansing product has recently been introduced and is being advertised on television as having remarkable cleansing qualities. The manufacturer believes that if a homemaker is selected at random, then the probability that she watches television and sees the commercial between the hours of 12 noon and 4 PM is 4/11. Furthermore, if she sees the commercial, then the probability that she buys the cleanser is 22/36. What is the probability that a homemaker selected at random will watch television *and* buy the product?

SOLUTION

We use Formula 6.4. We have

$$
\begin{aligned}
p(\text{watches TV and buys product}) &= p(\text{buys product}|\text{watches TV}) \cdot p(\text{watches TV}) \\
&= \frac{22}{36} \cdot \frac{4}{11} \\
&= \frac{88}{396} = \frac{2}{9}
\end{aligned}
$$

Thus, the probability that a homemaker selected at random watches television and buys the cleanser is 2/9.

In many cases it turns out that whether or not one event happens does not affect whether another will happen. For example, if two cards are drawn from a deck and the first card is replaced before the second card is drawn, the outcome on the first draw has nothing to do with the outcome on the second draw. Also, if two dice are rolled, the outcome for one die has nothing to do with the outcome for the second die. Such events are called **independent events.**

DEFINITION 6.2

Two events A and B are said to be **independent** if the occurrence of event B is in no way affected by the occurrence or non-occurrence of event A.

When dealing with independent events, we can simplify Formula 6.4. The following example will show this.

EXAMPLE 3

Two cards are drawn from a deck of 52 cards. Find the probability that both cards drawn are aces if the first card
a. is *not* replaced before the second card is drawn.
b. is replaced before the second card is drawn.

SOLUTION

a. Since the first card is not replaced, we use the multiplication rule. We have

p(both cards are aces) $= p$(2nd card is ace $|$ 1st card is ace) $\cdot p$(1st card is ace)

Notice that since the first card is not replaced, there are only 3 aces remaining out of a possible 51 cards. This is because the first card removed was an ace. Thus

$$p\text{(both cards are aces)} = \frac{3}{51} \cdot \frac{4}{52}$$

$$= \frac{12}{2652} = \frac{1}{221}$$

Thus, the probability that both cards are aces is 1/221.

b. Since the first card is replaced before the second card is drawn, then whether or not an ace appeared on the first card in no way affects what happens on the second draw. The events "ace on

second draw" and "ace on first draw" are independent. Thus
p(2nd card is ace | 1st card is ace) is exactly the same as
p(2nd card is ace). Therefore

p(both cards are aces) $= p$(2nd card is an ace | 1st card is ace) $\cdot p$(1st card is ace)

$$= p(\text{2nd card is ace}) \cdot p(\text{1st card is ace})$$

$$= \frac{4}{52} \cdot \frac{4}{52}$$

$$= \frac{16}{2704} = \frac{1}{169}$$

Hence the probability that both cards are aces in this case is 1/169.

Example 3 suggests that if two events A and B are independent, we can substitute $p(B)$ for $p(B|A)$ since B is in no way affected by what happens with A. We then get a special multiplication rule for independent events.

FORMULA 6.5

If A and B are independent events, then

$$p(A \text{ and } B) = p(A) \cdot p(B)$$

EXAMPLE 4

Two students who do not know each other wish to satisfy the language requirement at their school. The probability that Carlos chooses Spanish is 0.86 and the probability that Pedro chooses Spanish is 0.73. What is the probability that they both choose Spanish?

SOLUTION

Since both students do not know each other, the events "Carlos chooses Spanish" and "Pedro chooses Spanish" are independent. We therefore use Formula 6.5. We have

p(both choose Spanish) $= p$(Carlos chooses Spanish) $\cdot p$(Pedro chooses Spanish)

$$= (0.86)(0.73)$$

$$= 0.6278$$

Thus the probability that they both choose Spanish is approximately 0.63

EXAMPLE 5

If the probability of a college student having untreated gonorrhea is 0.15, what is the probability that two totally unrelated college students do *not* have untreated gonorrhea?

SOLUTION

These are independent events so we use Formula 6.5. The probability of a college student having untreated gonorrhea is 0.15. Thus the probability that the student does not have untreated gonorrhea is $1 - 0.15$, or 0.85. Therefore,

$$p(\text{both students no gonorrhea}) = p(\text{student 1 no gonorrhea}) \cdot p(\text{student 2 no gonorrhea})$$
$$= (0.85)(0.85)$$
$$= 0.7225$$

Hence, the probability that neither of two totally unrelated college students have untreated gonorrhea is approximately 0.72.

Comment The multiplication rule for independent events can be generalized for more than two independent events. We simply multiply all the respective probabilities. Thus if event A has probability 0.7 of occurring, event B has probability 0.6 of occurring, and event C has probability 0.5 of occurring, and if these events are independent, then the probability that all three occur is

$$p(A \text{ and } B \text{ and } C) = p(A) \cdot p(B) \cdot p(C)$$
$$= (0.7)(0.6)(0.5)$$
$$= 0.21$$

Therefore, the probability that all three occur is 0.21.

EXERCISES

1. The probability that Floyd will forget to put money in the meter when he parks is 0.62. The probability that he gets a ticket given that he forgot to put money in the meter is 0.48. What is the probability that he will forget to put money in the meter *and* that he will get a ticket?
2. The probability that Jack will be late to work is 0.37. The probability that his boss will find out given that he is late is 0.72. What is the probability that Jack will be late to work today *and* that his boss finds out?
3. A leading department store has had bad experiences with customers who pay by check. The probability that a customer will pay for some purchases with a check that "bounces" is 0.28. Furthermore, the probability that a customer will

pay by check is 0.78. Bill is about to pay for some purchases. What is the probability that Bill pays for them with a check that will bounce?

4. The probability that the truck drivers who are negotiating for a new contract with a bakery get a pay raise is 0.42. The probability that the price of bread will go up given that the truck drivers got a pay raise is 0.39. What is the probability that the truck drivers get a raise and that the price of bread goes up?

5. Peter is interested in becoming a secret service agent. The probability that he passes the physical examination is 0.89. The probability that he passes the written exam given that he has passed the physical exam is 0.62. What is the probability that Peter passes both exams?

6. The probability that they raise tuition next semester at Whipple University is 0.56. The probability that the student enrollment will drop by at least 5000 students if they raise tuition is 0.48. What is the probability that they raise tuition and that the student enrollment drops by at least 5000 students?

7. Peter and Mary took their driving tests today. The probability that Peter passes the test is 6/10 and the probability that Mary passes the test is 7/10. What is the probability that they both pass their driving tests?

8. An executive has two phones on his desk, each with a different number and neither of which is an extension of the other. The probability that phone 1 rings is 0.38 and the probability that phone 2 rings is 0.59. What is the probability that both phones ring?

9. The probability that Danny plays a guitar is 0.55, the probability that Manny plays the guitar is 0.43, and the probability that Fanny plays the guitar is 0.72. What is the probability that all three play the guitar? Assume independence.

10. In Exercise 9 what is the probability that Danny and Manny play the guitar and that Fanny *does not* play the guitar?

11. Two people who do not know each other have volunteered to search for a lost dog independently. The probability that the first volunteer finds the dog is 0.62 and the probability that the second volunteer finds the dog is 0.56. What is the probability that neither volunteer finds the lost dog?

12. At Los Gables airport two kinds of security checks are used to prevent any passenger from taking a bomb and/or a gun on the airplane. One is a visual check by a security guard and the second is a screening by a metal detector. The probability that the guard stops a person carrying a gun is 0.52 and the probability that the person is caught by the metal detector given that he was not stopped by the guard is 0.96. What is the probability that a person will *not* be stopped by the guard nor caught by the metal detector?

13. Clark Kent has just boarded a four-engine plane to attend a crime commission conference in Washington. Each engine operates independently and the probability of any engine failing is 0.19. If the plane can fly safely with only one engine working, what is the probability that the plane crashes due to engine failure?

14. Martha is a nature lover. Recently she snapped seven pictures of an egg about to hatch. If the probability that any one picture comes out clearly is 0.7, what is the probability that none of the seven pictures will be clear?

★6.4 **BAYES' FORMULA**

The conditional probability formula discussed in previous sections allows us to calculate the probability of some event, given that some other event has occurred. In many situations, however, we are given the outcome of an experiment and we are interested in determining the probability that the outcome happened because of a particular cause. The clipping on page 146 is an example of this. For these situations we need a formula.

EXAMPLE 1

Two boxes containing marbles are placed on a table. The boxes are labeled B_1 and B_2. Box B_1 contains 7 green marbles and 4 white marbles. Box B_2 contains 3 green marbles and 10 yellow marbles. Furthermore, the boxes are arranged so that the probability of selecting box B_1 is 1/3 and the probability of selecting box B_2 is 2/3. Trudy is blindfolded and asked to select a marble. She will win a color television set if she selects a green marble.
a. What is the probability that Trudy will win the television, that is, select a green marble?
b. If Trudy wins the color television, what is the probability that the green marble was selected from the first box?

SOLUTION

Trudy can win the color television by selecting a green marble from either box B_1 or box B_2. Let A represent the event of drawing a green marble, B_1 represent the event "the marble was taken from box B_1," and let B_2 represent the event "the marble was selected from box B_2." We use a tree diagram to illustrate the possible ways in which Trudy can win.

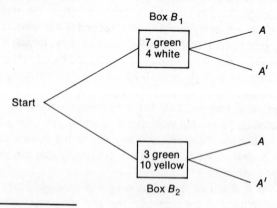

★A star indicates that the section requires more time and thought.

We are told that the probability of selecting box B_1 is 1/3 and the probability of selecting box B_2 is 2/3. Since box B_1 contains 7 green and 4 white marbles we have

$$p(\text{selecting green marble}|\text{box } B_1 \text{ is chosen}) = \frac{7}{11}$$

Symbolically,

$$p(A|B_1) = \frac{7}{11}$$

Using the multiplication formula (Formula 6.4 on page 164) we have

$p(\text{choosing box } B_1 \text{ and selecting green marble})$
$$= p(\text{selecting green marble}|B_1 \text{ is chosen}) \cdot p(\text{choosing box } B_1)$$

$$= \frac{7}{11} \cdot \frac{1}{3}$$

$$= \frac{7}{33}$$

Symbolically,

$$p(A \text{ and } B_1) = p(A|B_1) \cdot p(B_1) = \frac{7}{33}$$

Also, box B_2 contains 3 green and 10 yellow marbles, so that

$$p(\text{selecting green marble}|\text{box } B_2 \text{ is chosen}) = \frac{3}{13}$$

Thus,

$$p(A \text{ and } B_2) = p(A|B_2) \cdot p(B_2)$$

$$= \frac{3}{13} \cdot \frac{2}{3}$$

$$= \frac{2}{13}$$

Trudy can win in one of two mutually exclusive ways:
a. Box B_1 is selected and a green marble is chosen from it.
b. Box B_2 is selected and a green marble is chosen from it.

Since Trudy can select from only one box, these events are

mutually exclusive and we can then use Formula 6.1 (see page 148 to determine the probability of her winning). We have

$$p(\text{Trudy wins}) = p(\text{chooses box } B_1 \text{ and selects green marble})$$
$$+ \, p(\text{chooses box } B_2 \text{ and selects green marble})$$

Symbolically,

$$p(\text{Trudy wins}) = p(A \text{ and } B_1) + p(A \text{ and } B_2)$$
$$p(A) = p(A|B_1) \cdot p(B_1) + p(A|B_2) \cdot p(B_2)$$
$$= \frac{7}{33} + \frac{2}{13}$$
$$= \frac{91}{429} + \frac{66}{429}$$
$$= \frac{157}{429}$$

Thus, the probability that Trudy wins is 157/429.

b. According to the conditional probability formula, we have

$$p(\text{box } B_1 \text{ chosen} \,|\, \text{green marble selected})$$
$$= \frac{p(\text{chooses box } B_1 \text{ and selects a green marble})}{p(\text{green marble selected})}$$

$$p(B_1|A) = \frac{p(A \text{ and } B_1)}{p(A)}$$

$$= \frac{7/33}{157/429} \qquad \text{(We substitute the values obtained from part } a.\text{)}$$

$$= \frac{7}{33} \div \frac{157}{429}$$

$$= \frac{7}{33} \times \frac{429}{157}$$

$$= \frac{91}{157}$$

Thus the probability that box B_1 was chosen given that a green marble was selected is 91/157. In a similar manner we can calculate the probability that box B_2 was selected given that a green marble was chosen. We have

p(box B_2 chosen | green marble selected)

$$= \frac{p(\text{chooses box } B_2 \text{ and selects green marble})}{p(\text{green marble selected})}$$

$$p(B_2|A) = \frac{p(A \text{ and } B_2)}{p(A)}$$

$$= \frac{2/13}{157/429}$$

$$= \frac{2}{13} \div \frac{157}{429}$$

$$= \frac{2}{13} \times \frac{429}{157}$$

$$= \frac{66}{157}$$

Thus, the probability that box B_2 was chosen given that a green marble was selected is 66/157.

We can symbolize Example 1 in the following manner:

$$p(B_1|A) = \frac{p(A \text{ and } B_1)}{p(A)}$$

$$= \frac{p(A \text{ and } B_1)}{p(A \text{ and } B_1) + p(A \text{ and } B_2)}$$

$$= \frac{p(A|B_1) \cdot p(B_1)}{p(A|B_1) \cdot p(B_1) + p(A|B_2) \cdot p(B_2)}$$

Also

$$p(B_2|A) = \frac{p(A \text{ and } B_2)}{p(A)}$$

$$= \frac{p(A \text{ and } B_2)}{p(A \text{ and } B_1) + p(A \text{ and } B_2)}$$

$$= \frac{p(A|B_2) \cdot p(B_2)}{p(A|B_1) \cdot p(B_1) + p(A|B_2) \cdot p(B_2)}$$

The previous example can be generalized to more than two boxes. If we have three boxes, we get

$$p(B_1|A) = \frac{p(A|B_1) \cdot p(B_1)}{p(A|B_1) \cdot p(B_1) + p(A|B_2) \cdot p(B_2) + p(A|B_3) \cdot p(B_3)}$$

Similar formulas can be developed for $p(B_2|A)$ and $p(B_3|A)$.

More generally we have the following.

BAYES' RULE

Consider a sample space which is composed of the mutually exclusive events $A_1, A_2, A_3, \ldots, A_n$. Suppose each event has a non-zero probability of occurring and that one must definitely occur. If B is any event in the sample space, then

$$p(A_1|B) = \frac{p(B|A_1) \cdot p(A_1)}{p(B|A_1) \cdot p(A_1) + p(B|A_2) \cdot p(A_2) + \cdots + p(B|A_n) \cdot p(A_n)}$$

$$p(A_2|B) = \frac{p(B|A_2) \cdot p(A_2)}{p(B|A_1) \cdot p(A_1) + p(B|A_2) \cdot p(A_2) + \cdots + p(B|A_n) \cdot p(A_n)}$$

$$\vdots$$

$$p(A_n|B) = \frac{p(B|A_n) \cdot p(A_n)}{p(B|A_1) \cdot p(A_1) + p(B|A_2) \cdot p(A_2) + \cdots + p(B|A_n) \cdot p(A_n)}$$

Bayes' rule may seem rather complicated but it is easy to use as the following examples illustrate.

EXAMPLE 2

If $p(A|B) = 1/5$, $p(A|C) = 2/7$, $p(B) = 1/2$, and $p(C) = 1/2$, find
a. $p(B|A)$ and b. $p(C|A)$.

SOLUTION

a. We use Bayes' rule. We have

$$p(B|A) = \frac{p(A|B) \cdot p(B)}{p(A|B) \cdot p(B) + p(A|C) \cdot p(C)}$$

$$= \frac{(1/5)(1/2)}{(1/5)(1/2) + (2/7)(1/2)}$$

$$= \frac{1/10}{(1/10) + (1/7)}$$

$$= \frac{1/10}{17/70} = \frac{1}{10} \div \frac{17}{70}$$

$$= \frac{1}{10} \cdot \frac{70}{17} = \frac{7}{17}$$

Thus, $p(B|A) = 7/17$.

b. Again we use Bayes' formula. We have

$$p(C|A) = \frac{p(A|C) \cdot p(C)}{p(A|B) \cdot p(B) + p(A|C) \cdot p(C)}$$

$$= \frac{(2/7)(1/2)}{(1/5)(1/2) + (2/7)(1/2)} = \frac{1/7}{(1/10) + (1/7)} = \frac{1/7}{17/70}$$

$$= 10/17$$

Thus $p(C|A) = 10/17$.

EXAMPLE 3

A prisoner has just escaped from jail. There are three roads leading away from the jail. If the prisoner selects road A to make good her escape, the probability that she succeeds is 1/4. If she selects road B, the probability that she succeeds is 1/5. If she selects road C, the probability that she succeeds is 1/6. Furthermore, the probability that she selects each of these roads is the same. It is 1/3. If the prisoner succeeds in her escape, what is the probability that she made good her escape by using road B?

SOLUTION

We use Bayes' rule. We have

$$p(\text{used road } B \,|\, \text{succeeded}) = \frac{p(\text{succeeds}\,|\,\text{uses road } B) \cdot p(\text{uses road } B)}{\begin{array}{l} p(\text{succeeds}\,|\,\text{uses raod } A) \cdot p(\text{uses road } A) + p(\text{succeeds}\,|\,\text{uses} \\ \text{road } B) \cdot p(\text{uses road } B) + p(\text{succeeds}\,|\,\text{uses road } C) \cdot p(\text{uses} \\ \text{road } C) \end{array}}$$

$$= \frac{(1/5)(1/3)}{(1/4)(1/3) + (1/5)(1/3) + (1/6)(1/3)}$$

$$= \frac{1/15}{(1/12) + (1/15) + (1/18)}$$

$$= 12/37$$

Thus, the probability that she made good her escape by using road B is 12/37.

EXAMPLE 4

Professor Gordon, a faculty adviser, submits the following report about one of his students, Mildred. The probabilities that she registers for mathematics, ecology, history, and English next semester are 0.2, 0.5, 0.3, and 0.1, respectively. Furthermore, he estimates that the probability that she gets an A in each of these courses is 0.6, 0.2, 0.1, and 0.1, respectively. Professor Gordon does not see Mildred the following semester but is informed by the registrar that she received an A in one of her courses. He does not know which course. What is the probability that she got the A in mathematics?

SOLUTION

We use Bayes' formula. Let A represent the event "Mildred gets an A", B_1 represent the event she registers for math, B_2 represent the event she registers for ecology, B_3 represent the event she registers for history, and B_4 represent the event she registers for English. Then we are interested in $p(B_1|A)$. Using Bayes' rule we have

$$p(B_1|A) = \frac{p(A|B_1) \cdot p(B_1)}{p(A|B_1) \cdot p(B_1) + p(A|B_2) \cdot p(B_2) + p(A|B_3) \cdot p(B_3) + p(A|B_4) \cdot p(B_4)}$$

$$= \frac{(0.6)(0.2)}{(0.6)(0.2) + (0.2)(0.5) + (0.1)(0.3) + (0.1)(0.1)}$$

$$= \frac{0.12}{0.26} = \frac{12}{26} = \frac{6}{13}$$

Thus, the probability that she got the A in mathematics is 6/13.

EXERCISES

1. There are two boxes on a table. Box B_1 contains 10 red balls and 6 white balls. Box B_2 contains 4 red balls and 3 white balls. A ball is selected at random from one of these boxes and it is observed to be white. If the probability of selecting either of these boxes is 1/2, what is the probability that the white ball was selected from box B_2?

2. If $p(A|B) = 1/7$, $p(A|C) = 1/8$, $p(A|D) = 1/9$, $p(B) = 1/5$, $p(C) = 2/5$, and $p(D) = 2/5$, find
   a. $p(B|A)$.
   b. $p(C|A)$.
   c. $p(D|A)$.

3. A gambler is in Las Vegas. There are two machines into which he can insert money and possibly win the jackpot. The probability that he selects machine I is 2/5 and the probability that he selects machine II is 3/5. The probability that he wins at these machines is 1/9 and 1/15 respectively. If he wins the jackpot, what is the probability that it was from machine I?

4. A coin is tossed three times. If it is known that exactly one of these tosses showed tails, what is the probability that it was on the first toss?

5. Bill is in the bank. There are two tellers serving the customers. The probability that he goes to teller A is 0.6 and the probability that he goes to teller B is 0.4. The probabilities that he has to wait on line at least 10 minutes before being served by tellers A and B are 0.3 and 0.7 respectively. If he has waited at least 10 minutes, what is the probability that he is on teller A's line?

6. A psychologist is conducting experiments with rats. A rat can enter maze I, maze II, or maze III with probabilities 0.6, 0.3, and 0.1 respectively. Furthermore, the probability that the rat finds its way through each maze is 0.5, 0.4, and 0.1 respectively. If a rat actually finds its way through a maze, what is the probability that it went through maze II?

7. A manufacturer has just received two truckloads of televisions. Truckload I contains 6 defective and 29 good televisions. Truckload II contains 4 defective and 11 good televisions. An inspector randomly selects a television and notices that it is defective. Unfortunately, he does not remember from which truckload the sample came. What is the probability that the defective television came from truckload II?

SELF-STUDY GUIDE

In this chapter we discussed many different rules concerning the calculation of probabilities. Each formula given applies to different situations. Thus the addition rule allows us to determine the probability of event A or event B or both events happening. We distinguished between mutually exclusive and non-mutually exclusive events and their effect on the addition rule.

We then discussed conditional probability and how the probability of one event is affected by the occurrence or non-occurrence of a second event. This led us to the multiplication rule. When one event is in no way affected by the occurrence or non-occurrence of a second event, we have independent events and a simplified multiplication rule.

Finally we discussed Bayes' formula, which is used when we know the outcome of some experiment and are interested in determining the probability that it was caused by or is the result of some other event. In each case many applications of all the formulas introduced were given.

You should be able to identify each symbol in the following formulas, understand the relationships among the symbols expressed in each formula, understand the significance of each formula, and use the formulas in solving problems.

1. Addition rule: $p(A \text{ or } B) = p(A) + p(B)$ for mutually exclusive events
2. Addition rule: $p(A \text{ or } B) = p(A) + p(B) - p(A \text{ and } B)$ general case
3. Complement of event A: $p(A') = 1 - p(A)$
4. Conditional probability formula: $p(A|B) = \dfrac{p(A \text{ and } B)}{p(B)}$
5. Multiplication rule: $p(A \text{ and } B) = p(A|B) \cdot p(B)$
6. Multiplication rule: $p(A \text{ and } B) = p(A) \cdot p(B)$ for independent events
7. Bayes' rule:

$$p(A_n|B) = \frac{p(B|A_n) \cdot p(A_n)}{p(B|A_1) \cdot p(A_1) + p(B|A_2) \cdot p(A_2) + \cdots + p(B|A_n) \cdot p(A_n)}$$

You should now be able to demonstrate your knowledge of the following ideas presented in this chapter by giving definitions, descriptions, or specific examples. Page references are given for each term so that you can check your answer.

Mutually exclusive events (page 148) Conditional probability formula (page 161)
Complement of an event (page 151) Multiplication rule   (page 164)
Addition rule   (page 152) Independent events   (page 166)
Conditional probability (page 158) Bayes' rule   (page 174)

The tests of the following section will be more useful if you take them after you have studied the examples and solved the exercises given in this chapter.

MASTERY TESTS

Form *A*

1. Explain the difference between mutually exclusive events and independent events.
2. Which of the following events are mutually exclusive:
    a. getting pregnant and getting a headache?
    b. cheating on your math final and getting caught?
3. Which of the following events are independent:
    a. driving while intoxicated and getting into an accident?
    b. setting the alarm clock to wake you up in the morning and getting to class on time?
4. The probability that it rains tomorrow is 0.63. What is the probability that it does not rain?
5. True or false: If two events are mutually exclusive, their probabilities must add up to 1.

6. True or false: The multiplication rule is used when we are dealing with probability questions involving the word *and*.

7. True or false: The probability that Bill owns a Chevrolet is 0.62. The probability that he owns a Ford is 0.73. Therefore, the probability that he owns a Chevrolet or a Ford is 0.62 + 0.73, or 1.35.

8. The probability that your math teacher will be absent today is 0.7. The probability that your history teacher will be absent is 0.7. Therefore the probability that both will be absent is 0.49, which would mean that the probability that both will be present is 0.51. Do you agree with this reasoning? Explain your answer.

9. In a certain community the probability that an individual has a certain type of blood is as follows:

Blood Type	Probability
A	40%
B	12%
AB	3%
O	45%

   Robert is planning to marry Andrea. What is the probability that Robert has type A blood and Andrea has type O? Assume independence.

10. In test question 9 find the probability that Robert and Andrea both have the same blood type.

Form *B*

1. Three women are independently attempting to swim around the island of Manhattan in New York City. Their probabilities of succeeding are 0.3, 0.2, and 0.1. Find the probability that
   a. none of them succeed.
   b. all of them succeed.
   c. only one of them succeeds.
   d. at least one of them succeeds.

2. If only 2 people are in a bus, what is the probability that their birthdays are different?

3. If only 3 people are in a bus, what is the probability that none of them have the same birthday?

4. The probability that Phil attends the dance Friday night is 0.79, the probability that Bill attends the dance is 0.86, and the probability that Gill attends the dance is 0.53. What is the probability that all 3 attend the dance? Assume independence.

5. In test question 4 what is the probability that Phil and Bill attend and that Gill does not attend the dance?

Would you believe that we need only 23 people in a crowd to have a 50% probability that at least 2 of these people would have the same birthday? The probability increases to about 1, almost a certainty, when we have a crowd of 63 people.

6. A manufacturer has just received a shipment of 50 transistor radios. Unknown to him 8 are defective. He samples the shipment by selecting 2 radios without replacement. What is the probability that he selects
   a. 2 good radios?
   b. 2 defective radios?
   c. 1 defective radio and 1 good radio?

7. Dennis is stranded in the desert. He has 7 flares which he can use to send for help. If the probability is 0.3 that any flare will be noticed, what is the probability that his cries for help are *not* noticed?

8. A family that is known to have 2 children, not twins, is selected at random from among all families with 2 children. If it is known that there is a girl in the family, what is the probability that both children are girls?

9. Mr. and Mrs. Pascal have 2 girls and would like to have a boy. After taking a course in probability they decide to have a third child, reasoning that the probability that the third child is a boy is 7/8. There are 8 possible ways in which a family can consist of 3 children as shown in the following chart. Of these, 7 of them have at least one boy. Thus, they reason the probability that the third child is a boy is 7/8. Do you agree with this reasoning? Explain your answer.

	Child 1	Child 2	Child 3
Possibility 1	Boy	Boy	Boy
Possibility 2	Boy	Boy	Girl
Possibility 3	Boy	Girl	Boy
Possibility 4	Boy	Girl	Girl
Possibility 5	Girl	Boy	Boy
Possibility 6	Girl	Boy	Girl
Possibility 7	Girl	Girl	Boy
Possibility 8	Girl	Girl	Girl

10. In June of 1964 an elderly woman was mugged in San Pedro, California. In the vicinity of the crime a bearded black man sat waiting in a yellow car. Shortly after the crime was committed, a young white woman, wearing her blonde hair in a ponytail, was seen running from the scene of the crime and getting into the car which sped off. The police broadcast a description of the suspected muggers. Soon afterwards, a couple fitting the description was arrested and convicted of the crime. Although the evidence in the case was largely circumstantial, the prosecutor based his case on probability and the unlikeliness of another couple having such characteristics. He assumed the following probabilities.

Characteristic	Assumed probability
Drives yellow car	1/10
Black–white couple	1/1000
Black man	1/3
Man with beard	1/10
Blonde woman	1/4
Woman wears her hair in ponytail	1/10

He then multiplied the individual probabilities:

$$\frac{1}{10} \cdot \frac{1}{1000} \cdot \frac{1}{3} \cdot \frac{1}{10} \cdot \frac{1}{4} \cdot \frac{1}{10} = \frac{1}{12,000,000}$$

He claimed that the probability is 1/12,000,000 that another couple has such characteristics. The jury agreed and convicted the couple. The conviction was overturned by the California Supreme Court in 1968. The defense attorneys got some professional advice on probability. Serious errors were found in the prosecutor's probability calculations. Some of these involved assumptions about independent events. As a matter of fact, it was demonstrated that the probability is 0.41 that another couple with the same characteristics existed in the area once it was known that there was at least one such couple.

For a complete discussion of this probability case, read "Trial by Mathematics" which appeared in *Time* January 8, 1965, p. 42, and April 26, 1968, p. 41.

SUGGESTED READING

Freund, J. *Statistics, A First Course.* Englewood Cliffs, N.J.: Prentice-Hall, 1970.

Glass, G. V. and Stanley, J. C. *Statistical Methods in Education and Psychology.* Englewood Cliffs, N.J.: Prentice-Hall, 1970.

Guilford, J. P. and Fruchter, B. *Fundamental Statistics in Psychology and Education.* New York: McGraw-Hill, 1973.

Hoel, P. G. *Elementary Statistics.* New York: Wiley, 1966.

"Mathematics in the Modern World." Readings from *Scientific American,* article 22, San Francisco, Calif.: Freeman, 1968.

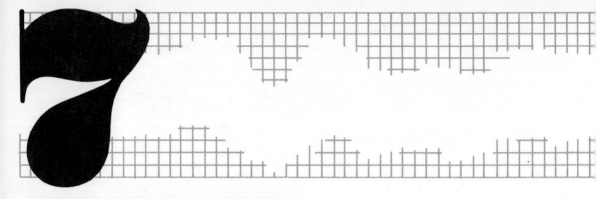

THE BINOMIAL DISTRIBUTION

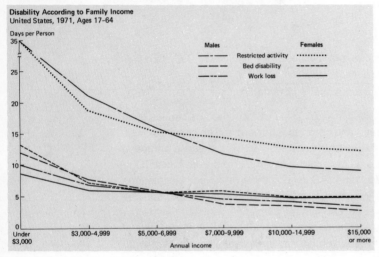

Disability According to Family Income
United States, 1971, Ages 17–64

Table courtesy of The Metropolitan Life
Insurance Company.

AMERICAN EXPERIENCE TABLE OF MORTALITY			
AGE IN YRS.	NUMBER ALIVE	AGE IN YRS.	NUMBER ALIVE
10	100,000	55	64,600
15	96,300	60	57,900
20	92,600	65	49,300
25	89,000	70	38,600
30	85,400	75	26,200
35	81,800	80	14,500
40	78,100	85	5,500
45	74,200	90	850
50	69,800	95	3

The graph indicates the results of a National Health Survey. At virtually every age and in every salary category, men experienced substantially lower disability rates than women. Moreover, the number of days of disability in 1971 decreased steadily with an advance in family income. Based upon such information, insurance companies set up various charts indicating the probability of the number of days that a disability will last depending upon family income.

The table gives the distribution of the number of people alive at various ages. Such information is used to set up the life insurance rates that we pay. They are based upon probability calculations.

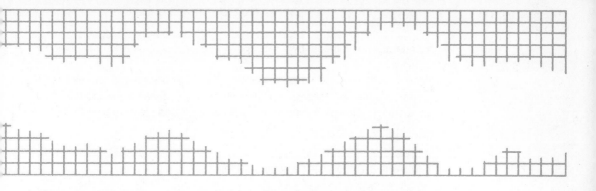

Preview

In this chapter we will discuss the following topics:

1. *Random Variable*
The outcome of some experiment that is of interest to us.

2. *Probability Functions*
These assign probabilities to the different values of the random variable.

3. *Mean and Variance of a Probability Distribution*
These give us the expected value of a probability function as well as the spread of the distribution.

4. *Bernoulli (Binomial) Experiments*
This is an experiment which has only two possible outcomes, success or failure.

5. *Binomial Distribution*
We discuss this very important probability distribution which gives us the probability of obtaining a specified number of successes when an experiment is performed *n* times.

6. *Binomial Distribution Properties*
We discuss the mean and standard deviation of the binomial distribution.

Introduction

In Chapter 2 we discussed frequency distributions of sets of data. Using these distributions we were able to analyze data more intelligently to determine which outcomes occurred most often, least often, etc. In Chapters 5 and 6 we discussed the various rules of probability and how they can be applied to many different situations. These rules enable us to predict how often something will happen in the long run. In this chapter we will combine these ideas.

In any given experiment there may be many different things of interest. For example, if a scientist decides to mate a white rat with a black rat, she may be interested in the number of offspring that are white, black, gray, etc. We will therefore have to define what is meant by a **random variable** and we will then discuss its **probability function.** Special emphasis will be given in this chapter to the binomial random variable and its probability distribution.

7.1 PROBABILITY FUNCTIONS

To understand what is meant by a probability function, let us analyze the following examples.

Valerie is a dentist and keeps accurate records on the number of cavities of each of her patients. Her records indicate that each patient has anywhere from 0 to 5 cavities. Based upon past experience, she has compiled the data given in Table 7.1.

TABLE 7.1

THE NUMBER OF CAVITIES AND THEIR PROBABILITIES

NUMBER OF CAVITIES	PROBABILITY
0	1/16
1	4/16
2	5/16
3	3/16
4	2/16
5	1/16
	Total $= 16/16 = 1$

Notice that each patient has anywhere from 0 to 5 cavities. Thus, the number of cavities that each patient has is somehow dependent upon chance, as the probabilities in the table indicate.

Eric is an elevator operator. The number of people who enter the elevator at exactly 9:00 A.M. varies from 0 to 10 people. The capacity of the elevator is 10 people. Eric has compiled the data given in Table 7.2.

TABLE 7.2

THE NUMBER OF PEOPLE ENTERING THE ELEVATOR AND THEIR PROBABILITIES

NUMBER OF PEOPLE ENTERING ELEVATOR	PROBABILITY
0	1/50
1	3/50
2	4/50
3	5/50
4	7/50
5	8/50
6	10/50
7	6/50
8	3/50
9	2/50
10	1/50
	Total $= 50/50 = 1$

In each of the two previous examples the values which the item of interest assumed, that is, the number of cavities or the number of people entering the elevator, were somehow dependent upon chance. We refer to such a quantity as a **random variable.**

The term random variable applies to many different situations. Thus, it may represent the number of people buying tickets to a movie, the number of typing mistakes that a secretary will make in typing a letter, the number of telephone calls received by the school switchboard during the month of September, the number of games that the Green Bay Packers will win next season, or the number of students that will enroll in a particular course, Music 161, to be offered for the first time in the spring.

The following examples further illustrate the idea of a random variable.

EXAMPLE 1

Three cards are selected, without replacement, from a deck of 52. The random variable may be the number of aces obtained. It would then have values of 0, 1, 2, or 3 depending upon the number of aces actually obtained.

EXAMPLE 2

Calvin drives his car over some nails. The random variable is the number of flat tires that Calvin gets. The values of the random variable are 0, 1, 2, 3, 4, corresponding to 0 flats, 1 flat, 2 flats, 3 flats, and 4 flats, respectively.

EXAMPLE 3

Let the number of people who will attend the next concert at the Hollywood Bowl be the random variable of interest. Then this random variable can assume values ranging from 0 to the seating capacity of the Hollywood Bowl.

EXAMPLE 4

A nurse is taking Chuck's blood pressure. Let the random variable be Chuck's systolic blood pressure. What values can the random variable assume?

EXAMPLE 5

Get on a scale and weigh yourself. Let the random variable be your weight. What values can the random variable assume?

185

Let us now return to the two examples discussed at the beginning of this section. You will notice that the probabilities associated with the different values of the random variable are indicated. Thus, Table 7.1 tells us that the probability of having one cavity is 4/16 and that the probability of having three cavities is 3/16. Similarly, Table 7.2 tells us that the probability of three people entering the elevator is 5/50.

When discussing a random variable, we are almost always interested in assigning probabilities to the various values of the random variable. For this reason we now discuss probability functions.

DEFINITION 7.1

A **probability function** is a correspondence which assigns probabilities to the values of a random variable.

EXAMPLE 6

If a pair of dice is rolled, the random variable that may be of interest to us is the number of dots on both of the dice together. When a pair of dice is rolled, there are 36 possible outcomes. These are shown in Figure 7.1. We can then set up the following chart:

Sum on Both Dice	Number of Different Ways in which Sum Can Be Obtained	Probability
2	1	1/36
3	2	2/36
4	3	3/36
5	4	4/36
6	5	5/36
7	6	6/36
8	5	5/36
9	4	4/36
10	3	3/36
11	2	2/36
12	1	1/36

$$\text{Total} = 36/36 = 1$$

EXAMPLE 7

An airplane has 4 engines, each of which operates independently of the other. If the random variable is the number of engines that are functioning properly, the random variable has values 0, 1, 2, 3, 4. For a particular airline we have the following probability function:

Number of Engines Functioning Properly	Probability
0	1/19
1	2/19
2	4/19
3	5/19
4	7/19
Total =	19/19 = 1

Notice that in Example 7, as well as in Examples 1 to 6, the sum of all the probabilities is 1. This will be true for any probability function. Also, the probability that the random variable assumes any one particular value is between 0 and 1. Again, this is true in every case. We state this formally as

RULE

a. The sum of all the probabilities of a probability function is always 1.

b. The probability that a random variable assumes any one value in particular is between 0 and 1. Zero means that it can never happen and 1 means that it will always have that value.

Comment Some authors distinguish between **probability functions** and **probability distributions.** In this book we will use these terms interchangeably.

Figure 7.1

Thirty-six possible outcomes when a pair of dice is rolled.

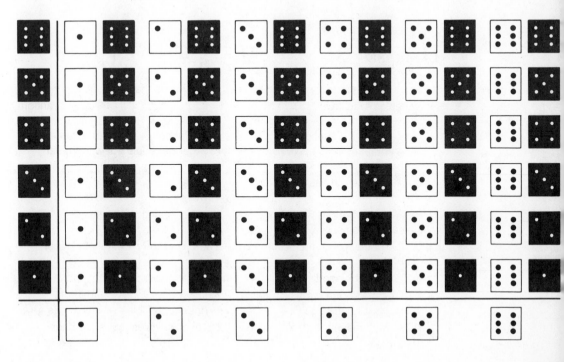

In all examples mentioned thus far, some of the random variables assumed many values and some took on only two values. The following random variables take on many values:

1. The number of flat tires that Calvin gets when he drives over nails.
2. The sum obtained when rolling a pair of dice. There are 11 possible values as indicated on page 186.
3. The number of people that are in an elevator.

The following random variables take on only two variables:

1. The possible outcome when one coin is flipped (head or tail).
2. The outcome of either having a cavity or not having a cavity (yes or no).
3. The sex of a newborn child (male or female).
4. The results of an exam (pass or fail only).

If a random variable has only two possible outcomes and if the probability of these outcomes remains the same for each trial regardless of what happened on any previous trial, the variable is called a **binomial variable** and its probability distribution is called a **binomial distribution.** Since the binomial distribution is so important in statistics, we will discuss it in great detail in Section 7.6.

EXERCISES

1. For each of the following situations, find the values that the indicated random variables may have.

Situation	Random Variable of Interest
a. A company which employs 227 people	Number of people out with the flu $0-227$
b. A motel which has 26 rooms	Number of rooms rented for the weekend $0-26$
c. A family that has 5 pet dogs or cats at home	Number of dogs at home $0-5$
d. A birthday party that occurs in February	The day of the month that it occurs $1-28$
e. You plant 4 seedlings	Number of seedlings that sprout $0-4$
f. A gas tank of a car which has a capacity of 22 gallons	Number of gallons (rounded off) needed to fill the tank with gas $0-22$
g. A marriage in the Arab country	Number of wives that a man has $1-?$

2. Can the following be a probability distribution for a random variable? If your answer is no, explain why.

Random Variable, x	Probability
0	0.31
1	0.08
2	0.22
3	0.17
4	0.23
5	0.02

no since prob. exceeds 1

1.03

3. Given the following probability distribution for the random variable x, what is the probability that the random variable has a value of 0?

Random Variable, x	Probability
0	? *.16*
1	0.23
2	0.08
3	0.04
4	0.32
5	0.17

.84

1.00
.84
.16

4. A family is known to have three children. Let x be the number of girls in the family. Find the probability distribution of x.

5. A coin is tossed twice. Let x be the number of heads that come up in both tosses of the coin. Find the probability distribution of x.

6. A pair of dice is thrown once. Let x be the number of 1s that come up on both dice together. Find the probability distribution of x. *1/36*

7. A man has a penny, a nickel, a dime, a quarter, a half-dollar, and a dollar piece in his pocket. Two coins are selected at random with the first coin replaced before the second coin is drawn. Let x denote the sum of the money obtained on both coins. Find the probability distribution of x.

x	p
0	1/8
1	3/8
2	3/8
3	1/8

7.2 THE MEAN OF A PROBABILITY DISTRIBUTION

Imagine that the traffic department of the City of Chelem is considering installing traffic signals at the intersection of Main Street and Broadway. The department's statisticians have kept accurate records over the past year on the number of accidents reported per day at this particularly dangerous intersection. They have submitted the report on the number of accidents per day and their respective probabilities shown in Table 7.3.

189

TABLE 7.3

REPORT ON ACCIDENTS AT INTERSECTION OF MAIN STREET AND
BROADWAY

NUMBER OF ACCIDENTS, x	PROBABILITY, $p(x)$	PRODUCT, $x \cdot p(x)$
0	1/32	$0 \cdot (1/32) = 0$
1	1/32	$1 \cdot (1/32) = 1/32$
2	9/32	$2 \cdot (9/32) = 18/32$
3	10/32	$3 \cdot (10/32) = 30/32$
4	8/32	$4 \cdot (8/32) = 32/32$
5	3/32	$5 \cdot (3/32) = 15/32$

Let us multiply each of these possible number of accidents given in Table 7.3 by
their respective probabilities. The results of these multiplications are shown in the
third column of Table 7.3. If we now add the products we get

$$0 + (1/32) + (18/32) + (30/32) + (32/32) + (15/32) = 96/32 = 3$$

This result is known as the **mean** of the distribution. It tells us that on the average
there are 3 accidents per day at this dangerous intersection.

Now consider the following. Suppose we intend to flip a coin 4 times. What is
the average number of heads that we can expect to get? To answer this question
we will first find the probability distribution. Let x represent the random variable "the
number of heads obtained in 4 flips of the coin." Since the coin is flipped 4 times,
we may get 0, 1, 2, 3, or 4 heads. Thus, the random variable x can have the values
0, 1, 2, 3, or 4. The probabilities associated with each of these values is indicated in
Table 7.4. You should verify that the probabilities given in this table are correct.

TABLE 7.4

NUMBER OF HEADS THAT CAN BE OBTAINED WHEN A COIN IS FLIPPED 4
TIMES

RANDOM VARIABLE, x	PROBABILITY, $p(x)$	PRODUCT, $x \cdot p(x)$
0	1/16	$0 \cdot (1/16) = 0$
1	4/16	$1 \cdot (4/16) = 4/16$
2	6/16	$2 \cdot (6/16) = 12/16$
3	4/16	$3 \cdot (4/16) = 12/16$
4	1/16	$4 \cdot (1/16) = 4/16$

We now multiply each possible outcome by its probability, the results of which
are shown in the third column of Table 7.4. If we now add these products, the result
is again called the mean of the probability distribution. In our case the mean is

$$0 + (4/16) + (12/16) + (12/16) + (4/16) = 32/16 = 2$$

This tells us that on the average we can expect to get 2 heads.

We generalize the results of the previous two examples as follows:

DEFINITION 7.2

The **mean** of a probability distribution is the number obtained by multiplying all the possible values of the random variable having this particular distribution by their respective probabilities and adding these products together.

FORMULA 7.1

The mean of a probability distribution is denoted by the Greek letter μ, read as mu. Thus,

$$\mu = \Sigma x \cdot p(x)$$

where this summation is taken over all the values which the random variable x can assume and the quantities $p(x)$ are the corresponding probabilities.

Comment   Many books refer to the mean of a probability distribution as its **mathematical expectation** or its **expected value.** We will use the word **mean.** The mean is similar to the mathematical expectation that was discussed in Section 5.5 of Chapter 5.

We illustrate the use of Formula 7.1 with several examples.

EXAMPLE 1

Matthew is a waiter. The following table gives the probabilities that customers will give tips of varying amounts of money.

Amounts of Money (in cents) x	30	35	40	45	50	55	60
Probability, p(x)	0.45	0.25	0.12	0.08	0.05	0.03	0.02

Find the mean for this distribution.

SOLUTION

Applying Formula 7.1 we have

$$\mu = \Sigma x \cdot p(x) = 30(0.45) + 35(0.25) + 40(0.12) + 45(0.08)$$
$$+ 50(0.05) + 55(0.03) + 60(0.02)$$
$$= 13.5 + 8.75 + 4.8 + 3.6 + 2.5 + 1.65 + 1.2$$
$$= 36$$

The mean is 36. Thus, Matthew can expect to receive an average tip of 36 cents.

EXAMPLE 2

Rosemary works for the Census Bureau in Washington. For a particular midwestern town the number of children per family and their respective probabilities is as follows:

Number of Children, x	0	1	2	3	4	5	6
Probability, $p(x)$	0.07	0.17	0.31	0.27	0.11	0.06	0.01

Find the mean for this distribution.

SOLUTION

Applying Formula 7.1 we have

$$\mu = \Sigma x \cdot p(x) = 0(.07) + 1(0.17) + 2(0.31) + 3(0.27) + 4(0.11)$$
$$+ 5(0.06) + 6(0.01)$$
$$= 0 + 0.17 + 0.62 + 0.81 + 0.44 + 0.30 + 0.06$$
$$= 2.4$$

The mean is 2.4. How can the average number of children per family be 2.4? Should it not be a whole number such as 2 or 3, not 2.4?

7.3 MEASURING CHANCE VARIATION

Suppose a manufacturer guarantees that a tire will last 20,000 miles under normal driving conditions. If a tire is selected at random and lasts only 12,000 miles, can the difference between what was expected and what actually happened be reasonably attributed to chance or is there something wrong with the claim?

Similarly, if a coin is flipped 100 times, we would expect the average number of heads to be 50. If a coin was actually flipped 100 times and resulted in only 25 heads, can we conclude that the difference between what was expected and what actually happened is to be attributed to chance or is it possible that the coin is loaded?

To answer these questions we need some method of measuring the variations of a random variable that are due to chance. Thus, we will discuss the variance and standard deviation of a probability distribution.

You will recall that in Chapter 3 on page 69 we discussed variation of a set of numbers. We now extend this idea to variation of a probability distribution. We let μ represent the mean, $x - \mu$ represent the difference of any number from the mean, and $(x - \mu)^2$ represent the square of the difference. The difference of a number from the mean is called the **deviation from the mean.** We multiply each of the squared deviations from the mean by their respective probabilities. The sum of these products is called the variance of the probability distribution. Formally we have

DEFINITION 7.3

The **variance** of a probability distribution is the number obtained by multiplying each of the squared deviations from the mean by their respective probabilities and adding these products.

FORMULA 7.2

The variance of a probability distribution is denoted by the Greek letter σ^2, read as sigma squared. Thus

$$\sigma^2 = \Sigma(x - \mu)^2 p(x)$$

where this summation is taken over all the values which random variable x can take on. The quantities $p(x)$ are the corresponding probabilities and $(x - \mu)^2$ is the square of the deviations from the mean.

DEFINITION 7.4

The **standard deviation** of a probability distribution is the square root of the variance of the probability distribution. We denote the standard deviation by the symbol σ(sigma). Thus

$$\sigma = \sqrt{\text{variance}}$$

EXAMPLE 1

A random variable has the following probability distribution:

x	0	1	2	3	4	5
$p(x)$	7/24	5/24	1/8	1/8	1/12	1/6

Find the mean, variance, and standard deviation of this distribution.

SOLUTION

We first find μ by using Formula 7.1 of Section 7.2 and then proceed to use Formula 7.2. We arrange the computations in the form of a chart as follows:

x	$p(x)$	$x \cdot p(x)$	$x - \mu$	$(x - \mu)^2$	$(x - \mu)^2 \cdot p(x)$
0	7/24	$0(7/24) = 0$	$0 - 2 = -2$	$(-2)^2 = 4$	$4(7/24) = 28/24$
1	5/24	$1(5/24) = 5/24$	$1 - 2 = -1$	$(-1)^2 = 1$	$1(5/24) = 5/24$
2	1/8	$2(1/8) = 2/8$	$2 - 2 = 0$	$0^2 = 0$	$0(1/8) = 0$
3	1/8	$3(1/8) = 3/8$	$3 - 2 = 1$	$1^2 = 1$	$1(1/8) = 1/8$
4	1/12	$4(1/12) = 4/12$	$4 - 2 = 2$	$2^2 = 4$	$4(1/12) = 4/12$
5	1/6	$5(1/6) = 5/6$	$5 - 2 = 3$	$3^3 = 9$	$9(1/6) = 9/6$

We then have

$\mu = \Sigma x \cdot p(x)$

$\quad = 0(7/24) + 1(5/24) + 2(1/8) + 3(1/8) + 4(1/12) + 5(1/6)$

$\quad = 0 + (5/24) + (2/8) + (3/8) + (4/12) + (5/6)$

$\quad = 0 + (5/24) + (6/24) + (9/24) + (8/24) + (20/24)$

$\quad = 48/24 = 2$

Also $\sigma^2 = \Sigma(x - \mu)^2 \cdot p(x)$

$\quad = (28/24) + (5/24) + 0 + (1/8) + (4/12) + (9/6)$

$\quad = 80/24 = 3.33$

Thus, the mean is 2, the variance is 3.33, and the standard
deviation is $\sqrt{3.33}$, or 1.82

EXAMPLE 2

A dress manufacturer claims that the probability that a customer will
buy a particular size dress is as follows:

Size, x	8	10	12	14	16	18
Probability, $p(x)$	0.11	0.21	0.28	0.17	0.13	0.10

Find the mean, variance, and standard deviation of this distribution.

SOLUTION

We first find μ by using Formula 7.2 and then arrange the data in
tabular form as follows:

x	$p(x)$	$x \cdot p(x)$	$x - \mu$	$(x - \mu)^2$	$(x - \mu)^2 \cdot p(x)$
8	0.11	0.88	$8 - 12.6 = -4.6$	21.16	$(21.16)(0.11) = 2.33$
10	0.21	2.10	$10 - 12.6 = -2.6$	6.76	$(6.76)(0.21) = 1.42$
12	0.28	3.36	$12 - 12.6 = -0.6$	0.36	$(0.36)(0.28) = 0.10$
14	0.17	2.38	$14 - 12.6 = 1.4$	1.96	$(1.96)(0.17) = 0.33$
16	0.13	2.08	$16 - 12.6 = 3.4$	11.56	$(11.56)(0.13) = 1.50$
18	0.10	1.80	$18 - 12.6 = 5.4$	29.16	$(29.16)(0.10) = 2.92$

$\mu = \Sigma x \cdot p(x) = 0.88 + 2.10 + 3.36 + 2.38 + 2.08 + 1.80 = 12.6$
$\sigma^2 = 2.33 + 1.42 + 0.10 + 0.33 + 1.50 + 2.92 = 8.6$

Thus the mean is 12.6, the variance is 8.6, and the standard deviation is $\sqrt{8.6}$, or 2.93.

Formula 7.2, like the formula for the variance of a set of numbers (see page 69), requires us to first compute the mean. In many cases we do not wish to do this. For such situations we can use an alternate formula to calculate the variance of a probability distribution.

FORMULA 7.3

The variance of a probability distribution is given by

$$\sigma^2 = \Sigma x^2 \cdot p(x) - [\Sigma x \cdot p(x)]^2$$

Formula 7.3 may seem strange but it is similar to Formula 3.5 on page 72. Let us see how Formula 7.3 is used.

EXAMPLE 3

Calculate the variance of the probability distribution given in Example 2 by using Formula 7.3.

SOLUTION

We arrange the data as follows:

x	$p(x)$	$x \cdot p(x)$	x^2	$x^2 \cdot p(x)$
8	0.11	0.88	64	$64(0.11) = 7.04$
10	0.21	2.10	100	$100(0.21) = 21.00$
12	0.28	3.36	144	$144(0.28) = 40.32$
14	0.17	2.38	196	$196(0.17) = 33.32$
16	0.13	2.08	256	$256(0.13) = 33.28$
18	0.10	1.80	324	$324(0.10) = 32.4$
		Total = 12.60		Total = 167.36

Using Formula 7.3, we find the variance is

$$\sigma^2 = \Sigma x^2 \cdot p(x) - [\Sigma x \cdot p(x)]^2$$
$$= 167.36 - (12.6)^2$$
$$= 167.36 - 158.76$$
$$= 8.6$$

The variance is 8.6. This is the same result that we got using Formula 7.2.

EXERCISES

1. Phil often forgets to put money in the parking meter when he parks. The probability that he will get different amounts of summons during the month of January is as follows:

Number of Summons, x	0	1	2	3	4	5	6	7
Probability, p(x)	0.07	0.19	0.22	0.15	0.08	0.12	0.09	0.08

Find the mean, variance, and standard deviation of this distribution.

2. Jennifer Robbins is a hunter. The number of animals that she will shoot on a weekend hunting trip and the associated probabilities is as follows:

Number of Animals Shot, x	Probability, p(x)
0	0.08
1	0.17
2	0.19
3	0.20
4	0.23
5	0.12
6	0.01

Find the mean, variance, and standard deviation for this distribution.

3. The following table gives the probabilities of the number of people who will go to the registrar's office at State University to request copies of their transcripts:

Number of Students Going to Registrar's Office, x	Probability, p(x)
5	0.08
6	0.13
7	0.18
8	0.20
9	0.19
10	0.18
11	0.04

Find the mean of this distribution.

4. A die is altered by painting an additional dot on the face that originally had 1 dot. Let x be the number that appears when this die is rolled. Find the probability distribution of x and then find μ, σ^2, and σ for this distribution.

5. In Danville the number of television sets that a family has and the corresponding probabilities are as follows:

Number of Television Sets, x	Probability, p(x)
0	0.09
1	0.38
2	0.29
3	0.12
4	0.08
5	0.04

Find the mean, variance, and standard deviation for this distribution.

6. The probability that a car in a drive-in movie will contain x people is given by the following probability distribution:

Number of People in Car, x	Probability, p(x)
1	0.23
2	0.32
3	0.19
4	0.14
5	0.08
6	0.04

Find the mean, variance, and standard deviation for this distribution.

7. Beverly Lewis works for the Passport Agency of the United States. The number of passports that she approves per hour and the associated probabilities is as follows:

Number of passports, x	1	2	3	4	5	6	7	8	9	10	11	12
Probability, p(x)	0.05	0.10	0.15	0.17	0.12	0.08	0.09	0.06	0.05	0.05	0.04	0.04

Find the mean for this distribution.

8. The following is the probability distribution for the number of daily requests for urine analysis that the laboratory of a large hospital receives:

Number of Requests, x	Probability
1	0.05
2	0.43
3	0.17
4	0.25
5	0.06
6	0.03
7	0.01

Find the mean of this distribution.

9. The following is the probability distribution for the number of prescriptions that a local pharmacy will fill on a typical day:

Number of Prescriptions, x	Probability, p(x)
10	0.01
11	0.05
12	0.13
13	0.18
14	0.26
15	0.18
16	0.12
17	0.06
18	0.01

.1
.55
1.56
2.34
3.64
2.7
1.92
1.02
.18
14.01

Find the mean of this distribution.

10. The following table gives the probabilities of the number of patients who will come to Dr. Jenkins, a dentist, on a given day:

Number of People, x	5	6	7	8	9	10	11
Probability, p(x)	0.06	0.18	0.22	0.20	0.16	0.08	0.10

Find the mean of this distribution.

7.4 THE BINOMIAL DISTRIBUTION

Consider the following probability problem.

EXAMPLE 1

Paula is about to take a five question true–false quiz. She is not prepared for the exam and decides to guess the answers without reading the questions.

Answer Sheet

Directions: For each question darken the appropriate space

1. [True] [False]
2. [True] [False]
3. [True] [False]
4. [True] [False]
5. [True] [False]

Find the probability that she gets

a. all the answers correct.

b. all the answers wrong.

c. three out of the five answers correct.

SOLUTION

Let us denote a correct answer by the letter c and a wrong answer
by the letter w. There are 2 possible outcomes for question 1, c or
w. Similarly, there are 2 possible outcomes for question 2
regardless of whether the first question was correct or incorrect.
There are 2 possible outcomes for each of the remaining questions
3, 4, and 5. Thus, there are 32 possible outcomes since

$$2 \times 2 \times 2 \times 2 \times 2 = 32$$

These outcomes we list as follows:

CCCCC	CWCCC	WCCCC	WWCCC
CCCWC	CWCWC	WCCWC	WWCWC
CCCCW	CWCCW	WCCCW	WWCCW
CCCWW	CWCWW	WCCWW	WWCWW
CCWCC	CWWCC	WCWCC	WWWCC
CCWCW	CWWCW	WCWCW	WWWCW
CCWWC	CWWWC	WCWWC	WWWWC
CCWWW	CWWWW	WCWWW	WWWWW

Once we have listed all the possible outcomes, we can construct a
chart similar to Table 7.5.

TABLE 7.5

NUMBER OF CORRECT ANSWERS

0 CORRECT	1 CORRECT	2 CORRECT	3 CORRECT	4 CORRECT	5 CORRECT
WWWWW	CWWWW	CCWWW	CCCWW	WCCCC	CCCCC
	WCWWW	CWCWW	CCWCW	CWCCC	
	WWCWW	CWWCW	CCWWC	CCWCC	
	WWWCW	CWWWC	CWCWC	CCCWC	
	WWWWC	WCCWW	CWCCW	CCCCW	
		WCWCW	CWWCC		
		WCWWC	WCCWC		
		WWCWC	WCCCW		
		WWCCW	WCWCC		
		WWWCC	WWCCC		

199

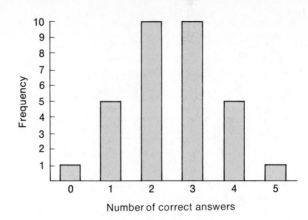

Figure 7.2

Frequency distribution of the number of correct answers obtained by guessing at five true-false questions.

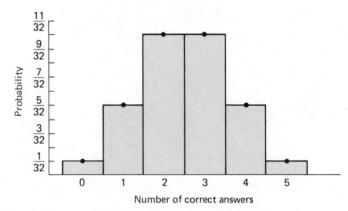

Figure 7.3

Histogram for the frequency distribution of Figure 7.2.

Now we can calculate the probability associated with each outcome. We have

$$p(0 \text{ correct}) = 1/32$$
$$p(1 \text{ correct}) = 5/32$$
$$p(2 \text{ correct}) = 10/32$$
$$p(3 \text{ correct}) = 10/32$$
$$p(4 \text{ correct}) = 5/32$$
$$p(5 \text{ correct}) = 1/32$$

We can picture these results in the form of a frequency distribution, as shown in Figure 7.2. The histogram for Figure 7.2 is shown in Figure 7.3.

We can now answer the question raised at the beginning of the problem. We have

$$p(\text{all correct answers}) = 1/32$$
$$p(\text{all wrong answers}) = 1/32$$
$$p(3 \text{ out of 5 correct answers}) = 10/32$$

Many experiments or probability problems result in outcomes that can be grouped into two categories, success or failure. For example, when a coin is flipped, there are two possible outcomes, heads or tails; when a hunter shoots at a target there are two possible outcomes, hit or miss; and when a baseball player is at bat the result is hit or out. We will consider a walk as a hit.

Statisticians apply this idea of success and failure to wide-ranging problems. For instance, if a quality control engineer is interested in determining the life of a typical light bulb in a large shipment, each time a bulb burns out he has a "success." Similarly, if we were interested in determining the probability of a family having ten boys, assuming they planned to have ten children, then each time a boy is born we have a "success."

As we mentioned in Section 7.1 (page 188), experiments which have only two possible outcomes are referred to as **binomial probability experiments** or **Bernoulli experiments** in honor of the mathematician Jacob Bernoulli (1654–1705) who studied them in detail. His contributions to probability theory are contained in his book *Ars Conjectandi* published after his death in 1713. This book also contains a reprint of an earlier treatise of Huygens. In 1657 the great Dutch mathematician Christian Huygens had written the first formal book on probability based upon the Pascal–Fermat correspondences discussed in Chapter 1 (page 10). Huygens also introduced the important ideas of mathematical expectation discussed in Chapter 5. All these ideas are contained in Bernoulli's book.

Binomial probability experiments are characterized by the following:

DEFINITION 7.5

A **binomial probability experiment** is an experiment which satisfies the following properties:

a. There are n independent repeated trials of the experiment.

b. Each trial results in one of two possible outcomes. We call one outcome a success and denote it by the letter S and the other outcome a failure, denoted by F.

c. The probability of success on a single trial equals p and remains the same from trial to trial. The probability of a failure equals $1 - p = q$. Symbolically,

$$p(\text{success}) = p \quad \text{and} \quad p(\text{failure}) = q = 1 - p$$

d. We are interested in the number of successes obtained in n trials of the experiment.

Comment   Although very few real-life situations satisfy all the above requirements, many of them can be thought of as satisfying these requirements. Thus, we can apply the binomial distribution to many different problems.

Since all binomial probability experiments are similar in nature and result in either success or failure for each trial of the experiment, we seek a formula for determining the probability of obtaining x successes out of n trials of the experiment, where the probability of success on any one trial is p and the probability of failure is q. To achieve this goal let us consider the following.

A coin is tossed 4 times. What is the probability of getting exactly one head? We could list all the possible outcomes and count the number of favorable ones. This is shown below with the favorable outcomes circled.

HHHH	HHTT	THHH	$\widehat{THTT}$
HHHT	HHTH	THHT	THTH
HTHH	$\widehat{HTTT}$	TTHH	TTTT
HTTH	HTHT	$\widehat{TTTH}$	$\widehat{TTHT}$

Thus, the probability of getting exactly one head is 4/16.

However, in many cases it is not possible or advisable to list all the possible outcomes. It is for this reason that we consider an alternate approach.

When a coin is tossed there are 2 possible outcomes, heads or tails. Thus,

$$p(\text{heads}) = 1/2 \quad \text{and} \quad p(\text{tails}) = 1/2$$

Since we are interested in getting only a head, we classify this outcome as a success and write

$$p(\text{head}) = p(\text{success}) = 1/2 \quad \text{and} \quad p(\text{tail}) = p(\text{failure}) = 1/2$$

Therefore,

$$p = 1/2, q = 1/2 \quad \text{and} \quad p + q = 1/2 + 1/2 = 1$$

Each toss is independent of what happened on the preceding toss. We are interested in obtaining 1 head in 4 tosses. One possible way in which this can happen, along with the corresponding probabilities, is as follows:

Outcome	head	tail	tail	tail
Success or Failure	Success	Failure	Failure	Failure
Probability	1/2	1/2	1/2	1/2
Symbolically	p	q	q	q

Since the probability of success is p and the probability of failure is q, we can summarize this as $p \cdot q^3$. Remember q^3 means $q \cdot q \cdot q$. We would then say that the probability of getting one head is

$$(1/2) \cdot (1/2)^3 = (1/2) \cdot (1/2) \cdot (1/2) \cdot (1/2) = 1/16$$

However, we have forgotten one thing. The head may occur on the second, third, or fourth toss:

| Different ways in which one head can be obtained when a coin is tossed 4 times | head tail tail tail | tail head tail tail | tail tail head tail | tail tail tail head |

There are then 4 ways in which we can get one head. Thus the 1/16 that we calculated before can occur in 4 different ways. Therefore,

$$p(\text{exactly 1 head}) = 4(1/16) = 4/16$$

Notice that this is exactly the same answer we obtained by listing all the possible outcomes.

Similarly, if we were interested in the probability of getting exactly 2 heads, we could consider one particular way in which this can happen:

Outcome	head	head	tail	tail
Success or Failure	Success	Success	Failure	Failure
Probability	1/2	1/2	1/2	1/2
Symbolically	p	p	q	q

The probability is thus

$$p^2 \cdot q^2 = (1/2)^2(1/2)^2 = (1/2)(1/2)(1/2)(1/2) = 1/16$$

Again, we must multiply this result by the number of ways that these 2 heads can occur in the 4 trials. This is the number of combinations of 4 things taken 2 at a time. We can use Formula 5.4 of Chapter 5 and get

$$_4C_2 = \frac{4!}{2!(4-2)!} = \frac{4!}{2!2!} = 6$$

Thus, the probability of getting 2 heads in 4 flips of a coin is

$$6(1/16) = 6/16 = 3/8$$

More generally, if we are interested in the probability of getting x success out of n trials of an experiment, then we consider one way in which this can happen. Here we have assumed that all the x success occur first and all the failures occur on the remaining $n - x$ trials.

Success or Failure	Success·Success·Success	$\cdots$	Failure·Failure·Failure
Probability	$\underbrace{p \cdot p \cdot p}_{x \text{ of them}}$	$\cdots$	$\underbrace{q \cdot q \cdot q}_{n - x \text{ of them}}$

This gives $p^x q^{n-x}$. We then multiply this result by the number of ways that exactly x successes can occur in n trials.

The number of ways that exactly x successes can occur in a set of n trials is given by

$$_nC_x = \frac{n!}{x!(n-x)!}$$

This leads us to the following **binomial distribution formula.**

FORMULA 7.4

Consider a binomial experiment which has two possible outcomes, success or failure. Let p(success) = p and p(failure) = q. If this experiment is performed n times, then the probability of getting x successes out of the n trials is

$$p(x \text{ successes}) = \frac{n!}{x!(n-x)!} \, p^x q^{n-x}$$

We illustrate the use of Formula 7.4 with numerous examples.

EXAMPLE 1

Justin Case is an insurance agent. The probability that he will sell a life insurance policy to a family that he visits is 0.6. If he plans to visit 5 families today, what is the probability that he will sell exactly 3 policies?

SOLUTION

Since he plans to visit 5 families, $n = 5$. We are interested in the probability of him selling exactly 3 policies. Thus $x = 3$. Also, p(success) = p = 0.6 and p(failure) = q = 0.4, since $p + q$ must equal 1. Applying Formula 7.4 we have

$$p(3 \text{ successes}) = p(x = 3) = \frac{5!}{3!(5-3)!}(0.6)^3(0.4)^2$$

$$= \frac{5 \cdot \overset{2}{\cancel{4}} \cdot \cancel{3} \cdot \cancel{2} \cdot \cancel{1}}{\cancel{3} \cdot \cancel{2} \cdot \cancel{1} \cdot \cancel{2} \cdot \cancel{1}}(0.6)(0.6)(0.6)(0.4)(0.4)$$

$$= 10(0.6)(0.6)(0.6)(0.4)(0.4)$$
$$= 0.3456, \text{ or } 0.35$$

Therefore the probability that Justin will sell exactly 3 policies is 0.35.

EXAMPLE 2

Charlie has just returned from a fishing trip on which he caught 5
fish. If the probability that Charlie catches a tuna fish is 3/4, what is
the probability that exactly 4 of the fish are tuna?

SOLUTION

Since Charlie has caught 5 fish, $n = 5$. Each time that Charlie
catches a tuna he has a success so that

$$p(\text{success}) = 3/4 \quad \text{and} \quad p(\text{failure}) = 1/4$$

We are interested in getting 4 successes. Thus $x = 4$. Applying
Formula 7.4, we have

$$p(\text{caught 4 tuna fish}) = p(x = 4) = \frac{5!}{4!1!}\left(\frac{3}{4}\right)^4\left(\frac{1}{4}\right)$$

$$= \frac{5\cdot\cancel{4}\cdot\cancel{3}\cdot\cancel{2}\cdot\cancel{1}}{\cancel{4}\cdot\cancel{3}\cdot\cancel{2}\cdot\cancel{1}\cdot\cancel{1}}\left(\frac{3}{4}\right)\left(\frac{3}{4}\right)\left(\frac{3}{4}\right)\left(\frac{3}{4}\right)\left(\frac{1}{4}\right)$$

$$= \frac{405}{1024}$$

Consequently, the probability that exactly 4 of the fish are tuna is
405/1024, or 0.40.

EXAMPLE 3

Mario is taking a multiple choice examination which consists of 5
questions. Each question has 4 possible answers. Mario guesses at
every answer. What is the probability that he passes the exam if he
needs at least 4 correct answers to pass?

SOLUTION

In order to pass, Mario needs to get at least 4 correct answers.
Thus, he passes if he gets 4 answers correct or 5 answers correct.
Each question has 4 possible answers so that $p(\text{correct}$
answer$) = 1/4$ and $p(\text{wrong answer}) = 3/4$. Also there are 5
questions, so $n = 5$. Therefore,

$$p(\text{4 answers correct}) = \frac{5!}{4!1!}\left(\frac{1}{4}\right)^4\left(\frac{3}{4}\right)$$

$$= \frac{5\cdot\cancel{4}\cdot\cancel{3}\cdot\cancel{2}\cdot\cancel{1}}{\cancel{4}\cdot\cancel{3}\cdot\cancel{2}\cdot\cancel{1}\cdot\cancel{1}}\left(\frac{1}{4}\right)\left(\frac{1}{4}\right)\left(\frac{1}{4}\right)\left(\frac{1}{4}\right)\left(\frac{3}{4}\right)$$

$$= \frac{15}{1024}$$

and

$$p(5 \text{ answers correct}) = \frac{5!}{0!5!}\left(\frac{1}{4}\right)^5\left(\frac{3}{4}\right)^0$$

$$= \frac{\cancel{5}\cdot\cancel{4}\cdot\cancel{3}\cdot\cancel{2}\cdot 1}{\cancel{1}\cdot\cancel{5}\cdot\cancel{4}\cdot\cancel{3}\cdot\cancel{2}\cdot\cancel{1}}\left(\frac{1}{4}\right)\left(\frac{1}{4}\right)\left(\frac{1}{4}\right)\left(\frac{1}{4}\right)\left(\frac{1}{4}\right)\cdot 1$$

$$= \frac{1}{1024}$$

(Any number to the 0 power is 1.)

Adding the two probabilities, we get

$$p(\text{at least 4 correct answers}) = p(4 \text{ answers correct}) + p(5 \text{ answers correct})$$

$$= \frac{15}{1024} + \frac{1}{1024}$$

$$= \frac{16}{1024} = \frac{1}{64}$$

Hence, the probability that Mario passes is 1/64.

EXAMPLE 4

A shipment of 100 tires from the Flatt Tire Corporation is known to contain 20 defective tires. Five tires are selected at random and each tire is replaced before the next tire is selected. What is the probability of getting *at most* 2 defective tires?

SOLUTION

We are interested in the probability of getting *at most* 2 defective tires. This means 0 defective tires, 1 defective tire, or 2 defective tires. Thus the probability of at most 2 defective tires equals

$$p(0 \text{ defective}) + p(1 \text{ defective}) + p(2 \text{ defective})$$

The probability of a defective tire is 20/100, or 1/5. Therefore, the probability of getting a non-defective tire is 4/5. Now

$$p(0 \text{ defective}) = \frac{5!}{0!5!}\left(\frac{1}{5}\right)^0\left(\frac{4}{5}\right)^5 = \frac{1024}{3125}$$

$$p(1 \text{ defective}) = \frac{5!}{1!4!}\left(\frac{1}{5}\right)^1\left(\frac{4}{5}\right)^4 = \frac{1280}{3125}$$

$$p(2 \text{ defective}) = \frac{5!}{2!3!}\left(\frac{1}{5}\right)^2\left(\frac{4}{5}\right)^3 = \frac{640}{3125}$$

Adding, we get

$$p(\text{at most 2 defectives}) = \frac{1024}{3125} + \frac{1280}{3125} + \frac{640}{3125} = \frac{2944}{3125}$$

Hence the probability of getting at most 2 defective tires is
2944/3125, or 0.94.

EXAMPLE 5

If the conditions are the same as in the previous problem except
that now 15 tires are selected, what is the probability of getting *at
least* 1 defective tire?

SOLUTION

We could proceed as we did in Example 4. Thus

$$p(\text{at least 1 defective}) = p(1 \text{ defective}) + p(2 \text{ defective}) + \cdots + p(15 \text{ defective})$$

However this involves a tremendous amount of computation. Recall
(see the Rule on page 187) that the sum of all the values of a
probability function must be 1. Thus

$$p(0 \text{ defective}) + p(1 \text{ defective}) + p(2 \text{ defective}) + \cdots + p(15 \text{ defective}) = 1$$

Therefore, if we subtract $p(0 \text{ defective})$ from both sides we have

$$p(1 \text{ defective}) + p(2 \text{ defective}) + \cdots + p(15 \text{ defective}) = 1 - p(0 \text{ defective})$$

Now

$$p(0 \text{ defective}) = \frac{15!}{0!15!} \left(\frac{1}{5}\right)^0 \left(\frac{4}{5}\right)^{15} = 0.035$$

Consequently, the probability of obtaining at least 1 defective tire is
$1 - 0.035$, or 0.965.

Comment Calculating binomial probabilities can sometimes be quite a time
consuming task. To make the job a little easier, we can use Table IV of the
Appendix which gives us the binomial probabilities for different values of $n, x,$ and
p. No computations are needed. We only need to know the values of $n, x,$ and p.

EXAMPLE 6

Given a binomial distribution with $n = 11$ and $p = 0.4$, use Table IV
of the Appendix to find the probability of getting

a. exactly 4 successes. .236

b. at most 3 successes. .297

c. 5 or more successes. .467

SOLUTION

We use Table IV of the Appendix with $n = 11$ and $p = 0.4$. We first
locate $n = 11$ and then move across the top of the table until we
reach the $p = 0.4$ column.

a. To find the probability of exactly 4 successes, we look for the
   value given in the table for $x = 4$. It is 0.236. Thus when $n = 11$
   and $p = 0.4$, the probability of exactly 4 successes is 0.236.

b. To find the probability of getting at most 3 successes, we look in
   the table for the values given for $x = 0$, $x = 1$, $x = 2$, and $x = 3$.
   These probabilities are 0.004, 0.027, 0.089, and 0.177
   respectively. We add these (Why?) and get

$$0.004 + 0.027 + 0.089 + 0.177 = 0.297$$

   Thus, the probability of at most 3 successes is 0.297.

c. To find the probability of 5 or more successes, we look in the
   chart for the values given for $x = 5$, $x = 6$, $x = 7$, $x = 8$, $x = 9$,
   $x = 10$, and $x = 11$. These probabilities are 0.221, 0.147, 0.070,
   0.023, 0.005, 0.001, and 0. We add these and get

$$0.221 + 0.147 + 0.070 + 0.023 + 0.005 + 0.001 + 0 = 0.467$$

   Thus, the probability of 5 or more successes is 0.467.

Comment In the previous example you will notice that there is no value
given in Table IV when $x = 11$. It is left blank. Whenever there is a blank in the
chart, this means that the probability is approximately 0. This is the reason that
we used 0 as the probability in our calculations.
Using Table IV we can draw the graphs of binomial probabilities as shown in
Figure 7.4. In each case $n = 5$.

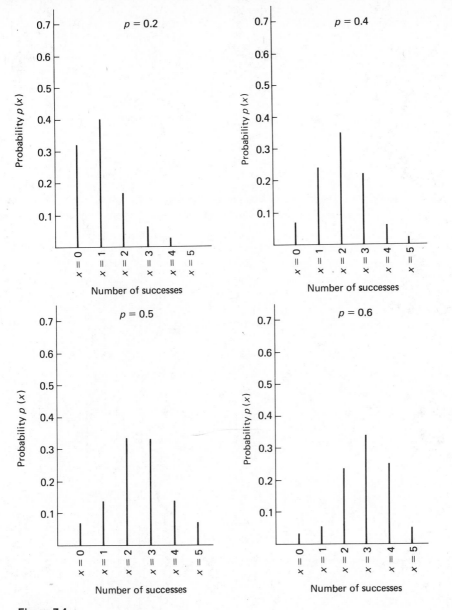

Figure 7.4

Graphs of binomial probabilities for $n = 5$.

For a given value of p, the distribution becomes symmetrical as the value of n gets larger. This is shown in Figure 7.5. We will have more to say about this in the next chapter.

**THE BINOMIAL
DISTRIBUTION**

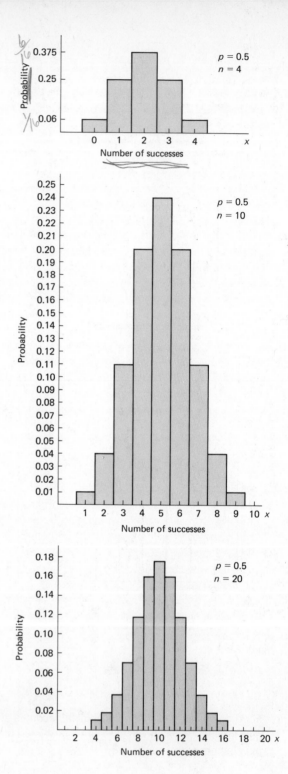

Figure 7.5

For a given value of p the distribution becomes symmetrical as the value of n gets larger.

1. The probability that a student at Hall University uses a particular brand of mouthwash is 2/5. If 6 students are selected at random, find the probability that
   a. exactly 3 of them use the mouthwash.
   b. all of them use the mouthwash.
   c. none of them use the mouthwash.
2. Many people jog to maintain their physical fitness. In a certain community the probability that a man and wife jog together is 0.35. If 5 men are selected at random, what is the probability that 2 of them jog with their wives?
3. The probability that a driver making a gas purchase will pay by credit card is 3/5. If 6 cars pull up to a gas station to buy gas, what is the probability that 4 of the drivers will pay for the gas by credit card?
4. The Grey Brothers are training elephants to perform a certain act. The probability that a trained elephant will perform the act properly is 4/5. If 7 elephants have been trained, find the probability that 3 of them perform the act properly.
5. It is known from past experience that, on the average, 3% of all traps produced by a company are defective. If 10 traps produced by the company are selected at random, find the probability that 6 of them are defective.
6. Fred, Ted, Ned, and Jed are inmates in a state penitentiary. The probability that any one of them escapes is 1/2. Assume independence. What is the probability that at least 2 of these prisoners will escape in a jailbreak?
7. On a grocer's shelf there are 40 loaves of bread. Ten of these are known to be stale. Gail purchases 4 loaves of bread from the grocer. Find the probability that
   a. all the loaves of bread purchased are fresh.
   b. half the number of the loaves of bread purchased are fresh.
   c. none of the loaves of bread purchased are fresh.
8. A coin is tossed in such a way that the probability of it coming up heads is 0.7. This coin is tossed 5 times. What is the probability of getting
   a. at least 3 heads?
   b. at most 3 heads?
9. From past experience the Community Savings Bank knows that the probability that a person making a deposit will cash a check is 0.60. If 8 people are standing on line to make their transactions, what is the probability that
   a. at least 4 of them will cash a check?
   b. at most 4 of them will cash a check?
10. The probability that a family in Doversville has an unlisted phone number is 0.02. If 9 families are selected at random, find the probability that at most 2 of them have an unlisted number.

11. Forty percent of all people in Boondock have Type A blood. If 10 people are selected at random, find the probability that at least 3 of them have type A blood.

12. What is the probability that a family of 5 children consists of 3 boys and 2 girls? Assume that a child is as likely to be a boy as it is a girl.

13. A corporation claims that 97% of its trains arrive on time. If 10 trains are selected at random, what is the probability that at least 8 of them will arrive on time?

14. Suppose it is known that 1 out of every 20 novels published by a company makes the best seller list. If the publisher has selected 10 new books to be published, what is the probability that 2 of them will make the best seller list?

7.5 THE MEAN AND STANDARD DEVIATION OF THE BINOMIAL DISTRIBUTION

Consider the binomial distribution given on page 200, which is also repeated here. Recall that x represents the number of correct answers that Paula obtained on an exam of 5 questions where the probability of a correct answer is 1/2.

x	$p(x)$	$x \cdot p(x)$	x^2	$x^2 \cdot p(x)$
0	1/32	0	0	0
1	5/32	5/32	1	5/32
2	10/32	20/32	4	40/32
3	10/32	30/32	9	90/32
4	5/32	20/32	16	80/32
5	1/32	5/32	25	25/32
		Total = 80/32		Total = 240/32
		= 5/2		= 15/2

Let us calculate the mean and variance for this distribution. Using Formula 7.1 of Section 7.3 we get

$$\mu = \Sigma x \cdot p(x) = 0 + (5/32) + (20/32) + (30/32) + (20/32) + (5/32) = 80/32 = 5/2$$

To calculate the variance we use Formula 7.3 of Section 7.2. From the preceding table we have

$$\sigma^2 = \Sigma x^2 \cdot p(x) - [\Sigma x \cdot p(x)]^2$$
$$= 15/2 - (5/2)^2$$
$$= 30/4 - 25/4 = 5/4$$

Thus $\mu = 5/2$, or 2.5, and $\sigma^2 = 5/4$, or 1.25.

Notice that if we multiply the total number of exam questions, which is 5, by the probability of a correct answer, which is 1/2, we get $5(1/2) = 2.5$. This is exactly the same answer we get for the mean by applying Formula 7.1. We might be tempted to conclude that $\mu = np$. This is indeed the case.

Similarly, if we multiply the total number of questions with the probability of a correct answer and with the probability of a wrong answer we get $5(1/2)(1/2) = 1.25$.

Here we might conclude that $\sigma^2 = npq$. Again this is indeed the case. We can generalize these ideas in the following:

FORMULA 7.5

The **mean of a binomial distribution**, μ, is found by multiplying the total number of trials with the probability of success on each trial. If there are n trials of the experiment and if the probability of success on each trial is p, then

$$\mu = np$$

FORMULA 7.6

The **variance of a binomial distribution** is given by

$$\sigma^2 = npq$$

The standard deviation σ is the square root of the variance. Thus

$$\sigma = \sqrt{\text{variance}}$$

The proofs of Formulas 7.5 and 7.6 can be found in many textbooks on mathematical statistics.

EXAMPLE 1

A die is rolled 600 times. Find the mean and standard deviation of the number of 1s that show.

SOLUTION

This is a binomial distribution. Since the die is rolled 600 times, $n = 600$. Also there are 6 possible outcomes so that $p = 1/6$ and $q = 5/6$. Thus, using Formulas 7.5 and 7.6 we have

$$\mu = np$$
$$= 600(1/6) = 100$$

and
$$\sigma^2 = npq$$
$$= 600(1/6)(5/6) = 83.33$$

213

Therefore, the mean is 100 and the standard deviation, which is the square root of the variance, is $\sqrt{83.33}$, or 9.13. This tells us that if this experiment were to be repeated many times, we could expect an average of 100 1s per trial with a standard deviation of $\sqrt{83.33}$, or 9.13.

EXAMPLE 2

A large mail order department store finds that approximately 17% of all purchases are returned for credit. If the store sells 100,000 different items this year, about how many items will be returned? Find the standard deviation.

SOLUTION

This is a binomial distribution. Since 100,000 items were sold, $n = 100,000$. Also the probability that a customer will return the item is 0.17 so that $p = 0.17$. Therefore the probability that the customer will not return the item is 0.83. Thus, using Formulas 7.5 and 7.6 we have

$$
\begin{aligned}
\mu &= np \\
&= (100,000)(0.17) \\
&= 17,000
\end{aligned}
$$

and

$$
\begin{aligned}
\sigma &= \sqrt{npq} \\
&= \sqrt{(100,000)(0.17)(0.83)} \\
&= \sqrt{14110} \\
&= 118.79
\end{aligned}
$$

Thus, the department store can expect about 17,000 items to be returned with a standard deviation of 118.79.

Formulas 7.5 and 7.6 will be applied in greater detail in Chapter 8.

EXERCISES

1. Calculate the mean, variance, and standard deviation of a binomial distribution where $n = 300$ and $p = 1/5$.
2. A restaurant owner finds that approximately 8% of the people who make reservations do not show. Three hundred people have made reservations on a

certain weekend. How many people can be expected to show up? Find the standard deviation.

3. A camp finds that about 5% of its campers are dissatisfied with the food. If 400 campers have registered for the summer season, about how many campers can be expected to be dissatisfied with the food? Find the standard deviation.

4. A physician knows that the probability is 0.30 that a patient will recover from a particular disease. If the physician is treating 100 patients with this disease, about how many people can be expected to recover? Find the standard deviation.

5. Seventy-five percent of all people who buy a particular camera also buy the carrying case for the camera. If two hundred cameras are to be sold this year, about how many carrying cases will be sold? Find the standard deviation.

6. Seventeen percent of all fire alarms in Fun City are false. If fire department officials expect 10,000 fire alarms next year, about how many will be false? Find the standard deviation.

7. The dropout rate at a certain school is 18% per semester. If 36,000 students have enrolled, about how many students can be expected to withdraw from school?

8. The probability that the sanitary conditions in a restaurant are not acceptable to the board of health authorities is 0.12. If 200 restaurants are selected at random, about how many of them will have unsanitary conditions? Find the standard deviation.

9. The Board of Directors of the Dreck Corporation believes that about 65% of the company stockholders will favor a stock split. If there are 10,000 stockholders, about how many can be expected to actually favor the stock split?

10. Two percent of all calculators produced by the Scientific Corporation must be returned to the manufacturer. If the company sold 50,000 calculators last year, about how many will be returned to the manufacturer?

SELF-STUDY GUIDE

In this chapter we discussed how the ideas of probability can be combined with frequency distributions. Specifically we introduced the idea of a random variable and its probability distribution. These enable an experimenter to analyze outcomes of experiments and to speak about the probability of different outcomes.

We then discussed the mean and variance of a probability distribution. These allow us to determine the expected number of favorable outcomes of an experiment.

Although we did not emphasize the point, all the random variables discussed and their associated probabilities were mutually exclusive. Thus, we were able to add probabilities by the addition rule for probabilities. Also, the events were independent. This allowed us to multiply probabilities by the multiplication rule for probabilities of independent events.

We discussed one particular distribution in detail, the binomial distribution, since it is one of the most widely used distributions in statistics. In addition to the binomial distribution formula itself, which allows us to calculate the probability of getting a specified number of successes in repeated trials of an experiment, formulas for calculating its mean, variance, and standard deviation were given. These formulas were applied to numerous examples. Because of its importance we will discuss the binomial distribution further in Chapter 8.

You should be able to identify each symbol in the following formulas, understand the relationships among the symbols expressed in each formula, understand the significance of each formula, and use the formulas in solving problems.

1. Mean of a Probability Distribution $\mu = \Sigma x \cdot p(x)$
2. Variance of a Probability Distribution $\sigma^2 = \Sigma(x - \mu)^2 \cdot p(x)$
3. Variance of a Probability Distribution $\sigma^2 = \Sigma x^2 \cdot p(x) - [\Sigma x \cdot p/x)]^2$
4. Standard Deviation of a Probability Distribution $\sigma = \sqrt{\text{variance}}$
5. $p(x$ successes out of n trials$) = \dfrac{n!}{x!(n-x)!} p^x q^{n-x}$
6. Mean of the Binomial Distribution $\mu = np$
7. Variance of the Binomial Distribution $\sigma^2 = npq$

You should now be able to demonstrate your knowledge of the following ideas presented in this chapter by giving definitions, descriptions, or specific examples. Page references are given for each term so that you can check your answer.

Random variable (page 185)
Probability function (page 186)
Probability distribution (page 187)
Binomial variable (page 188)
Binomial distribution (page 188)
Mean (page 191)
Mathematical expectation (page 191)
Expected Value (page 191)
Deviation from the mean (page 192)

Variance (page 193)
Standard deviation (page 193)
Binomial probability experiment (page 201)
Bernoulli experiment (page 201)
Binomial distribution formula (page 204)

The tests of the following section will be more useful if you take them after you have studied the examples and solved the exercises given in this chapter.

Form A

1. Consider the following probability distribution. Find the probability that $x = 1$.

x	p(x)	
0	1/3	.33
1	?	
2	1/6	.17

2. If for a particular binomial distribution $n = 100$ and $p = 1/2$, what is the value of μ?

3. What is σ for the distribution given in question 2?

4. The term grade that Irving will receive in his statistics course is a random variable. Find the values of this random variable.

5. The number of teeth that an adult has is a random variable. What are the values of this random variable?

6. In an experiment a coin is tossed 400 times and the number of tails which appears is recorded. If the coin is fair, find the mean of this distribution.
   a. 10 *b.* 200 *c.* 400 *d.* 100 *e.* none of these

7. What is the standard deviation of the distribution described in question 6?
   a. 10 *b.* 200 *c.* 400 *d.* 100 *e.* none of these

8. Consider the following probability distribution. Find the probability that $x = 3$.

x	p(x)
0	1/10
1	1/5
2	1/8
3	?
4	3/20

9. A family is planning to have 3 children. What is the probability that the family will consist of exactly 2 girls and 1 boy?
   a. 2/3 *b.* 3/8 *c.* 1/4 *d.* 1/8 *e.* none of these

10. Is the following distribution a probability distribution? Explain your answer.

x	p(x)
0	1/11
1	1/10
2	1/9
3	1/8
4	2/7
5	1/6
6	3/5
7	1/3
8	1/2

217

11. Jimmy has just planted 20 trees. If the probability that any one tree sprouts is 0.3, find the probability that exactly 11 of them will sprout.

a. $\dfrac{20!}{11!9!}(0.3)^{11}(0.7)^9$

c. $\dfrac{20!}{11!}(0.3)^{11}(0.7)^9$

e. none of these

b. $\dfrac{20!}{11!9!}(0.7)^{11}(0.3)^9$

d. $\dfrac{20!}{9!}(0.3)^{11}(0.7)^9$

Form B

1. A box contains the following four cards: the 5 of hearts, the 3 of clubs, the 2 of spades, and the 7 of diamonds. Two cards are drawn from the box with the first card replaced before the second card is drawn. Let x denote the sum of the numbers obtained on the two cards. Find the probability distribution of x.

2. Suppose that 80% of the students who register for a statistics course actually pass the course. Find the probability that in a class of 15 students at least 3 will fail the course. Use Table IV of the Appendix.

3. The probability that Alice has a cavity in any one tooth is 1/10. If Alice has only 30 teeth, find the probability that she has
   a. at most one cavity   b. no cavities

*4. Show that Formulas 7.2 and 7.3 are equivalent, that is, derive Formula 7.3 from Formula 7.2.

5. Calculate the mean and variance for x, the number of dots that show when rolling an honest die.

6. Weather records in Petersville show that, on the average, it will snow on 6 of the 30 days in April. Assuming a binomial distribution with each day of April treated as an independent trial, find the probability that it will snow on at most 4 days in April. Are we really justified in using the binomial distribution?

7. In Chapter 3 we mentioned that the sum of differences from the mean, that is $\Sigma(x - \mu)$, is 0. Yet in many of the probability distributions discussed in this chapter, for example, the one discussed on page 194, the sum of the differences around the mean is not 0. Can you explain why?

*8. In the baseball world series the series ends when any one of the teams win 4 games. Assume that the New York Yankees are one of the teams playing in the series. Let the probability that the Yankees win any one game be 1/2. Assume that this probability remains the same from game to game. Find the probability that the world series will end
   a. in 4 games   c. in 6 games
   b. in 5 games   d. in 7 games

9. A certain blood disease occurs in 10% of all white females. If 8 females are randomly selected what is the probability that
   a. four of them have the disease.
   b. at most two of them have the disease.

10. A coin is bent so that the probability of heads coming up is 0.6. If this coin is tossed 10 times, find the probability of getting at least 7 heads.

"Computers and Computation." Readings from *Scientific American,* article 16, San Francisco, Calif.: Freeman, 1971.

Hoel, Paul G., Port, Sidney C., and Stone, Charles J. *Introduction to Probability Theory*. Boston: Houghton Mifflin, 1972.

Mendenhall, William. *Introduction to Probability and Statistics.* 3rd ed. North Scituate, Mass.: Duxbury Press, 1971.

Meyer, P. L. *Introductory Probability and Statistical Applications.* Reading, Mass.: Addison-Wesley, 1965.

Mosteller, Frederick, Rourke, R. E. K., and Thomas, G. B., Jr. *Probability with Statistical Applications*. 2nd ed. Reading, Mass.: Addison-Wesley, 1970.

National Bureau of Standards. *Tables of the Binomial Probability Distribution.* Applied Mathematics Series 6. U.S. Department of Commerce, Washington, 1949.

Runyon, R. P. and Haber, Audrey. *Fundamentals of Behavioral Statistics.* Reading, Mass.: Addison-Wesley, 1967.

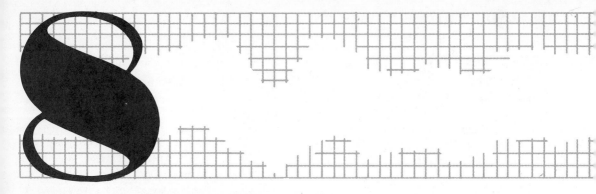

THE NORMAL DISTRIBUTION

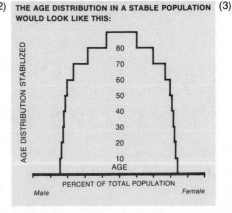

(1)   **Distribution of IQ's in 2904 Children**

THE 1970 AGE DISTRIBUTION OF OUR POPULATION (2)
LOOKS LIKE THIS:

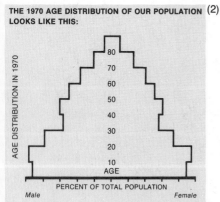

THE AGE DISTRIBUTION IN A STABLE POPULATION (3)
WOULD LOOK LIKE THIS:

Graphs 2 and 3:
From *ZPG Has Not
Been Reached,*
Zero Population
Growth, Inc.,
Washington, D.C.

The first graph indicates the distribution of IQ scores of 2904 children. Notice the type of frequency distribution that is illustrated. This is approximately a normal distribution.

Now look at the second graph. Is the age distribution of our population normally distributed? To achieve zero population growth, many people believe that the age distribution should be given by the third graph.

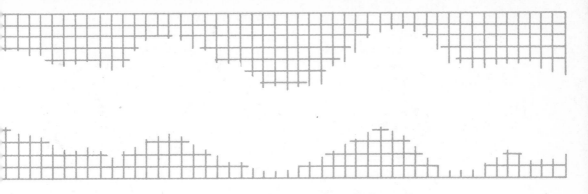

Preview

In this chapter we will discuss the following topics:

1. *The Normal Probability Distribution*
We discuss in detail a probability distribution which is bell-shaped or mound-shaped. It is called the normal distribution.

2. *The Standard Normal Distribution*
To be able to apply the normal distribution to many different situations, we must convert z–scores into standard scores.

3. *Applications of the Normal Distribution*
We indicate how we can apply the normal distribution to calculate probabilities. We also discuss statistical quality control charts used in industry as an additional application.

4. *Approximating the Binomial Distribution*
We point out how the normal distribution can be used to simplify lengthy computations involving the binomial distribution.

5. *Historical Background*
We briefly discuss some of the mathematicians who used the normal distribution.

Introduction

Until now the random variables that we discussed assumed only the limited values of 0, 1, 2, ... Thus when a coin is flipped many times, the number of heads that comes up is 0, 1, 2 . . . . Similarly, if a student is guessing answers on a multiple choice test, the number of correct answers can be 0, 1, 2, 3, ... Also, the number of defective bulbs in a shipment of 100 bulbs is 0, 1, 2, 3, . . . , 100. Since these random variables can assume only the values 0, 1, 2, . . . , they are called **discrete variables.**

As opposed to the preceding examples, consider the following random variables: the length of a page of this book, the height of your pet dog, the temperature at noon on New Year's Day, or the weight of a bag of sugar on your grocer's shelf. Since each of these variables can assume an infinite number of values on a measuring scale, they are called **continuous variables.** Thus, the weight of a bag of sugar can be 5 pounds, 5.1 pounds, 5.161 pounds, 5.16158 pounds, 5.161581 pounds, and so forth, depending upon the accuracy of the scale. Similarly, the temperature at noon on

New Year's Day may be 38°, 38.2°, 38.216°, and so forth.

Among the many different kinds of distributions of random variables that are used by statisticians, the normal distribution is by far the most important. This type of distribution was first discovered by the English mathematician Abraham De Moivre (1667–1754). De Moivre spent many hours with London gamblers. In his *Annuities Upon Lives,* which played an important role in the history of actuarial mathematics, and his *Doctrine of Chances,* which is a manual for gamblers, he essentially developed the first treatment of the normal probability curve which is important in the study of statistics. De Moivre also developed a formula, known as Stirling's formula, which is used for approximating factorials of large numbers.

A rather interesting story is told of De Moivre's death. According to the story De Moivre was ill and each day he noticed that he slept a quarter of an hour longer than on the preceding day. Using progressions he computed that he would die in his sleep on the day after he slept 23 hours and 45 minutes. On the day following a sleep of 24 hours De Moivre died.

Many years later the French mathematician Pierre-Simon Laplace (1749–1827) applied the normal distribution to astronomy and other practical problems. The normal distribution was also used extensively by the German mathematician Carl Friedrich Gauss (1777–1855) in his studies of physics and astronomy. Gauss is considered by many as the greatest mathematician of the nineteenth century. At the age of three he is alleged to have detected an error in his father's bookkeeping records.

The normal distribution is sometimes known as the **bell-shaped** or **Gaussian distribution** in honor of Gauss who studied it extensively.

Although there are other distributions of continuous variables that are important in statistics, the normal distribution is by far the most important. In this chapter we will discuss in detail the nature of the normal distribution, its properties, and its uses.

8.1   THE GENERAL NORMAL CURVE

Refer to the histogram given on page 39 of Chapter 2. Such a curve was called the **normal curve.** Experience has taught us that for many distributions drawn from large populations the histograms approximate what is known as a normal or bell-shaped curve as shown in Figure 8.1.

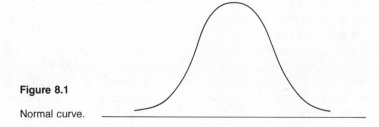

Figure 8.1

Normal curve.

Heights and weights of people, IQ scores, waist sizes, or even life expectancy of cars, to name but a few, are all examples of distributions whose histograms approach a normal curve when the samples taken are from large populations. When the graph of a frequency distribution resembles the bell-shaped curve shown in Figure 8.1, the graph is called a **normal curve** and its frequency distribution is known as a **normal distribution.** The word normal is simply a name for this particular distribution. It does not indicate that this distribution is more typical than any other.

Since the normal distribution has wide-ranging applications, we need a careful description of what a normal curve is and some of its properties.

As mentioned before, the graph of a normal distribution is a bell-shaped curve. It extends in both directions. Although the curve gets closer and closer to the horizontal axis, it never really crosses it, no matter how far it is extended. The normal distribution is a probability distribution satisfying the following properties:

1. The mean is at the center of the distribution and the curve is symmetrical about the mean. This tells us that we can fold the curve along the dotted line shown in Figure 8.2 and either portion of the curve will correspond with the other portion.

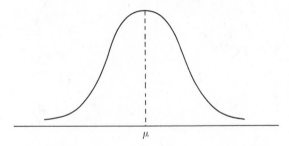

Figure 8.2

2. Its graph is the bell-shaped curve, most often referred to as the normal curve.

3. The mean equals the median.

4. The scores that make up the normal distribution tend to cluster around the middle with very few values more than three standard deviations away from the mean on either side.

Normal distributions can come in different sizes and shapes. Some are tall and skinny or flat and spread out as shown in Figure 8.3. However, for a given mean and a given standard deviation, there is one and only one normal distribution. The normal distribution is completely specified once we know its mean and standard deviation.

We mentioned in Chapter 2 (see page 21) that the area under a particular rectangle of a histogram gives us the probability of obtaining values within that rectangle. We can generalize this idea to any distribution. We say that the area under the curve between any two points *a* and *b* gives us the probability that the random variable having this particular continuous distribution will assume values between *a* and *b*. This idea is very important since calculating probabilities for the normal distribution will depend upon the areas under the curve. Also since the total probability of a random variable must be 1 (see page 187), the total area under its probability curve must also be 1.

8.2 THE STANDARD NORMAL CURVE

A normal distribution is completely specified by its mean and standard deviation. Thus, although all normal distributions are basically mound-shaped, different means and different standard deviations will describe different bell-shaped curves. However, it is possible to convert each of these different normal distributions into one standardized form. You may be wondering, why bother? The answer is rather simple.

Since areas under a normal distribution are related to probability, we can use special normal distribution tables for calculating probabilities. Such a table is given in the Appendix at the end of this book. However, since the mean and standard deviation can be any values, it would seem that we need an endless number of tables. Fortunately, this is not the case. We only need one standardized table. Thus, the area under the curve between 40 and 60 of a normal distribution with a mean of 50 and a standard deviation of 10 will be the same as the area between 70

Figure 8.3

Different normal distributions.

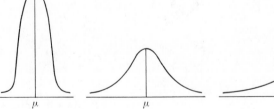

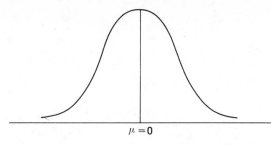

Figure 8.4

Curve of a typical standard
distribution.

$\mu = 0$

and 80 of another normally distributed variable with mean 75 and standard deviation 5. They are both within one standard deviation unit from the mean. It is for this reason that statisticians use a standard normal distribution.

DEFINITION 8.1

A **standardized normal distribution** is a normal distribution with a mean of 0 and a standard deviation of 1. The curve of a typical standard distribution is shown in Figure 8.4.

Table V in the Appendix gives us the areas of a standard normal distribution between $z = 0$ and $z = 3.09$. We read this table as follows. The first two digits of the z-score are under the column headed by z, the third digit heads the other columns. Thus, to find the area from $z = 0$ to $z = 2.43$ we first look under z to 2.4 and then read across from $z = 2.4$ to the column headed by 0.03. The area is 0.4925, or 49.25%.

Similarly, to find the area from $z = 0$ to $z = 1.69$ we first look under $z = 1.6$ and then read across from $z = 1.6$ to the column headed by 0.09. The area is 0.4545.

EXAMPLE 1

Find the area between $z = 0$ and $z = 1$ in a standard normal curve.

SOLUTION

We first draw a sketch as shown in Figure 8.5. Then using Table V for $z = 1.00$ we find that the area between $z = 0$ and $z = 1$ is 0.3413. This means that the probability of a score with this normal distribution falling between $z = 0$ and $z = 1$ is 0.3413.

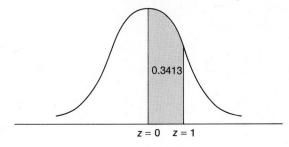

0.3413

$z = 0$ $z = 1$

Figure 8.5

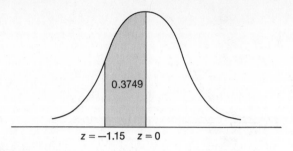

Figure 8.6

$z = -1.15 \quad z = 0$

EXAMPLE 2

Find the area between $z = -1.15$ and $z = 0$ in a standard normal curve.

SOLUTION

We first draw a sketch as shown in Figure 8.6. Then using Table V we look up the area between $z = 0$ and $z = 1.15$. The area is 0.3749, not -0.3749. A negative value of z just tells us that the value is to the left of the mean. The area under the curve (and the resulting probability) is *always* a positive number. Thus, the probability of getting a z–score between 0 and -1.15 is 0.3749.

EXAMPLE 3

Find the area between $z = -1.63$ and $z = 2.22$ in a standard normal curve.

SOLUTION

We draw the sketch shown in Figure 8.7. Since Table V gives the area only from $z = 0$ on, we first look under the normal curve from $z = 0$ to $z = 1.63$. We get 0.4484. Then we look up the area between $z = 0$ and $z = 2.22$. We get 0.4868. Finally we add these two together and get

$$0.4484 + 0.4868 = 0.9352$$

Thus, the probability that a z–score is between $z = -1.63$ and $z = 2.22$ is 0.9352.

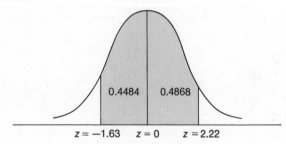

Figure 8.7

$z = -1.63 \quad z = 0 \qquad z = 2.22$

By following a procedure similar to that used in Example 3, you should verify the following:

1. The probability that a z–score falls within 1 standard deviation of the mean on either side, that is, between $z = -1$ and $z = 1$, is approximately 68%.

2. The probability that a z–score falls within 2 standard deviations of the mean, that is, between $z = -2$ and $z = 2$, is approximately 95%.

3. The probability that a z–score falls within 3 standard deviations of the mean is 99.7%.

Thus, approximately 99.7% of the z–scores fall within $z = -3$ and $z = 3$.

In many cases we have to find areas between two given values of z or areas to the right or left of some value of z. Finding these areas is an easy task provided we remember that the area under the entire normal distribution is 1. Thus, since the normal distribution is symmetrical about $z = 0$, we conclude that the area to the right of $z = 0$ and the area to the left of $z = 0$ are both equal to 0.5000.

EXAMPLE 4

Find the area between $z = 0.87$ and $z = 2.57$ in a standard normal distribution.

SOLUTION

We cannot look this up directly since the chart starts at 0, not at 0.87. However, we can look up the area between $z = 0$ and $z = 2.57$ and get 0.4949 and then look up the area between $z = 0$ and $z = 0.87$ and get 0.3078. We then take the difference between the two and get

$$0.4949 - 0.3078 = 0.1871$$

See Figure 8.8.

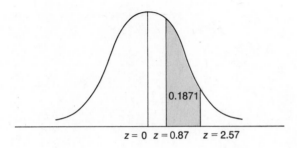

0.1871

$z = 0$ $z = 0.87$ $z = 2.57$ **Figure 8.8**

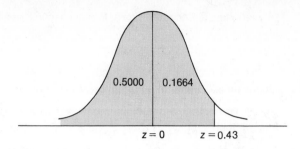

Figure 8.9

0.5000   0.1664

$z = 0$ $z = 0.43$

EXAMPLE 5

Find the probability of getting a z–value less than 0.43 in a standard normal distribution.

SOLUTION

The probability of getting a z–value less than 0.43 really means the area under the curve to the left of $z = 0.43$. This represents the shaded portion of Figure 8.9. We look up the area from $z = 0$ to $z = 0.43$ and get 0.1664 and add this to 0.5000 and get

$$0.5000 + 0.1664 = 0.6664$$

Thus the probability of getting a z–value less than 0.43 is 0.6664.

EXAMPLE 6

Find the probability of getting a z–value in a standard normal distribution which is
a. greater than -2.47.
b. greater than 1.82.
c. less than -1.53.

SOLUTION

a. Using Table V we first find the area between $z = 0$ and $z = 2.47$. See Figure 8.10. We get 0.4932. Then we add this to 0.5000

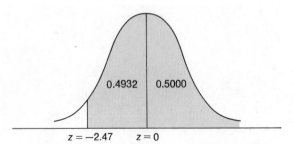

Figure 8.10

0.4932   0.5000

$z = -2.47$ $z = 0$

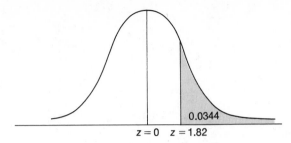

0.0344

$z = 0$ $z = 1.82$

Figure 8.11

which is the area to the right of $z = 0$ and get

$$0.4932 + 0.5000 = 0.9932$$

b. Here we are interested in finding the area to the right of
$z = 1.82$. See Figure 8.11. We find the area from $z = 0$ to
$z = 1.82$. It is 0.4656. Since we are interested in the area to the
right of $z = 1.82$, we must *subtract* 0.4656 from 0.5000 which
represents the *entire* area to the right of $z = 0$. We get

$$0.5000 - 0.4656 = 0.344$$

Thus, the probability that a z–score is greater than 1.82 is
0.0344.

c. Here we are interested in the area to the left of $z = -1.53$. See
Figure 8.12. We calculate the area between $z = 0$ and $z = 1.53$.
It is 0.4370. We subtract this from 0.5000. Our result is

$$0.5000 - 0.4370 = 0.0630$$

Thus, the probability that a z–score is less than $z = -1.53$ is
0.0630.

In the preceding examples we interpreted the area under the normal curve as a
probability. If we know the probability of an event, we can look at the probability
chart and find the z–value that corresponds to this probability.

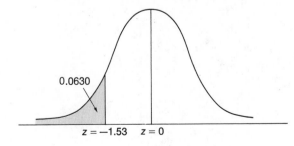

0.0630

$z = -1.53$ $z = 0$

Figure 8.12

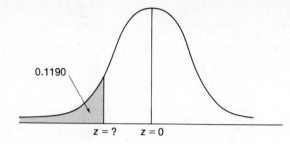

Figure 8.13

$z = ?$ $z = 0$

EXAMPLE 7

If the probability of getting less than a certain z–value is 0.1190, what is the z–value?

SOLUTION

We first draw the sketch shown in Figure 8.13. Since the probability is 0.1190, which is less than 0.5000, we know that the z–value must be to the left of the mean. We subtract 0.1190 from 0.5000 and get

$$0.5000 - 0.1190 = 0.3810$$

See Figure 8.13. This means that the area between $z = 0$ and some z–value is 0.3810. Table V tells us that the z–value is 1.18. However, this is to the left of the mean. Therefore, our answer is $z = -1.18$.

EXAMPLE 8

If the probability of getting a z–value larger than a certain z–value is 0.0129, what is the z–value?

SOLUTION

In this case we are told that the area to the right of some z–value is 0.0129. This z–value must be on the right side. If it were on the left side, the area would have to be at least 0.5000. Why? See Figure 8.14. Thus, we subtract 0.0129 from 0.5000 and get 0.4871. Then we look up this probability in Table V and find the z–value that gives this probability. It is $z = 2.23$. If $z = 2.23$, the probability of getting a z–value greater than 2.23 is 0.0129.

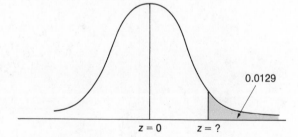

Figure 8.14

$z = 0$ $z = ?$

If we are given a normal distribution whose mean is different from 0 and whose
standard deviation is different from 1, we can convert this normal distribution into a
standardized normal distribution by converting each of its scores into standard
scores. To accomplish this we use Formula 4.2 of Chapter 4 which we will now call
Formula 8.1:

FORMULA 8.1 $\qquad z = \dfrac{x - \mu}{\sigma}$

Expressing the scores of a normal distribution as standard scores allows us to
calculate different probabilities, as the following examples will show.

EXAMPLE 9

In a normal distribution, $\mu = 25$ and $\sigma = 5$. What is the probability
of obtaining a value
a. greater than 30?
b. less than 15?

SOLUTION

a. We use Formula 8.1. We have $\mu = 25$, $x = 30$, and $\sigma = 5$, so
that

$$z = \frac{x - \mu}{\sigma} = \frac{30 - 25}{5} = \frac{5}{5} = 1$$

See Figure 8.15. Thus we are really interested in the area to the
right of $z = 1$ of a standardized normal curve. The area from
$z = 0$ to $z = 1$ is 0.3413. The area to the right of $z = 1$ is then

$$0.5000 - 0.3413 = 0.1587$$

Therefore, the probability of obtaining a value greater than 30 is
0.1587.

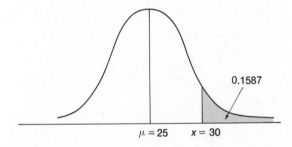

0.1587

$\mu = 25$ $\qquad x = 30$

Figure 8.15

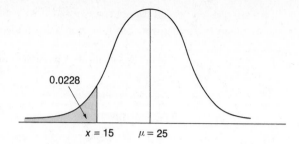

0.0228

Figure 8.16   x = 15   μ = 25

b. We use Formula 8.1. We have $\mu = 25$, $x = 15$, and $\sigma = 5$ so that

$$z = \frac{x - \mu}{\sigma} = \frac{15 - 25}{5} = \frac{-10}{5} = -2$$

See Figure 8.16. Thus, we are interested in the area to the left of $z = -2$. The area from $z = 0$ to $z = -2$ is 0.4772. Thus the area to the left of $z = -2$ is

$$0.5000 - 0.4772 = 0.0228$$

The probability of obtaining a value less than 15 is therefore 0.0228.

EXAMPLE 10

Find the percentile rank of 20 in a normal distribution with $\mu = 15$ and $\sigma = 2.3$.

SOLUTION

The problem is to find the area to the left of 20 in a normal distribution with $\mu = 15$ and $\sigma = 2.3$. We use Formula 8.1 with $\mu = 15$, $x = 20$, and $\sigma = 2.3$ so that

$$z = \frac{x - \mu}{\sigma} = \frac{20 - 15}{2.3} = \frac{5}{2.3} = 2.17$$

The area between $z = 0$ and $z = 2.17$ is 0.4850. See Figure 8.17. Thus, the area to the left of 20 is

$$0.5000 + 0.4850 = 0.9850$$

The percentile rank of 20 is therefore 98.5.

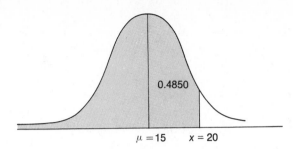

0.4850

$\mu = 15$ $x = 20$

Figure 8.17

go over

EXAMPLE 11

In a certain school, heights of students are normally distributed with
$\mu = 63$ inches and $\sigma = 2$ inches. If Sam is in the 90th percentile,
find his height.

SOLUTION

Since Sam is in the 90th percentile, this means that 90% of the
class is shorter than he. So, the problem here is to find a z-value
which has 90 percent of the area to the left of z. Therefore, we look
in the area portion of Table V to find a z-value which has 0.4000 of
the area to its left. See Figure 8.18. We use 0.4000, not 0.9000,
since 0.5000 of this is to the left of $z = 0$. The closest entry is
0.3997 which corresponds to $z = 1.28$. Now we convert this score
into a raw score by using Formula 4.3 on page 91. We have

$$x = \mu + z\sigma$$
$$= 63 + 1.28(2)$$
$$= 63 + 2.56$$
$$= 65.56$$

Thus, Sam's height is approximately 65.56 inches.

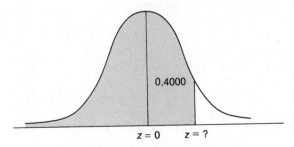

0.4000

$z = 0$ $z = ?$

Figure 8.18

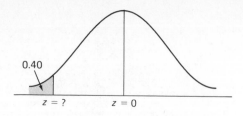

Figure 8.19

EXAMPLE 12

Same information as Example 11 of this section except Bill's percentile rank is 40. Find his height.

SOLUTION

Since Bill's percentile rank is 40, this means that 40% of the class is shorter than he. The problem here is to find a z–value which has 40 percent of the area to the left of z. See Figure 8.19. Since we are given the area to the left of z, we must subtract 0.4000 from 0.5000.

$$0.5000 - 0.4000 = 0.1000$$

Then we look in the area portion of Table V to find a z–value which has 0.1000 as the area between $z = 0$ and that z–value. The closest entry is 0.0987 which corresponds to $z = 0.25$. Since this z–value is to the left of the mean, we have $z = -0.25$. Now when we convert this score into a raw score we have

$$x = \mu + z\sigma$$
$$= 63 + (-0.25)2$$
$$= 63 - 0.50 = 62.50$$

Therefore, Bill's height is approximately 62.50 inches.

EXERCISES

1. In a standard normal distribution, find the area which lies
   a. between $z = 0$ and $z = 1.47$.
   b. between $z = -0.56$ and $z = 0$.
   c. to the right of $z = 1.93$.
   d. to the left of $z = -0.78$.
   e. to the right of $z = 2.31$.
   f. between $z = -1.56$ and $z = 1.69$.
   g. between $z = -1.82$ and $z = 2.56$.
   h. between $z = -0.92$ and $z = -1.66$.

2. Find the percentage of z–scores in a standard normal distribution that are
   a. above $z = -2.46$.
   b. below $z = 0.86$.
   c. between $z = 1.24$ and $z = 2.09$.
   d. above $z = 2.51$.
   e. above $z = 3.94$.
   f. between $z = -1.42$ and $z = -0.93$.
   g. between $z = -1.38$ and $z = 1.09$.
   h. between $z = -2.91$ and $z = -1.82$.

3. Find the z–score in a normal distribution which cuts off the bottom
   a. 28%.
   b. 18%.
   c. 3%.
   d. 1%.

4. Find z if the area under a standard normal curve
   a. between $z = 0$ and z is 0.4913.
   b. to the left of z is 0.9793.
   c. to the left of z is 0.2912.
   d. to the right of z is 0.0188.
   e. between $z = 1$ and z is 0.1441.
   f. between $z = 2$ and z is 0.0211.

5. Find the percentage of z–scores in a normal distribution with $\mu = 75$ and $\sigma = 10$ that are
   a. between 60 and 80.
   b. greater than 83.
   c. between 60 and 70.
   d. between 40 and 65.

6. In a normal distribution with $\mu = 40$ and $\sigma = 8$, find the percentile rank of a score of
   a. 50.
   b. 35.
   c. 53.
   d. 29.

7. In a certain girl's school the weights of students are normally distributed such that $\mu = 120$ lbs. and $\sigma = 12$ lbs. Find the weight of
   a. Caroline if her percentile rank is 55.
   b. Alexis if she is in the 10th percentile.
   c. Norma if she is in the 95th percentile.

8. The following is a standard normal curve with various z–values marked on it. If this curve also represents a normal distribution with $\mu = 24$ and $\sigma = 5$, replace the z–values with raw scores.

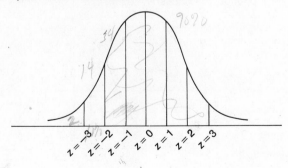

235

*9. A normal distribution has mean $\mu = 60$ and unknown standard deviation σ. However, it is known that 40% of the area lies to the right of 80. Find σ.

*10. A normal distribution has an unknown mean, μ, with a standard deviation of $\sigma = 19.8$. However, it is known that the probability that a z-score is less than 60 is 0.7123. Find μ.

8.3 SOME APPLICATIONS

We mentioned earlier that the importance of the normal distribution lies in its wide ranging applications. In this section we will apply the normal distribution to some concrete examples.

EXAMPLE 1

From past experience it has been found that the weight of a newborn infant at a maternity hospital is normally distributed with mean 7½ pounds (which equals 120 ounces) and standard deviation 21 ounces. If a newborn baby is selected at random, what is the probability that the infant weighs less than 4 pounds 15 ounces (which equals 79 ounces)?

μ 120

SOLUTION

We use Formula 8.1 of Section 8.2. Here $\mu = 120$, $\sigma = 21$, and $x = 79$ so that

$$z = \frac{x - \mu}{\sigma} = \frac{79 - 120}{21} = \frac{-41}{21}, \text{ or } -1.95$$

Thus, we are interested in the area to the left of $z = -1.95$. The area from $z = 0$ to $z = -1.95$ is 0.4744 so that the area to the left of $z = -1.95$ is $0.5000 - 0.04744$, or 0.0256. See Figure 8.20. Therefore, the probability that a randomly selected baby weighs less than 4 pounds 15 ounces is 0.0256.

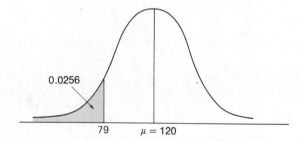

0.0256

Figure 8.20

79    $\mu = 120$

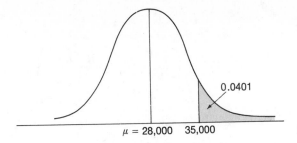

0.0401

$\mu = 28{,}000$ 35,000

Figure 8.21

EXAMPLE 2

The Flatt Tire Corporation claims that the useful life of its tires is normally distributed with a mean life of 28,000 miles and with a standard deviation of 4000 miles. What percentage of the tires are expected to last more than 35,000 miles?

SOLUTION

Here $\mu = 28{,}000$, $\sigma = 4000$, and $x = 35{,}000$. Using Formula 8.1 we get

$$z = \frac{x - \mu}{\sigma} = \frac{35{,}000 - 28{,}000}{4000} = 1.75$$

See Figure 8.21. We are interested in the area to the right of $z = 1.75$. The area between $z = 0$ and $z = 1.75$ is 0.4599, so the area to the right of $z = 1.75$ is 0.0401. Thus approximately 4% of the tires can be expected to last more than 35,000 miles.

EXAMPLE 3

In a recent study it was found that the number of hours that a typical ten-year-old child in a certain town watches television per week is normally distributed with a mean of 12 hours and with a standard deviation of 1.5 hours. If Gary is a typical ten-year-old child in this town, what is the probability that he watches between 9 and 14 hours of television per week?

SOLUTION

We first find the probability that Gary will watch television between 12 and 14 hours and add to this the probability that he will watch television between 9 and 12 hours per week. Using Formula 8.1 we have

$$z = \frac{x - \mu}{\sigma} = \frac{14 - 12}{1.5} = 1.33$$

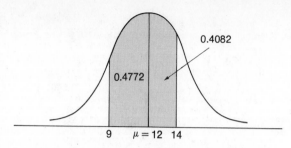

Figure 8.22

The area between $z = 0$ and $z = 1.33$ is 0.4082. See Figure 8.22.
Similarly,

$$z = \frac{x - \mu}{\sigma} = \frac{9 - 12}{1.5} = \frac{-3}{1.5} = -2$$

The area between $z = 0$ and $z = -2$ is 0.4772. See Figure 8.22.
Adding these two probabilities we get $0.4082 + 0.4772 = 0.8854$.
Thus, the probability that Gary watches between 9 and 14 hours of
television per week is 0.8854.

EXAMPLE 4

Daisy discovers that the amount of time it takes her to drive to work
is normally distributed with a mean of 35 minutes and a standard
deviation of 7 minutes. At what time should Daisy leave her home
so that she has a 95% chance of arriving at work by 9 A.M.?

SOLUTION

We first find a z–value which has 95% of the area to the left of z.
See Figure 8.23. Thus we look in the area portion of Table V to
find a z–value which has 0.4500 of the area to its left. (Remember
0.5000 of the area is to the left of $z = 0$.) From Table V we find that
z is midway between 1.64 and 1.65. We will use 1.65. We then find
the raw score corresponding to $\mu = 35$, $\sigma = 7$, and $z = 1.65$. We
have

$$\begin{aligned}
x &= \mu + z\sigma \\
&= 35 + (1.65)7 \\
&= 35 + 11.55 \\
&= 46.55
\end{aligned}$$

So, if Daisy leaves her house 46.55 minutes before 9 A.M., she will
arrive on time about 95% of the time. She should leave her home at
8:13 A.M.

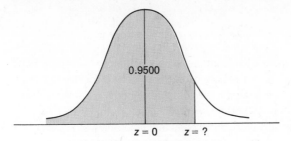

z = 0   z = ?   **Figure 8.23**

*EXAMPLE 5

In one study a major television corporation finds that the life of a color television tube is normally distributed with a standard deviation of 1.53 years. If 7% of these tubes last more than 6.9944 years, find the mean life of a television tube.

SOLUTION

We are told that 7% of the tubes last more than 6.9944 years. See Figure 8.24. Thus, approximately 43% of the tubes last between μ and 6.994 years. This means that on a standardized normal distribution the area between $z = 0$ and z is 0.43. Using Table V we find that 0.43 of the area is between $z = 0$ and $z = 1.48$. Then

$$x = \mu + z\sigma$$
$$6.9944 = \mu + (1.48)(1.53)$$
$$6.9944 = \mu + 2.2644$$
$$6.9944 - 2.2644 = \mu$$
$$4.73 = \mu$$

Thus, the mean life of a television tube is 4.73 years.

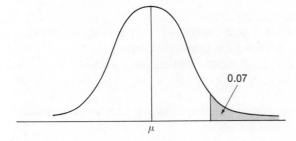

0.07

μ **Figure 8.24**

EXERCISES

1. The College Entrance Examination scores are approximately normally distributed with a mean of 500 and a standard deviation of 100 for graduates of McDonald High School. Find the probability that a randomly selected student graduate scored lower than 450 on this exam. $.3085$

2. The life of a washing machine produced by one major company is known to be normally distributed with a mean life of 5.3 years and a standard deviation of 1.93 years. Find the probability that a randomly selected machine will last more than 6 years.

3. The XYZ Corporation sells photocopying machines. The number of copies that a machine will give before requiring new toner is normally distributed with a mean of 20,000 copies and a standard deviation of 3500 copies. Find the probability that a typical machine will give between 15,000 to 25,000 copies before new toner is needed.

4. The amount of liquor consumed at a Friday night party is normally distributed with a mean of 50 pints and a standard deviation of 7 pints. What is the probability that less than 40 pints will be consumed at this gathering?

5. The average yearly salary of workers of the Ace Construction Company is $15,000 with a standard deviation of $1800. Bill Jackson is an employee of the company. If salaries are normally distributed, what is the probability that he earns more than $17,000 per year?

6. The average time needed for a rat to find its way through a maze is normally distributed with a mean of 12 minutes and a standard deviation of 3.2 minutes. Find the probability that a randomly selected rat will find its way through the maze in

   a. less than 8 minutes.

   b. not less than 11 minutes.

7. The distribution of lengths for the career of a major league ball player is approximately normal with a mean of 8 years and a standard deviation of 4 years. Find the probability that a randomly selected player will have a playing career of more than 10 years.

8. Mortgage statistics collected by a bank indicate that the number of years that the average new homeowner will occupy the house before moving or selling is normally distributed with a mean of 6.3 years and a standard deviation of 2.31 years. If a homeowner is selected at random, what is the probability that the owner will sell the house after 3 years?

9. Professor Milford has found that the grades on a final exam are normally distributed with a mean of 62 and a standard deviation of 12.

   a. If the passing grade is 52, what percent of the class will fail?

   b. If Professor Milford wants only 80% of the class to pass, what should be the passing grade?

   c. If Professor Milford wants only 8% of the class to get A's, what grade must a student have in order to get an A?

10. Phil finds that the number of hours he sleeps per night is normally distributed with a mean of 4.3 hours and a standard deviation of 0.37 hours. If a night is selected at random, what is the probability that Phil will sleep
    a. at most 6 hours?
    b. at least 4 hours?
    c. between 3.2 and 5.8 hours?

11. A school is administering a scholarship exam. Assume that the length of time required to complete the test is normally distributed with a mean of 50 minutes and a standard deviation of 17 minutes. How much time must be allowed by the proctors if the school wishes to assure enough time for only 85% of the applicants to complete the exam?

12. It is known that IQ scores of workers in a large factory are normally distributed with a mean of 109 and a standard deviation of 11. Furthermore, it is also known that certain factory jobs which demand only a minimum IQ of 102 bore people who have IQ scores over 123. On the basis of IQ, what percentage of the company's employees can be used for this particular job?

13. The life of a pressure cooker is normally distributed with a mean of 4.7 years and a standard deviation of 0.9 years. The manufacturer will replace any defective pressure cooker free of charge while under the guarantee. For how many years should the manufacturer guarantee the pressure cookers if not more than 5% of them are to be replaced?

14. In a certain doctor's office the waiting time is normally distributed with a mean of 25 minutes and a standard deviation of 5.3 minutes. Laura has just walked into the doctor's office. Find the probability that she will have to wait at least 18 minutes before seeing the doctor.

15. According to weather bureau records, the number of degree-days (fuel consumption units) for a large northeastern U.S. city during a typical winter season is normally distributed with a mean of 3900 and a standard deviation of 400. Find the probability that next winter will be a mild winter and that as a result 4000 degree days will be recorded.

8.4 THE NORMAL CURVE APPROXIMATION TO THE BINOMIAL DISTRIBUTION

An important application of the normal distribution is the approximation of the binomial distribution. To see why such an approximation is needed, suppose we were interested in determining the probability of getting 9 heads when a coin is flipped 20 times. This is a binomial distribution problem where $n = 20$, $x = 9$, $p = 1/2$, and $q = 1/2$. We use Formula 7.4 of Chapter 7 to determine this probability. We get

$$p(9 \text{ heads}) = \frac{20!}{9!11!}\left(\frac{1}{2}\right)^9\left(\frac{1}{2}\right)^{11}$$

$$= \frac{167,960}{1,048,576} = 0.1602$$

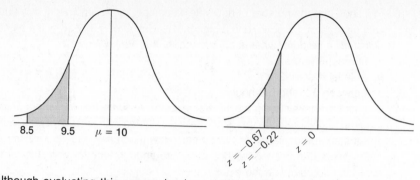

Figure 8.25

Although evaluating this expression is not difficult, the calculations involved are time consuming. If we actually compute the answer, we get a value of 0.1602. Thus, the probability of getting 9 heads when tossing a coin 20 times is 0.1602.

It turns out that we can obtain a fairly good approximation to the binomial distribution by using the normal curve. To accomplish this, we need the mean and standard deviation of the binomial distribution. Recall that for a binomial distribution the mean is $\mu = np$ and that the standard deviation is $\sigma = \sqrt{npq}$. These formulas were given in Section 7.5 of Chapter 7. Applying them to our example we get

$$\mu = np = 20(1/2) = 10$$

and

$$\sigma = \sqrt{npq} = \sqrt{20(1/2)(1/2)} = \sqrt{5}^1 = 2.236$$

We can now approximate this binomial distribution by a normal curve with $\mu = 10$ and $\sigma = 2.24$. We get this approximation by calculating the area under the normal curve between $x = 8.5$ and $x = 9.5$. See Figure 8.25. Any time we use the normal curve as an approximation for the binomial, we must calculate probabilities using an extra 0.5 either added to or subtracted from the number. A complete justification of this requires a knowledge of calculus and is beyond the scope of this text.

Returning to our example we have

$$z = \frac{x - \mu}{\sigma} = \frac{8.5 - 10}{2.24} = -0.67$$

and

$$z = \frac{x - \mu}{\sigma} = \frac{9.5 - 10}{2.24} = -0.22$$

The area between $z = 0$ and $z = -0.67$ is 0.2486 and the area between $z = 0$ and $z = -0.22$ is 0.0871 so that the area between $z = -0.67$ and $z = -0.22$ is $0.2486 - 0.0871 = 0.1615$.

---

[1]A knowledge of how to compute square roots is not assumed. These values can be obtained from the square root table given in the Appendix.

Using the binomial distribution, we find the probability of getting 9 heads is 0.1602. Using the normal curve approximation to the binomial distribution, we find the probability of getting 9 heads in 20 tosses of a coin is 0.1615. Although the answers differ slightly, the answer we get using the normal curve approximation is accurate enough for most applied problems. Furthermore, it is considerably easier to calculate.

More generally, if we were interested in the probability of getting 13 heads in 20 tosses of a coin, we can approximate this by calculating the area between $x = 12.5$ and $x = 13.5$.

Comment Any time you approximate a binomial probability with the normal distribution, depending upon the situation make sure to add or subtract 0.5 from the number.

The normal curve approximation to the binomial distribution is especially helpful when we must calculate the probability of many different values. The following examples will illustrate its usefulness.

EXAMPLE 1

Melissa is a nurse at Maternity Hospital. From past experience she determines that the probability that a newborn child is a boy is 1/2. (In the United States today, the probability that a newborn child is a boy is 0.53, not 0.50.) What is the probability that among 100 newborn babies, there are at least 60 boys?

SOLUTION

We can determine the probability *exactly* by using the binomial distribution or we can get an *approximation* by using the normal curve approximation. To determine the answer exactly, we say that a newborn child is either a boy or a girl with equal probability. Thus, $p = 1/2$ and $q = 1/2$. The probability that there are at least 60 boys means that we must calculate the probability of having 60 boys, 61 boys, . . . , and finally the probability of having 100 boys. Using the binomial distribution formula, we have

$$p(\text{at least 60 boys}) = p(60 \text{ boys}) + p(61 \text{ boys}) + \cdots + p(100 \text{ boys})$$
$$= \frac{100!}{60!(40)!}\left(\frac{1}{2}\right)^{60}\left(\frac{1}{2}\right)^{40} + \frac{100!}{61!39!}\left(\frac{1}{2}\right)^{61}\left(\frac{1}{2}\right)^{39} + \cdots + \frac{100!}{100!0!}\left(\frac{1}{2}\right)^{100}\left(\frac{1}{2}\right)^{0}$$

Calculating these probabilities requires some lengthy computations. However, the same answer can be obtained more quickly by using

243

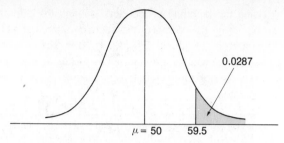

Figure 8.26

$\mu = 50 \qquad 59.5$

the normal curve approximation. We first determine the mean and standard deviation:

$$\mu = np = 100(1/2) = 50$$
$$\sigma = \sqrt{npq} = \sqrt{100(1/2)(1/2)} = \sqrt{25} = 5$$

Then we find the area to the right of 59.5 as shown in Figure 8.26. We use 59.5 rather than 60.5 in our approximation since we want to include 60 boys in our calculations. If the problem had specified more than 60 boys, we would have used 60.5 rather than 59.5 since more than 60 means do not include 60.

In our case we have

$$z = \frac{x - \mu}{\sigma} = \frac{59.5 - 50}{5} = \frac{9.5}{5} = 1.9$$

From Table V we find that the area to the right of $z = 1.9$ is $0.5000 - 0.4713$, or 0.0287.

Thus, the probability that among 100 newborn children there are at least 60 boys is 0.0287.

EXAMPLE 2

Refer to Example 1 of this section. Find the probability that there will be between 45 and 60 boys (not including these numbers) among 100 newborn babies at Maternity Hospital.

SOLUTION

The probability is approximated by the area under the normal curve between 45.5 and 59.5. We do not use 44.5 or 60.5 since 45 and 60 are not to be included. To find the area between 45.5 and 50 we have

$$z = \frac{x - \mu}{\sigma} = \frac{45.5 - 50}{5} = -0.90$$

The area is thus 0.3159.

Also to find the area between 50 and 59.5, we have

$$z = \frac{x - \mu}{\sigma} = \frac{59.5 - 50}{5} = 1.9$$

This area is 0.4713.

Adding these two areas we get

$$0.3159 + 0.4713 = 0.7872$$

Therefore, the probability that there are between 45 and 60 boys among 100 newborn babies at Maternity Hospital is 0.7872.

EXAMPLE 3

A large television network is considering canceling its weekly 7:30 P.M. comedy show because of a decrease in the show's viewing audience. The network decides to randomly phone 5000 viewers and to cancel the show if fewer than 1900 viewers are actually watching the show. What is the probability that the show will be canceled if $p = .4$
a. only 40% of all television viewers actually watch the comedy show?
b. only 39% of all television viewers actually watch the comedy show? $p = .39$

SOLUTION

a. Since a randomly selected television viewer that is phoned either watches the show or does not watch the show, we can consider this as a binomial distribution with $n = 5000$ and $p = 0.40$. We first calculate μ and σ:

$$\mu = 5000(0.40) = 2000$$
$$\sigma = \sqrt{5000(0.40)(0.60)} = \sqrt{1200} = 34.64$$

Since the show will be canceled only if fewer than 1900 people watch it, we are interested in the probability of having 0, 1, 2, . . ., 1899 viewers. Using a normal curve approximation, we calculate the area to the left of 1899.5.

245

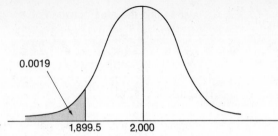

0.0019

Figure 8.27   1,899.5   2,000

We have

$$z = \frac{x - \mu}{\sigma} = \frac{1899.5 - 2000}{34.64} = -2.90$$

The area to the left of $z = -2.90$ is

$$0.5000 - 0.4981 = 0.0019$$

See Figure 8.27. Thus, the probability that the show is canceled
is 0.0019.

b. In this case the values of μ and σ are different, since the value
    of p is 0.39. We have

$$\mu = 5000(0.39) = 1950$$
$$\sigma = \sqrt{5000(0.39)(0.61)} = \sqrt{1189.5} = 34.49$$

Since the show will be canceled if fewer than 1900 people watch
it, we calculate the area to the left of 1899.5.
We have

$$z = \frac{1899.5 - 1950}{34.49} = -1.46$$

The area to the left of $z = -1.46$ is

$$0.5000 - 0.4279 = 0.0721$$

See Figure 8.28. Thus, the probability that the show is canceled
is 0.0721.

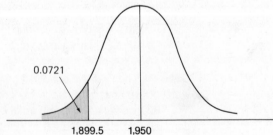

0.0721

Figure 8.28   1,899.5   1,950

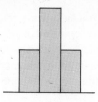

Binomial distribution
with $n = 2$ and $p = \frac{1}{2}$

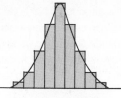

Binomial distribution
with $n = 10$ and $p = \frac{1}{2}$

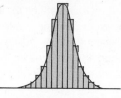

Binomial distribution
with $n = 25$ and $p = \frac{1}{2}$

Figure 8.29

Histogram for a binomial
distribution approaches that of
a normal distribution as n
gets larger.

Comment The degree of accuracy of the normal curve approximation to the binomial distribution depends upon the values of n and p. Figure 8.29 indicates how fast the histogram for a binomial distribution approaches that of a normal distribution as n gets larger. As a rule, the normal curve approximation can be used with fairly accurate results when *both* np and nq are greater than 5.

EXERCISES

1. It is claimed that 60% of all married men in a particular city have life insurance policies. If 500 married men are randomly selected, find the probability that at most 320 of them have life insurance policies.

2. If 65% of all students attending State University have type O blood, what is the probability that a random sample of 700 students will contain 600 or more students with type O blood?

3. If 55% of all people in Bayerville speak 2 languages, what is the probability that a random sample of 80 people will contain at most 44 people who speak 2 languages?

4. Thirty-seven percent of all calculators sold by a leading department store are not fully charged when originally purchased. If 200 calculators were sold during the month of January, find the probability that between 120 to 140 of the calculators were fully charged.

5. It is claimed that 1 out of every 5 people in York City is receiving some form of government financial aid. If 5000 people are selected at random, what is the probability that at most 950 of these people are receiving some form of government aid?

6. Eighty percent of all turkeys sold by Farmer McDonald are injected with a certain growth hormone. In a random sample of 400 turkeys, find the probability that at most 300 of them will have been injected with this growth hormone.

7. Ninety-five percent of all calls for assistance to a service station on a cold winter day are for cars that will not start because of a dead battery. If the service station receives 50 calls for assistance, find the probability that 46 of these calls for assistance are for cars that will not start because of a dead battery.

8. The Miller Car Rental Agency finds that 9% of all people making reservations for a car do not actually show up for the car. If the car rental agency has

accepted reservations for 90 cars although it has only 84 cars available, what is the probability that it will have a car for each person that has reserved one?

9. Sid is taking a true–false examination consisting of 60 questions. Since he has not studied for the exam, he decides to guess at the answers. If Sid needs at least 40 correct answers to pass, what is the probability that he passes the exam?

10. A certain birth control device claims to be 94% effective. If 200 women are using this device, find the probability that not more than 8 pregnancies will occur among these women.

11. Seven percent of all people who borrow books from the library fail to return them on time. If 450 people borrowed books from the library today, find the probability that 25 of these people will not return the books on time.

12. A large department store has found that 19% of all purchases are returned for one reason or another. If 1000 purchases were made, what is the probability that at most 170 of these will be returned for credit?

8.5 APPLICATION TO STATISTICAL QUALITY CONTROL[2]

In recent years numerous articles and books have been written on how statistical quality controls operate. This is an important branch of applied statistics. What are quality control charts? To answer this question we must consider the mass production process.

Industrial experience shows that most production processes can be thought of as normally distributed. So, when a manufacturer adjusts his machines to fill a jar with 10 ounces of coffee, although not all the jars will actually weigh 10 ounces, the weight of a typical jar will be very close to 10 ounces. When too many jars weigh more than 10 ounces, the manufacturer will lose money. When too many jars weigh less than 10 ounces, he will lose customers. He is therefore interested in maintaining the weight of the jars as close to 10 ounces as possible.

If the production process behaves in the manner described above, the weight of a typical jar of coffee is either acceptable or not acceptable. Thus, it can be thought of as a binomial variable. We can then use the normal approximation to the binomial distribution.

Rather than weigh each individual jar of coffee, the manufacturer can use **quality control charts.** This is a simple graphical method which has been found to be highly useful in the solution to problems of this type.

Figure 8.30 is a typical quality control chart. The horizontal line represents the time scale. The vertical line has three markings: μ, $\mu + 3\sigma$, and $\mu - 3\sigma$.

The middle line is thought of as the mean of the production process although in reality it is usually the mean weight of past daily samples. The two other lines serve as control limits for the daily production process. These lines have been spaced 3

---

[2]This section can be skipped without any loss of continuity.

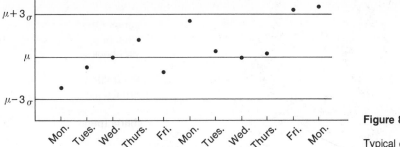

Figure 8.30

Typical quality control chart.

standard deviation units above and below the mean. From the normal distribution table, we find that approximately 99.7% of the area should be between $\mu - 3\sigma$ and $\mu + 3\sigma$. Thus, the probability that the weight of a typical jar of coffee falls outside the control bands is only 0.003. This is a relatively small probability. Therefore, if a sample of sufficient size is taken and the average weight is outside the control bands, the manufacturer can then assume that the production process is not operating properly and that immediate adjustment is necessary to avoid losing money or customers.

Figure 8.30 indicates that the production process went out of control on a Friday.

SELF-STUDY GUIDE

In this chapter we discussed the difference between a discrete random variable and a continuous random variable. We studied the normal distribution in detail since it is the most important probability distribution of a continuous random variable. Because of their usefulness, normal curve area charts have been constructed. These charts allow us to calculate the area under the standard normal curve. Thus, we can determine the probability that a random variable will fall within a specified range.

Not only can these charts be used to calculate probabilities for variables that are normally distributed, but they can also be used to obtain a fairly good approximation to binomial probabilities. This is especially helpful when we must calculate the probability that a binomial random variable assumes many different values. Numerous applications of these ideas were given.

The normal distribution was then applied to the construction of quality control charts which are so important in many industrial processes. Today statistical quality control is an important branch of applied statistics. Without discussing them in detail, we indicated the usefulness of these quality control charts.

You should be able to identify each symbol in the following formulas, understand the relationships among the symbols expressed in each formula, understand the significance of each formula, and use the formulas in solving problems.

1. $z = \dfrac{x - \mu}{\sigma}$ 2. $x = \mu + z\sigma$

You should now be able to demonstrate your knowledge of the following ideas presented in this chapter by giving definitions, descriptions, or specific examples. Page references are given for each term so that you can check your answer.

Discrete variable (page 221)   Normal distribution (page 223)
Continuous variable (page 221)   Standardized normal distribution (page
Bell-shaped distribution (page 222)   225)
Gaussian distribution (page 222)   Quality control charts (page 248)
Normal curve (page 222)

The tests of the following section will be more useful if you take them after you have studied the examples and solved the exercises given in this chapter.

MASTERY TESTS

Form A

1. The normal distribution is symmetric about the _____.
2. The total percent of area under a normal curve is _____.
3. In a normal distribution the mode is at
    a. $z = -2$ b. $z = -1$ c. $z = 0$ d. $z = 1$ e. none of these

   For questions 4 through 6 consider a distribution of the weights of college students which is normally distributed with a mean of 150 pounds and a standard deviation of 5.

4. The percent of the weights above 155 pounds is
    a. 16 b. 14 c. 5 d. 2 e. 34
5. The percent of the weights below 145 pounds is
    a. 34 b. 68 c. 16 d. 84 e. 5
6. The probability that a person's weight is between 145 and 155 pounds is
    a. 0.50 b. 1.00 c. 0.68 d. 0.66 e. 0.34
7. The proportion of area under the normal curve from $z = 0$ to $z = 0.5$ is 0.1915. What is the proportion of area under the normal curve from $z = 0.5$ to $z = 1$?
    a. 0.3413 b. 0.1915 c. 0.1498 d. 0.3830 e. none of these
8. Which of the following z–scores for a normal distribution has the smallest probability of occurrence?
    a. 0.23 b. −1.2 c. 4.5 d. 0 e. −3.1
9. The standard deviation of a distribution of z–scores is
    a. 1 b. 0 c. −1 d. 4 e. none of these
10. The area under a normal curve from $z = -1$ to $z = 0.75$ is
    a. 0.1587 b. 0.2266 c. 0.3413 d. 0.6417 e. none of these

Form B

1. The amount of soda dispensed by a soft drink vending machine in a school cafeteria is normally distributed with a mean of 8 ounces and a standard deviation of 0.9 ounce.

   a. What percentage of the cups will contain more than 8.2 ounces?

   b. What is the probability that a cup will overflow if it can hold exactly 9 ounces?

2. Bill's time for running 100 yards has been found to be normally distributed with a mean of 11.2 seconds and a standard deviation of 0.14 seconds. Find the probability that Bill runs the 100 yards in less than 9.8 seconds.

3. The mean grade on the last statistics exam was 75 with a standard deviation of 10. The top 10% of the students will get an A. What is the lowest grade that a student can get and still receive an A?

4. An airline company finds that 95% of the people who make reservations actually keep them. If the airline company accepts reservations from 1000 people, what is the probability that 950 people actually show up?

5. Transistors are manufactured by a certain process that gives them a mean lifetime of 600 hours and a standard deviation of 45 hours. A random sample of transistors is taken from this group. What percent will have a mean lifetime of between 580 and 620 hours?

6. The graduating averages at a Staten Island high school were normally distributed with a mean of 80 and a standard deviation of 7.8. What is the probability that a graduating senior had an average below 68?

7. In a bowling match the scores were normally distributed with a mean of 250 and a standard deviation of 25. What percent of the bowlers scored below 225?

8. A manufacturer claims that "4 out of 5 dentists recommend sugarless gum for their patients who chew gum." Assuming that this claim is true, find the probability that in a randomly selected group of 60 dentists, 48 or more will recommend sugarless gum for their patients who chew gum.

9. It is known that 35% of all summons issued by a particular police precinct are for parking violations. If 200 tickets will be issued during the month of January, what is the probability that at least 50 of them will be for parking violations?

10. If 50% of all girls at Staten Island Community College are blondes, what is the probability that a survey of 100 girls will contain 65 or more blondes?

11. A psychological introvert–extrovert test yields scores which are normally distributed with a mean of 80 and a standard deviation of 3. People scoring in the top 10% will be classified as extroverts. What score is the cut-off point for an extrovert?

SUGGESTED READING

Hoel, Paul G. *Elementary Statistics*. 4th ed. chap. 4. New York: Wiley, 1976.

Huntsberger, D. V. *Elements of Statistical Inference*. 3rd ed. Boston, Mass.: Allyn and Bacon, 1973.

Mendenhall, W. *Introduction to Probability and Statistics*. 3rd ed. Belmont, Calif.: Duxbury Press, 1971.

Walpole, R. E. *Introduction to Statistics*. New York: MacMillan, 1968.

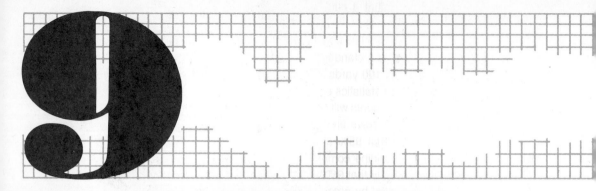

LINEAR CORRELATION
AND REGRESSION

LIVE BIRTHS AND MARRIAGES IN THE UNITED STATES — 1910–1974

Year	Number of births (in thousands)	Number of marriages (in thousands)	Year	Number of births (in thousands)	Number of marriages (in thousands)
1910	2777	948	1963	4098	1654
1915	2965	1008	1964	4027	1725
1920	2950	1274	1965	3760	1800
1925	2909	1188	1966	3606	1857
1930	2618	1127	1967	3521	1927
1935	2377	1327	1968	3502	2069
1940	2559	1596	1969	3600	2145
1945	2858	1613	1970	3731	2159
1950	3632	1667	1971	3559	2196
1955	4104	1531	1972	3256	2269
1960	4258	1523	1973	3141	2277
1962	4167	1577			

Source: *Vital Statistics of the United States, 1975*, U.S. National Center for Health Statistics.

In recent years the number of marriages in the United States has been increasing. Based upon the data in the table can we say that there is a relationship between the number of births and the number of marriages in the United States? Will more marriages result in more births? Although this table would seem to indicate that the two variables are related or that there is a negative correlation between them, other factors have to be considered. (As used in this chapter, the word *variable* will mean any measurable quantity that may change from situation to situation. For example, the number of marriages is a variable since it changes from year to year.)

Now look at the table on the distribution of employment by sex. Is it true that there is a relationship between the percent of males or females in the professional or technical areas and in the sales force area? Will a greater percent of males in one occupation result in a smaller percentage of males in another occupation?

These tables and the mathematical ideas to be discussed in this chapter allows one to predict the behavior of one variable from a knowledge of the other. It does not give any cause-and-effect relationship.

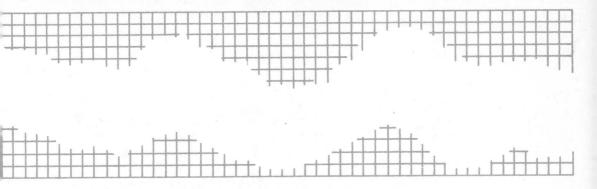

Distribution of Employment by Sex

	1970		1980*		1985*	
Occupation	Male	Female	Male	Female	Male	Female
Professional, technical and kindred workers	54.3%	45.7%	54.7%	45.3%	57.4%	42.6%
Computer programmers	77.3	22.7	80.0	20.0	81.3	18.7
Designers	76.5	23.5	72.0	28.0	69.7	30.3
Drafters	92.0	8.0	89.5	10.5	88.2	11.8
Electrical and electronic engineering technicans	94.3	5.7	93.3	6.7	92.0	7.2
Engineering and science technicians						
(not elsewhere classified)	82.3	17.7	79.6	20.4	78.2	21.8
Mechanical engineering technicians	97.1	2.9	94.2	5.8	92.7	7.3
Other technicians, except health	89.2	10.8	87.0	13.0	85.9	14.1
Personnel and labor relations workers	68.8	31.2	70.5	29.5	71.3	28.7
Recreation workers	59.6	40.4	66.5	33.5	69.0	31.0
Registered nurses	2.6	97.4	2.7	97.3	2.8	97.2
Therapists	36.4	63.6	36.0	64.0	35.8	64.2
Tool programmers, numerical control	84.9	15.1	69.8	30.2	62.3	37.7
Managers and administrators, except farm	84.7	15.3	86.2	13.8	85.4	14.6
Bank officials and financial managers	82.6	17.4	78.6	21.4	76.5	23.5
Buyers and shippers, farm products	97.9	2.1	96.9	3.1	96.4	3.6
Buyers, wholesale and retail trade	70.6	29.4	76.7	23.3	79.7	20.3
Managers and administrators						
(not elsewhere classified)	88.4	11.6	89.5	10.5	89.0	11.0
Managers and superintendents, building	59.3	40.7	62.2	37.8	63.7	36.3
Restaurant, cafeteria and bar managers	66.2	33.8	64.9	35.1	64.2	35.8
Sales managers and department heads, retail	75.9	24.1	75.2	24.8	74.8	25.2
Salesworkers	61.8	38.2	59.4	40.6	58.6	41.4
Insurance agents, brokers and underwriters	87.5	12.5	84.6	15.4	83.1	16.9
Real estate agents and brokers	67.7	32.3	68.2	31.8	55.1	44.9
Sales representatives, manufacturing	91.5	8.5	93.4	6.6	94.3	5.7
Sales representatives, wholesale	93.6	6.4	91.3	8.7	90.1	9.9
Salesclerks, retail trade	35.2	64.8	33.7	66.3	32.9	67.1
Salesworkers, retail trade	87.2	12.8	85.1	14.9	84.0	16.0
Salespeople, service and construction	65.9	34.1	55.7	44.3	51.4	48.6
Stock and bond sales agents	91.4	8.6	88.8	11.2	87.5	12.5
Total, All Occupations Studied	**64.1**	**36.9**	**61.0**	**39.0**	**59.5**	**40.5**
Total, All U.S. Occupations	**62.3**	**37.7**	**61.2**	**38.8**	**61.0**	**39.0**

*Projected   *Source: Department of Labor; 1960 and 1970 Census of Population; Conference Board*

Preview

In this chapter we will discuss the following topics:

1. *Historical Discussion*

We briefly discuss some of the pioneers in the field of correlation and regression.

2. *Scatter Diagrams*

To get some idea of whether two variables are related, we draw two lines, one vertical and one horizontal. Then we place dots in various places corresponding to the given data.

3. *Linear Correlation*

This is a relationship that tells us whether an increase in the value of one variable will cause a change in the average value of the other variable.

4. *Coefficient of Linear Correlation*

This measures the strength of the linear relationship that exists between two variables.

5. *Linear Regression*

The regression line is a mathematical expression which allows us to predict the value of one of the variables if the value of the other variable is known.

6. *Method of Least Squares*

This is the method that we use to determine the estimated regression line to be used in prediction.

7. *Standard Error of the Estimate*

This tells us how well the least square prediction equation describes the relationship between two variables.

Introduction

Up to this point we have been discussing the many statistical procedures for analyzing a single variable. However, when dealing with the problems of applied statistics in education, psychology, sociology, etc., we may be interested in determining whether a relationship exists between two or more variables. For example, if college officials have just administered a vocational aptitude test to 1000 entering freshmen, they may be interested in knowing whether there is any relationship between the math aptitude scores and the business aptitude scores. Do students who score well on the math section of the aptitude exam also do well on the business part? On the other hand, is it true that a student who scored poorly on the math part will necessarily score poorly on the business part? Similarly, the college officials may be interested in determining if there is a relationship between high school averages and college performance.

Questions of this nature frequently arise when we have many variables and are interested in determining relationships between these sets of scores. In this chapter we will learn how to compute a number which measures the relationship between two sets of scores. This number is called the **correlation coefficient.** The English mathematician Karl Pearson (1857–1936) studied it in great detail and to some extent so did another English mathematician, Sir Francis Galton (1822–1911).

Sir Francis Galton, a cousin of Charles Darwin, undertook a detailed study of human characteristics. He was interested in determining whether a relationship exists between the heights of fathers and the heights of their sons. Do tall parents have tall children? Do intelligent parents or successful parents have intelligent or successful children? In *Natural Inheritance* Galton introduced the idea referred to today as correlation. This mathematical idea allows us to measure the closeness of the relationship between two variables. Galton found that there exists a very close relationship between the heights of fathers and the heights of their sons. On the question of whether intelligent parents have intelligent children, it has been found that the correlation is 0.55. As we shall see, this means that it is not necessarily true that intelligent parents have intelligent children. In many cases children will score higher or lower than their parents. Figure 9.1 shows how the correlations for IQ range from 0.28 between parents and adopted children to 0.97 for identical twins reared together.

The precise mathematical measure of correlation as we use it today was actually formulated by Karl Pearson.

If there is a high correlation between two variables, we may be interested in representing this correspondence by some form of an equation. So we will discuss the **method of least squares.** The statistical method of least squares was developed by Adrien-Marie Legendre (1752–1833). Although Legendre is best known for his work in geometry, he also did important work in statistics. He developed the method of least squares. This method is used when we want to find the regression equation.

Finally, we will discuss how this equation can be used to make predictions and we will discuss the reliability of these predictions.

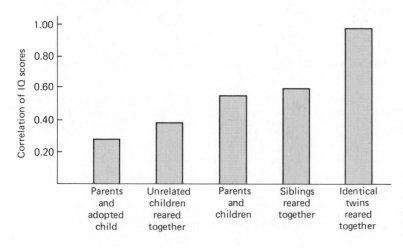

Figure 9.1

Correlations between intelligence of parents and that of their children.

9.1 SCATTER DIAGRAMS

To help us understand what is meant by a correlation coefficient, let us consider a guidance counselor who has just received the scores of an aptitude test administered to ten students. See Table 9.1.

TABLE 9.1

DIFFERENT APTITUDE SCORES RECEIVED BY TEN STUDENTS

STUDENT	MATH APTITUDE	BUSINESS APTITUDE	LANGUAGE APTITUDE	MUSIC APTITUDE
A	52	48	26	22
B	49	49	53	23
C	26	27	48	57
D	28	24	31	54
E	63	59	67	13
F	44	40	75	20
G	70	72	31	9
H	32	31	22	50
I	49	50	11	17
J	51	49	19	24

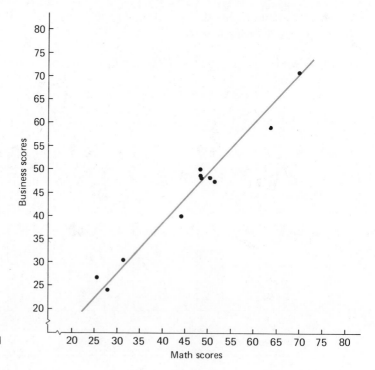

Figure 9.2

Scatter diagram for the math and business scores.

The counselor may be interested in determining if there is any correlation among these sets of scores. For example, the counselor may wish to know whether a student who scores well on the math aptitude part of the exam will also score well on the business aptitude part. She can analyze the situation pictorially by means of a **scatter diagram.**

To make a scatter diagram, we draw two lines, one vertical and one horizontal. On the horizontal line we indicate the math scores, and on the vertical line we indicate the business scores. Although we could put the math scores on the vertical line, we have purposely labeled the math scores on the horizontal line. This is done because we are interested in predicting the scores of the business aptitude part on the basis of the math scores. After both axes, that is, both lines are labeled, we use one dot to represent each person's score. The dot is placed directly above the person's math score and directly to the right of the business score. Thus the dot for Student A's score is placed directly above the 52 score on the math axis and to the right of 48 on the business axis. Similarly, the dot for Student B's score is placed directly above the 49 score on the math axis and to the right of 49 on the business axis. The same procedure is used to locate all the dots of Figure 9.2.

You will notice that these dots form an approximate straight line. When this happens we say that there is a **linear correlation** between the two variables.

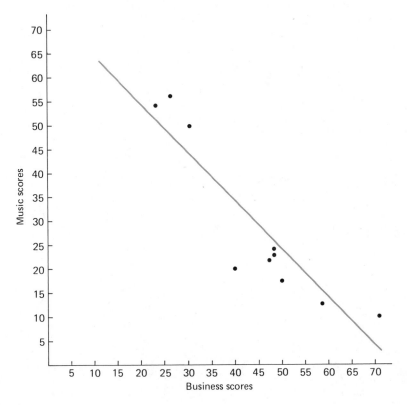

Figure 9.3

Scatter diagram for the business and music scores.

Notice also that the higher the math score, the higher the business score. The line moves in a direction which is from lower left to upper right. When this happens, we say that there is a **positive correlation** between the math scores and the business scores. This means that a student with a higher math score will have a higher business score.

Now let us draw the scatter diagram for the business aptitude scores and the music aptitude scores. It is shown in Figure 9.3. In this case you will notice that the higher the business score, the lower the music score. Again the dots arrange themselves in the form of a line, but this time the line moves in a direction which is from upper left to lower right. When this happens we say that there is a **negative correlation** between the business scores and the music scores. This means that a student with a high business score will have a low music score.

Now let us draw the scatter diagram for the language scores and music scores. It is shown in Figure 9.4. In this case the dots do not form a straight line. When this happens we say that there is little or no correlation between the language scores and the music scores.

Comment Although we will concern ourselves with only linear, that is, a straight line, correlation, the dots may suggest different types of curves. These are studied in detail by statisticians. Several examples of such scatter diagrams are given in Figure 9.5. In this text we will analyze only linear correlation.

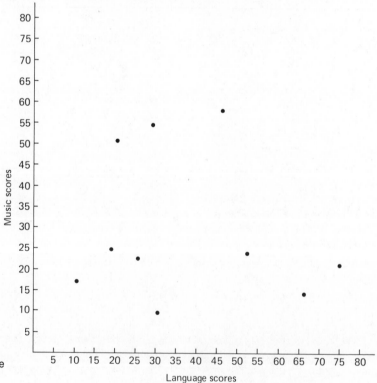

Figure 9.4

Scatter diagram for the language and music scores.

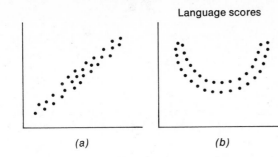

Language scores

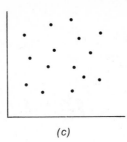

(a)   (b)   (c)

Figure 9.5

Scatter diagrams that
suggest (a) a linear
relationship, (b) a
curvilinear relationship,
(c) no relationship.

9.2 THE COEFFICIENT OF CORRELATION

Once we have determined that there is a linear correlation between two variables, we may be interested in determining the strength of the linear relationship. Karl Pearson developed a **coefficient of linear correlation** which measures the strength of a relationship between two variables. The value of the coefficient of linear correlation is calculated by means of Formula 9.1.

FORMULA 9.1

The coefficient of linear correlation is given by

$$r = \frac{n(\Sigma xy) - (\Sigma x)(\Sigma y)}{\sqrt{n(\Sigma x^2) - (\Sigma x)^2} \ \sqrt{n(\Sigma y^2) - (\Sigma y)^2}}$$

where x = label for one of the variables
          y = label for the other variable
and n = number of pairs of scores.

In Formula 9.1 the coefficient of correlation will always have a value between -1 and $+1$. A value of $+1$ means perfect positive correlation and corresponds to the situation where all the dots lie exactly on a straight line. A value of -1 means perfect negative correlation and again corresponds to the situation where all the points lie exactly on a straight line. Correlation is considered high when it is close to $+1$ or -1 and low when it is close to 0. If the coefficient of linear correlation is zero, we say that there is no linear correlation. These possibilities are indicated in Figures 9.6 and 9.7.

Although Formula 9.1 looks complicated, it is rather easy to use. The only new symbol that appears is Σxy. This value is found by multiplying the corresponding values of x and y and then adding all the products. The following examples will illustrate the procedure.

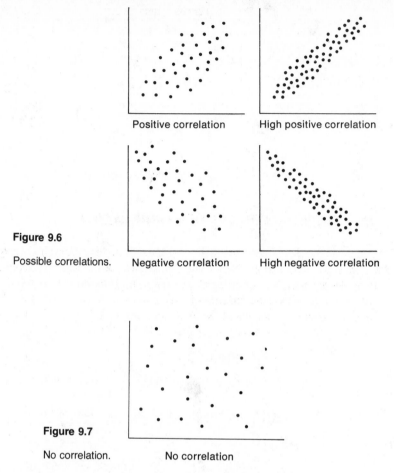

Positive correlation   High positive correlation

Figure 9.6

Possible correlations.   Negative correlation   High negative correlation

Figure 9.7

No correlation.   No correlation

EXAMPLE 1

Find the correlation coefficient for the data of Figure 9.2 (see page 256), which follows:

Math Score	52	49	26	28	63	44	70	32	49	51
Business Score	48	49	27	24	59	40	72	31	50	49

SOLUTION

We first let x represent the math score and y represent the business score. Then we arrange the data in tabular form as follows and apply Formula 9.1.

x (math)	y (business)	x²	y²	xy
52	48	2704	2304	2496
49	49	2401	2401	2401
26	27	676	729	702
28	24	784	576	672
63	59	3969	3481	3717
44	40	1936	1600	1760
70	72	4900	5184	5040
32	31	1024	961	992
49	50	2401	2500	2450
51	49	2601	2401	2499
$\Sigma x = 464$	$\Sigma y = 449$	$\Sigma x^2 = 23{,}396$	$\Sigma y^2 = 22{,}137$	$\Sigma xy = 22{,}729$

$$\text{Then } r = \frac{n(\Sigma xy) - (\Sigma x)(\Sigma y)}{\sqrt{n(\Sigma x)^2 - (\Sigma x)^2}\,\sqrt{n(\Sigma y^2) - (\Sigma y)^2}}$$

$$= \frac{10(22{,}729) - (464)(449)}{\sqrt{10(23{,}396) - (464)^2}\,\sqrt{10(22{,}137) - (449)^2}}$$

$$= \frac{18{,}954}{\sqrt{18{,}664}\,\sqrt{19{,}769}}$$

$$= \frac{18{,}954}{(136.62)(140.60)}$$

$$= \frac{18{,}954}{19{,}208.77} = 0.9867$$

Thus, the coefficient of correlation is 0.9867. Since this value is close to +1, we say that there is a high degree of positive correlation. Figure 9.2 also indicated the same result.

EXAMPLE 2

Find the coefficient of correlation for the data of Figure 9.4 (see page 258), which follows:

Language Score	26	53	48	31	67	75	31	22	11	19
Music Score	22	23	57	54	13	20	9	50	17	24

SOLUTION

We first let x represent the language score and y represent the music score. Then we arrange the data in tabular form as follows and apply Formula 9.1.

x (language)	y (music)	x^2	y^2	xy
26	22	676	484	572
53	23	2809	529	1219
48	57	2304	3249	2736
31	54	961	2916	1674
67	13	4489	169	871
75	20	5625	400	1500
31	9	961	81	279
22	50	484	2500	1100
11	17	121	289	187
19	24	361	576	456
$\Sigma x = 383$	$\Sigma y = 289$	$\Sigma x^2 = 18{,}791$	$y^2 = 11{,}193$	$\Sigma xy = 10{,}594$

Now we apply Formula 9.1. We have

$$r = \frac{n(\Sigma xy) - (\Sigma x)(\Sigma y)}{\sqrt{n(\Sigma x^2) - (\Sigma x^2)} \sqrt{n(\Sigma y^2) - (\Sigma y^2)}}$$

$$= \frac{10(10{,}594) - (383)(289)}{\sqrt{10(18{,}791) - (383)^2} \sqrt{10(11{,}193) - (289)^2}}$$

$$= \frac{-4747}{\sqrt{41{,}221} \sqrt{28{,}409}} = \frac{-4747}{(203.03)(168.55)}$$

$$= \frac{-4747}{34{,}220.71} = -0.1387$$

Thus, the coefficient of correlation is -0.1387. Since this value is close to zero, there is little correlation. Figure 9.4 indicated the same result.

In Chapter 3 we considered the effect of coding on the mean and standard deviation (see Section 3.4, page 73). We found that by adding (or subtracting) the same number to (or from) each term of a distribution, the calculations needed to compute the mean and standard deviation were simplified. Since the computations needed to calculate the coefficient of correlation are lengthy, we begin to wonder if the data also can be coded.

How will the correlation between x and y be affected if x is coded by adding the same number to (or subtracting the same number from) each score? How is y affected?

Fortunately, it turns out that the correlation coefficient is unaffected by adding or subtracting a number to either x or y or both. Thus, x can be coded in one way — perhaps by adding or subtracting a number — and y can be coded in another way — say by multiplying by a number. In either case the value of the correlation coefficient is unaffected. Of greater importance is the fact that if we code before calculating the value of r, we do not have to uncode our results.

Let us code the data of Example 2 of this section and see how coding simplifies the computations involved.

EXAMPLE 3

By coding the data, find the coefficient of correlation for the data of
Example 2 of this section.

SOLUTION

We will code the data by subtracting 38 from each x value and 29
from each y value. Our new distribution of test scores then becomes

Language Score	-12	15	10	-7	29	37	-7	-16	-27	-19
Music Score	-7	-6	28	25	-16	-9	-20	21	-12	-5

Now we calculate r from the coded data. We have the following:

x (language)	y (music)	x^2	y^2	xy
-12	-7	144	49	84
15	-6	225	36	-90
10	28	100	784	280
-7	25	49	625	-175
29	-16	841	256	-464
37	-9	1369	81	-333
-7	-20	49	400	140
-16	21	256	441	-336
-27	-12	729	144	324
-19	-5	361	25	95
$\Sigma x = 3$	$\Sigma y = -1$	$\Sigma x^2 = 4123$	$\Sigma y^2 = 2841$	$\Sigma xy = -475$

Then

$$r = \frac{n(\Sigma xy) - (\Sigma x)(\Sigma y)}{\sqrt{n(\Sigma x^2) - (\Sigma x)^2}\ \sqrt{n(\Sigma y^2) - (\Sigma y^2)}}$$

$$= \frac{10(-475) - (3)(-1)}{\sqrt{10(4123) - (3)^2}\ \sqrt{10(2841) - (-1)^2}}$$

$$= \frac{-4750 + 3}{\sqrt{41,230 - 9}\ \sqrt{28,410 - 1}} \quad \text{(Since a negative number multiplied with a negative number gives a positive number.)}$$

$$= \frac{-4747}{\sqrt{41,221}\ \sqrt{28,409}} = \frac{-4747}{(203.03)(168.55)} = -0.1387$$

Notice that the value of r obtained by coding and the value of r
obtained by working with the original uncoded data is exactly the
same. Coding simplifies computations if the values with which we
code are chosen carefully.

EXERCISES

1. Draw a scatter diagram for the following data. What type of relationship is suggested?

State	Total Number of Reported Crimes (per 100,000 population) for the Calendar Year 1973[1]	State and Local Police Expenditures (in millions of dollars) for the Calendar Year 1972[2]
Connecticut	3664	90
Maine	2544	18
Massachusetts	4521	184
New Hampshire	2329	16
Rhode Island	4678	25
New Jersey	4083	269
New York	4307	1051
California	6305	787
Idaho	3458	15
Oregon	5297	54
Washington	5090	84
Nebraska	2811	30

[1]Source: Uniform Crime Reports for the United States. U.S. Federal Bureau of Investigation, 1974.
[2]Source: Expenditure and Employment Data for the Criminal Justice System. U.S. Law Enforcement Assistance Administration, 1974.

2. For each of the following indicate whether you would expect a positive correlation, a negative correlation, or zero correlation:
    a. the age of your car and its value
    b. the number of female suicides in the United States over the past 20 years and the number of male suicides
    c. length of a jail term and the incidence of crime
    d. cheating on exams and the use of hard drugs by a college student
    e. the number of bank robberies and the unemployment rate
    f. a person's birthday and date of death
    g. speed limits and traffic accidents
    h. the weight of a baby at birth and the length of a baby at birth
    i. the number of cigarettes smoked and the incidence of cancer
    j. the amount of pollution in the air and the incidence of respiratory disease
    k. overweight and the incidence of heart attacks
3. An experiment was conducted to determine if there is any correlation between height and shoe size. Five people were randomly selected and their heights in inches and shoe sizes recorded in the following chart:

Height, x	64	66	68	71	72
Shoe Size, y	7½	8½	8	9	11

Draw a scatter diagram for the data and then compute the coefficient of correlation.

4. Professor Kelly suspects that two of his students, Bill and Jill, have been cheating together on exams, since their grades are quite similar. They received the following grades on the first six quizzes:

Bill, x	80	64	77	85	80	90
Jill, y	80	71	75	88	85	91

a. Draw a scatter diagram for the data and then compute the coefficient of correlation.

b. Are Professor Kelly's suspicions justified?

5. The following chart indicates the number of banks in the United States closed permanently or temporarily because of financial difficulties and the amount of money on deposit at the time of failure:

Year or Period	Number of Banks Closed, x	Amount of Money on Deposit (in millions), y
1951–1955	17	$58
1956–1960	19	41
1961–1965	28	99
1966–1970	11	50
1971	3	5
1972	2	57
1973	3	21

Source: *Federal Reserve Bulletin.* Board of Governors of the Federal Reserve System, 1974.

Draw a scatter diagram for the data and then compute the coefficient of correlation.

6. A large drug manufacturing company wishes its employees to be forced to work overtime. The union claims that the more hours an employee works, the greater the risk of an accident because of fatigue. To support its claim, the union gathered the following statistics on the average number of hours worked per week by an employee and the average number of accidents per week:

Number of Hours Worked, x	34	36	38	40	42	44	46	48
Number of Accidents, y	2.5	2.1	2.6	3.4	6.7	9.1	5.9	7.8

a. Draw a scatter diagram for the data and then compute the coefficient of correlation.

b. Does the union claim seem to be justified?

7. The president of the Dingdong Corporation is interested in knowing if there is any correlation between the number of health insurance policies and the

number of life insurance policies sold by his sales force. Over a five week period the following number of policies were sold:

Number of Health Insurance Policies Sold, x	24	32	23	16	22
Number of Life Insurance Policies Sold, y	18	16	27	14	19

Draw a scatter diagram for the data and then compute the coefficient of correlation.

8. The following chart summarizes some data on the United States Postal Service:

Year	Number of Post Offices (includes Puerto Rico and all outlying areas), x	Number of Pieces of Mail Handled (in millions), y
1967	33,624	71,873
1968	32,260	79,517
1969	32,064	82,005
1970	32,002	84,882
1971	31,947	86,983
1972	31,686	87,156
1973	31,385	89,683

Source: Annual Report of the Postmaster General U.S. Postal Service, 1975.

Draw a scatter diagram for the data and then compute the coefficient of correlation for the number of post offices and the number of pieces of mail handled. (*Hint:* Code the data by subtracting 32,500 from each x and 83,000 from each y.)

9. Trudy keeps track of her son's grades in math and science. He received the following grades on six of his most recent exams:

Math Grade, x	78	81	62	91	84	55
Science Grade, y	62	82	91	73	98	66

Draw a scatter diagram for the data and then compute the coefficient of correlation.

10. A fire department official believes that as the temperature decreases, the number of fires increases. To support this claim, he gathered the following statistics:

Temperature, x	40°	35°	30°	25°	20°	15°	10°	5°
Number of Fires, y	28	32	35	39	51	55	56	66

a. Draw a scatter diagram for the data and then compute the coefficient of correlation.

b. Does this data support the fire official's claim?

11. The following chart indicates the number of hours that each student in Professor Isaac's class studied for the final exam and their final exam grade:

Student	Number of Hours Studied, x	Final Exam Grade, y
Mary Lou	8	94
Sue	7	93
Drew	9	85
Hugh	6	71
Anabell	10	96
Mabel	2	60
Clarene	13	97
Abe	11	94

a. Draw a scatter diagram for the data and then compute the coefficient of correlation.

b. Did studying improve a student's grade?

9.3 THE RELIABILITY OF r

Although the coefficient of correlation is usually the first number that is calculated when we are given several sets of scores, great care must be used in how we interpret the results. It can undoubtedly be said that among all the statistical measures discussed in this book, the correlation coefficient is the one that is most misused. One reason for this misuse is the assumption that because the two variables are related, a change in one will result in a change in the other.

Many people have applied a positive correlation coefficient to prove a cause-and-effect relationship that may not even exist. To illustrate the point, it has been shown that there is a high positive correlation between teacher's salaries and the use of drugs on campus. Does this mean that the more money a teacher earns, the higher will be the use of drugs on campus?

Frequently, two variables may appear to have a high positive correlation even though they are not directly associated with each other. There may be a third variable which is highly correlated to these two variables.

There is another important consideration that is often overlooked. When r is computed on the basis of sample data, we may get a strong correlation, positive or negative, which is due purely to chance, not to some relationship that exists between x and y. For example, if x represents the amount of snowfall and y represents the number of hours that Joe studied on five consecutive days, we may have the following results:

Amount of Snow (in inches), x	1	4	2	6	3
Number of Hours Studied, y	2	6	3	4	4

The value of r in this case is 0.63. Can we conclude that if it snows more, then Joe studies more?

Fortunately, a chart has been constructed which allows us to interpret the value of the correlation coefficient correctly. This is Table VI in the Appendix. This chart allows us to determine the significance of a particular value of r. We use this table in the following way:

1. First compute the value of r using Formula 9.1.

2. Then look in the chart for the appropriate r value corresponding to some given n where n is the number of pairs of scores.

3. The value of r is *not* statistically significant if it is between $-r_{0.025}$ and $r_{0.025}$ for a particular value of n.

The subscript, that is, the little numbers, attached to r is called the **level of significance.** If we use this chart and use the $r_{0.025}$ values of the chart as our guideline, we will be correct in assuming that there is no statistical correlation between x and y approximately 95% of the time.

Table VI also gives us the values of $r_{0.005}$. We use these chart values when we want to be correct 99% of the time. In this book we will use the $r_{0.005}$ values only.

Returning to our example, we have $n = 5$ and $r = 0.63$. From Table VI we have $r_{0.025} = 0.878$. Thus, r will *not* be statistically significant if it is between -0.878 and $+0.878$. Since $r = 0.63$ is between -0.878 and $+0.878$, we conclude that the correlation is due purely to chance. We cannot say that if it snows more, then Joe will necessarily study more.

Similarly, in Example 1 of Section 9.2 (see page 260) we found that $r = 0.9867$. There were 10 scores, so $n = 10$. The chart values tell us that if r is between -0.632 and $+0.632$, there is *no* significant statistical correlation. Since the value of r that we obtained is greater than $+0.632$, we conclude that there is a *definite* positive correlation between x and y. Thus, we are justified in claiming that there is a relationship between the math aptitude scores and the business aptitude scores.

EXERCISES

1. Refer to Exercise 3, page 264. Determine if r is significant.
2. Refer to Exercise 4, page 265. Determine if r is significant.
3. Refer to Exercise 5, page 265. Determine if r is significant.
4. Refer to Exercise 6, page 265. Determine if r is significant.
5. Refer to Exercise 7, page 265. Determine if r is significant.
6. Refer to Exercise 8, page 266. Determine if r is significant.
7. Refer to Exercise 9, page 266. Determine if r is significant.
8. Refer to Exercise 10, page 266. Determine if r is significant.
9. Refer to Exercise 11, page 267. Determine if r is significant.

9.4 LINEAR REGRESSION

Let us return to the example discussed in Section 9.1. In that example a guidance counselor was interested in determining whether a relationship existed between the different aptitudes tested. Once a relationship, in the form of an equation, can be found between two variables, the counselor can use this relationship to *predict* the value of one of the variables if the value of the other variable is known. Thus the counselor may be interested in predicting how well a student will do on the business portion of the test if she knows the student's score on the math part.

Also, the counselor might be analyzing whether any correlation exists between high school averages and college grade point averages. Her intention would be to try to find some relationship which will *predict* a college student's academic success from a knowledge of high school average alone.

Comment It should be noted that the correlation coefficient merely determines whether two variables are related, but it does not specify how. Thus the correlation coefficient cannot be used to solve prediction problems.

When given a prediction problem, we first locate all the dots on a scatter diagram as we did in Section 9.1. Then we try to fit a straight line to the data in such a way that it best represents the relationship between the two variables. Such a fitted line is called an **estimated regression line.** Once we have such a line, we try to find an equation which will determine this line. We can then use this equation to predict the value of *Y* corresponding to a given value of *X*.

Fitting a line to a set of numbers is by no means an easy task. Nevertheless, methods have been designed to handle such prediction problems. These methods are known as **regression methods.** In this book we discuss **linear regression** only. This means that we will try to fit a straight line to a set of numbers.

Comment Occasionally the dots are so scattered that a straight line cannot be fitted to the set of numbers. The statistician may then try to fit a **curve** to the set of numbers. This is shown in Figure 9.8. We would then have **curvilinear regression.**

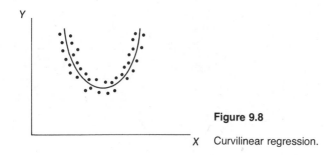

Figure 9.8

Curvilinear regression.

The following examples will illustrate how a knowledge of the regression line enables us to predict the value of Y for any given value of X. It is standard notation to call the variable to be predicted the **dependent variable** and to denote it by Y. The known variable is called the **independent variable** and is denoted by X.

EXAMPLE 1

A guidance counselor notices that there is a strong positive correlation between math scores and business scores. Based upon a random sample of many students, she draws the scatter diagram shown in Figure 9.9. To this scatter diagram we have drawn a straight line that best represents the relationship between the two variables. This line enables us to predict the value of Y for any given value of X. For example, if a student scored 35 on the math portion of the exam, then this line predicts that the student will score about 59 on the business part of the exam. This may be seen by first finding 35 on the horizontal axis, X, and then moving straight up until you hit the estimated regression line. Finally you move directly to the left to see where you cross the vertical axis, Y. This is indicated by the dotted line of Figure 9.10. Similarly, we can predict that a student whose math score is 27 will score 51 on the business portion of the exam.

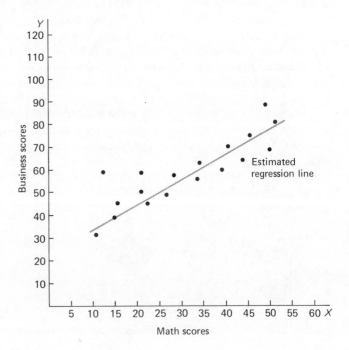

Figure 9.9

Math scores

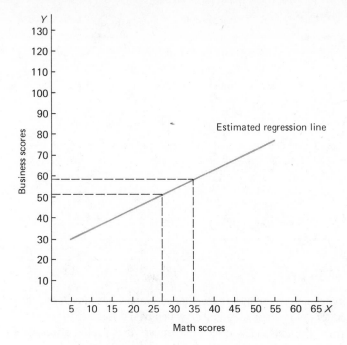

Math scores

Figure 9.10

EXAMPLE 2

A certain organization claims to be able to predict a person's height if given the person's weight. It has collected the following data for 10 people:

Weight (lbs), X	140	135	146	160	142	157	138	164	159	150
Height (inches), Y	63	61	68	72	66	65	64	73	70	69

If a person weighs 155 pounds, what is his predicted height?

SOLUTION

We first draw the scatter diagram as shown in Figure 9.11. Then we draw a straight line which best represents the relationship between the two variables. This line now enables us to predict a person's height if we know the person's weight. The estimated regression line predicts that a person who weighs 155 pounds will be about 70 inches tall.

How does one draw an estimated regression line? Since there is no set procedure, different people are likely to draw different regression lines. So, although we may speak of finding a straight line that best fits the data, how is one to know when the best fit has been achieved? There are, in fact, several reasonable ways in

271

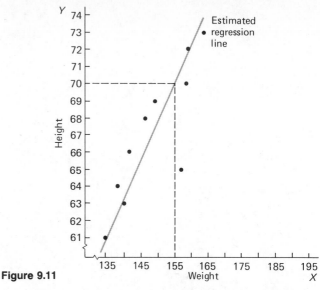

Figure 9.11

which best fit can be defined. For this reason statisticians have developed a mathematical method for determining an equation which best describes the linear relationship between two variables. The method is known as the **least squares method** and is discussed in detail in the next section.

9.5 THE METHOD OF LEAST SQUARES

Whenever we draw an estimated regression line, not all points will lie on the regression line. Some will be above it and some will be below. The difference between any point and the corresponding point on the regression line is called the **deviation** from the line. It represents the difference in value between what we predicted and what actually happened. See Figure 9.12. The **least squares method** determines the estimated regression line in such a way that the sum of the squares of these deviations is as small as possible. Although a background in calculus is needed to understand how we obtain the formula for the least square regression line, it is very easy to use the formula.

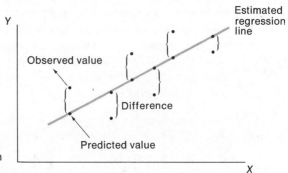

Figure 9.12

Difference between Y values
and the estimated regression
line.

FORMULA 9.2

The equation of the **estimated regression line** is

$$Y_{predicted} = m_Y + b(X - m_X)$$

where m_X and m_Y are the sample means of X and Y respectively, and

$$b = \frac{n(\Sigma XY) - (\Sigma X)(\Sigma Y)}{n(\Sigma X^2) - (\Sigma X)^2}$$

where n is the number of pairs of scores.

Let us use Formula 9.2 to find the regression equation connecting two variables.

EXAMPLE 1

Fifteen students were asked to indicate how many hours they studied before taking their statistics examination. Their responses were then matched with their grades on the exam which had a maximum score of 100.

Hours, X	0.50	0.75	1.00	1.25	1.50	1.75	2.00	2.25
Scores, Y	57	64	59	68	74	76	79	83

Hours, X	2.50	2.75	3.00	3.25	3.50	3.75	4.00
Scores, Y	85	86	88	89	90	94	96

a. Find the regression equation which will predict a student's score if we know how many hours the student studied.

b. If a student studied 0.25 hours, what is the student's predicted grade?

SOLUTION

a. To enable us to perform the computations, we arrange the data in the form of a chart:

X	Y	X^2	XY
0.50	57	0.2500	28.5
0.75	64	0.5625	48
1.00	59	1.0000	59
1.25	68	1.5625	85
1.50	74	2.2500	111
1.75	76	3.0625	133
2.00	79	4.0000	158
2.25	83	5.0625	186.75
2.50	85	6.2500	212.5
2.75	86	7.5625	236.5
3.00	88	9.0000	264
3.25	89	10.5625	289.25
3.50	90	12.2500	315
3.75	94	14.0625	352.5
4.00	96	16.0000	384
$\Sigma X = 33.75$	$\Sigma Y = 1188$	$\Sigma X^2 = 93.4375$	$\Sigma XY = 2863$

From the chart we have, $n = 15$, and

$$m_X = \frac{\Sigma X}{n} = \frac{33.75}{15} = 2.25$$

$$m_Y = \frac{\Sigma Y}{n} = \frac{1188}{15} = 79.2$$

$$b = \frac{n(\Sigma XY) - (\Sigma X)(\Sigma Y)}{n(\Sigma X^2) - (\Sigma X)^2}$$

$$b = \frac{15(2863) - (33.75)(1188)}{15(93.4375) - (33.75)^2}$$

$$b = \frac{42,945 - 40,095}{1401.5625 - 1139.0625}$$

$$b = \frac{2850}{262.5}$$

$$b = 10.86$$

Thus,

$$Y_{predicted} = m_Y + b(X - m_X)$$
$$= 79.2 + 10.86(X - 2.25)$$

The predicted regression line then is

$$Y = 79.2 + 10.86(X - 2.25)$$

b. For $X = 0.25$, we get

$$Y_{predicted} = 79.2 + 10.86(0.25 - 2.25)$$
$$= 79.2 + 10.86(-2)$$
$$= 79.2 - 21.72$$
$$= 57.48$$

Thus, the predicted grade of a student who studies 0.25 hours is approximately 57.

EXAMPLE 2

A west coast publishing company keeps accurate records of its monthly expenditure for advertising and its total monthly sales. For the first ten months of 1976, the records showed the following:

Advertising (in thousands), X	43	44	36	38	47	40	41	54	37	46
Sales (in millions), Y	74	76	60	68	79	70	71	94	65	78

Note that units are in dollars.
a. Find the least squares prediction equation appropriate for the data.
b. If the company plans to spend $60,000 for advertising next month, what is its predicted sales? Assume that all other factors can be neglected.

SOLUTION

a. We arrange the data in the form of a chart:

X	Y	X^2	XY
43	74	1849	3182
44	76	1936	3344
36	60	1296	2160
38	68	1444	2584
47	79	2209	3713
40	70	1600	2800
41	71	1681	2911
54	94	2916	5076
37	65	1369	2405
46	78	2116	3588
$\Sigma X = 426$	$\Sigma Y = 735$	$\Sigma X^2 = 18{,}416$	$\Sigma XY = 31{,}763$

From the chart we have $n = 10$ and

$$m_X = \frac{\Sigma X}{n} = \frac{426}{10} = 42.6$$

$$m_Y = \frac{\Sigma Y}{n} = \frac{735}{10} = 73.5$$

$$b = \frac{n(\Sigma XY) - (\Sigma X)(\Sigma y)}{n(\Sigma X^2) - (\Sigma X)^2}$$

$$= \frac{10(31{,}763) - (426)(735)}{10(18{,}416) - (426)^2}$$

$$= \frac{317{,}630 - 313{,}110}{184{,}160 - 181{,}476}$$

$$= \frac{4520}{2684} = 1.68$$

Thus,

$$Y_{predicted} = m_Y + b(X - m_X)$$
$$= 73.5 + 1.68(X - 42.6)$$

The predicted regression line is

$$Y = 73.5 + 1.68(X - 42.6)$$

b. For $X = 60$, not 60,000 since X is in thousands of dollars, we get

$$Y_{predicted} = 73.5 + 1.68(60 - 42.6)$$
$$= 73.5 + 1.68(17.4)$$
$$= 73.5 + 29.23$$
$$= 102.73$$

Thus, if the company spends $60,000 next month for advertising, its predicted sales are $102.73 million assuming all other factors can be neglected.

EXERCISES

1. As we indicated on page 255, Galton believed that there exists a very close relationship between the heights of fathers and the heights of their sons. To test this claim, a scientist selects seven men at random and records their heights and the heights of their sons (in inches):

Father, X	66	68	71	72	69	69	73
Son, Y	63	66	70	74	70	68	73

a. Find the least squares prediction equation.

b. If a father is 74 inches tall, what is the predicted height of his son?

2. In a certain community the number of deer licenses issued and the number of deer shot during the hunting season is as follows:

Year	Number of Licenses Issued, X	Number of Deer Shot, Y
1971	13	60
1972	10	50
1973	12	55
1974	16	70
1975	14	65
1976	17	75

a. Find the least squares prediction equation.

b. In order to preserve the deer population, the community plans to issue only 10 hunting permits next year. Neglecting all other factors, what is the predicted number of deer that will be shot next year?

3. Mary has determined that the number of miles per gallon of gas that she gets depends on the weight of the objects in the trunk of her car:

Weight of Objects in Trunk (in pounds), X	Miles per Gallon, Y
30	19
50	18
80	16
110	14
120	12
140	11

Find the least squares prediction equation.

4. In a certain school district the absentee rate for kindergarten classes depends upon the average outside temperature reading as follows:

Outside Temperature Reading (in degrees), X	Absentee Rate (in percent), Y
30	6
25	8
20	11
15	15
10	19
5	23
0	29

Find the least squares prediction equation.

5. The Blue Cab Company sends different numbers of cabs to the airport depending upon the number of arriving flights:

Number of Arriving Flights, X	Number of Cabs Sent to the Airport, Y
10	50
12	55
13	60
14	65
16	70
17	75

a. Find the least squares prediction equation.
b. If 20 flights are scheduled for tomorrow, what is the expected number of cabs to be sent to the airport?

6. Five secretaries of a large insurance company were asked to indicate how many pocketbooks and pairs of shoes they owned. Their answers were as follows:

Name	Number of Pocketbooks Owned, X	Number of Pairs of Shoes Owned, Y
Sylvia	9	11
Helen	6	8
Hilda	5	6
Fay	7	9
Renée	4	7

a. Find the least squares prediction equation.
b. If Lynn owns 8 pocketbooks, what is the predicted number of pairs of shoes that she owns?

7. A large corporation has seven branch offices. It has gathered the following information concerning the number of secretaries and the number of employees, excluding secretaries, at each of its branch offices:

City	Number of Secretaries, X	Number of Employees, Y
Pittsburgh	6	60
Los Angeles	12	84
Chicago	5	49
Dallas	15	101
New Orleans	7	68
Kansas City	10	75
Cleveland	9	71

Find the least squares prediction equation.

8. The number of pints of blood in storage on any given day at Willow Hospital depends upon the number of operations scheduled for that day:

Number of Operations Scheduled, X	Number of Pints of Blood in Storage, Y
10	78
12	91
13	108
15	155
16	201
17	271

a. Find the least squares prediction equation.

b. If 18 operations are scheduled for tomorrow, find the predicted number of pints of blood in storage.

9. Martha operates a grocery store. The number of cans of tuna fish that she can sell depends upon its price as indicated:

Price (in cents), X	Number of Cans Sold, Y
50	62
53	53
58	46
62	38
69	29

Find the least squares prediction equation.

10. Wilt is on the college basketball team. The following chart indicates the number of points that Wilt scored in the last six games and also the number of hours per week that Wilt practiced before each game:

Number of Practice Hours, X	Number of Points Scored, Y
8	29
6	25
4	20
9	33
8	28
7	27

Find the least squares prediction equation.

9.6 STANDARD ERROR OF THE ESTIMATE

In Section 9.5 we discussed the least squares regression line which predicts a value of Y when X has a particular value. Quite often it turns out that the predicted value of Y and the observed value of Y are different. If the correlation is low, these differences will be large. Only when the correlation is high can we expect the predicted values to be close to the observed values.

We then need some way of measuring the difference between the predicted and observed values. Statisticians have devised a method for measuring these average differences. This is the **standard error of the estimate.** The standard error of the estimate really indicates how well the least squares prediction equation describes the relationship between the two variables.

To determine the standard error of the estimate, we first calculate the predicted value of Y for each X and then compute the difference between the observed value and the predicted value. We then square these differences and divide the sum of these squares by $n - 2$. The square root of the result is called the standard error of the estimate.

FORMULA 9.3

The **standard error of the estimate** is denoted by S_{YX} and is defined as

$$\sqrt{\frac{\Sigma(Y - Y_p)^2}{n - 2}}$$

where Y_p is the predicted value, Y is the observed value, and n is the number of pairs of scores.

EXAMPLE 1

Find the standard error of the estimate for the least squares regression equation of Example 1 of Section 9.5 on page 273.

SOLUTION

The least squares regression equation was

$$Y_{predicted} = 79.2 + 10.86(X - 2.25)$$

Using this equation, we find the predicted value of Y corresponding to each value of X. We arrange our computations in the form of a chart:

X	Y	$Y_{predicted}$	$Y - Y_{predicted}$	$(Y - Y_p)^2$
0.50	57	60.2	−3.2	10.24
0.75	64	62.91	1.09	1.19
1.00	59	65.63	−6.63	43.96
1.25	68	68.34	−0.34	0.12
1.50	74	71.06	2.94	8.64
1.75	76	73.77	2.23	4.97
2.00	79	76.49	2.51	6.30
2.25	83	79.2	3.8	14.44
2.50	85	81.92	3.08	9.49
2.75	86	84.63	1.37	1.88
3.00	88	87.35	0.65	0.42
3.25	89	90.06	−1.06	1.12
3.50	90	92.78	−2.78	7.73
3.75	94	95.49	−1.49	2.22
4.00	96	98.21	−2.21	4.88

Total = 117.60

Applying Formula 9.3, we get

$$S_{XY} = \sqrt{\frac{\Sigma(Y - Y_p)^2}{n - 2}} = \sqrt{\frac{117.60}{15 - 2}} = \sqrt{\frac{117.60}{13}} = \sqrt{9.05} = 3.01$$

Thus, the standard error of the estimate is 3.01.

EXAMPLE 2

Find the standard error of the estimate for the least squares regression equation of Example 2 of Section 9.5 on page 275.

SOLUTION

The least squares regression equation was

$$Y_{predicted} = 73.5 + 1.68(X - 42.6)$$

Using this equation, we find the predicted value of Y corresponding to each value of X. We arrange our computations in the form of a chart:

X	Y	Y_p	$Y - Y_p$	$(Y - Y_p)^2$
43	74	74.17	−.17	0.03
44	76	75.85	0.15	0.02
36	60	62.41	−2.41	5.81
38	68	65.77	2.23	4.97
47	79	80.89	−1.89	3.57
40	70	69.13	0.87	0.76
41	71	70.81	0.19	0.04
54	94	92.65	1.35	1.82
37	65	64.09	0.91	0.83
46	78	79.21	−1.21	1.46

Total = 19.31

Applying Formula 9.3, we get

$$S_{XY} = \sqrt{\frac{(Y - Y_p)^2}{n - 2}}$$

$$= \sqrt{\frac{19.31}{8}} = \sqrt{2.41} = 1.55$$

Thus, the standard error of the estimate is 1.55.

The goodness of fit of the least squares regression line is determined by the value of the standard error of the estimate. A small value of S_{XY} indicates that the predicted and observed values of Y are fairly close. This means that the regression equation is a good description of the relationship between the two variables. On the other hand, a large value of S_{XY} indicates that there is a large difference between the predicted and observed values of Y. When this happens, the relationship between X and Y as given by the least squares equation is not a good indication of the relationship between the two variables. Only when the standard error of the estimate is zero can we say for sure that the least squares regression equation is a perfect description of the relationship between X and Y.

EXERCISES

For each of the following, refer to the exercise indicated and calculate the standard error of the estimate:
1. Exercise 1, page 276.
2. Exercise 2, page 277.
3. Exercise 3, page 277.
4. Exercise 4, page 277.
5. Exercise 5, page 278.
6. Exercise 6, page 278.
7. Exercise 7, page 278.
8. Exercise 8, page 278.
9. Exercise 9, page 279.
10. Exercise 10, page 279.

SELF-STUDY GUIDE

In this chapter we analyzed the relationship between two variables. Scatter diagrams were drawn which help us understand this relationship. We discussed the concept of correlation coefficients which tell us the extent to which two variables are related. Correlation coefficients vary between the values of -1 and $+1$. A value of $+1$ or -1 represents a perfect linear relationship between the two variables. A correlation coefficient of 0 means that there is no linear relationship between the two variables.

We indicated how to test whether a value of r is significant or not. Furthermore, we mentioned that even when there is an indication of positive correlation between two variables, great care must be shown in how we interpret this relationship.

Once we determine that there is a significant linear correlation between two variables, we find the least squares equation which expresses this relation mathematically.

Finally we discussed the standard error of the estimate. This is a way of measuring how good the estimated least squares regression line really fits the data. The smaller the value of S_{XY}, the better the estimate.

You should be able to identify each symbol in the following formulas, understand the relationships among the symbols expressed in each formula, understand the significance of each formula, and use the formulas in solving problems.

1. Coefficient of linear correlation

$$r = \frac{n(\Sigma xy) - (\Sigma x)(\Sigma y)}{\sqrt{n(\Sigma x^2) - (\Sigma x)^2} \; \sqrt{n(\Sigma y^2) - (\Sigma y)^2}}$$

2. Estimated regression line.

$$Y_{\text{predicted}} = m_Y + b(X - m_X)$$

where

$$b = \frac{n(\Sigma XY) - (\Sigma X)(\Sigma Y)}{n\Sigma X^2 - (\Sigma X)^2}$$

3. Standard error of the estimate

$$S_{XY} = \sqrt{\frac{\Sigma(Y - Y_p)^2}{n - 2}}$$

You should now be able to demonstrate your knowledge of the following ideas presented in this chapter by giving definitions, descriptions, or specific examples. Page references are given for each term so that you can check your answer.

Correlation coefficient (page 255)
Scatter diagram (page 257)
Linear correlation (page 257)
Positive correlation (page 258)
Negative correlation (page 258)
Coefficient of linear correlation
  (page 259)
Reliability of r (page 267)
Level of significance (page 268)

Estimated regression line (page 269)
Linear regression (page 269)
Curvilinear regression (page 269)
Dependent variable (page 270)
Independent variable (page 270)
Deviation (page 272)
Least squares method (page 272)
Standard error of the estimate
  (page 279)

The tests of the following section will be more useful if you take them after you have studied the examples and solved the exercises given in this chapter.

MASTERY TESTS

Form *A*

Questions 1–3 are based on the following graphs:

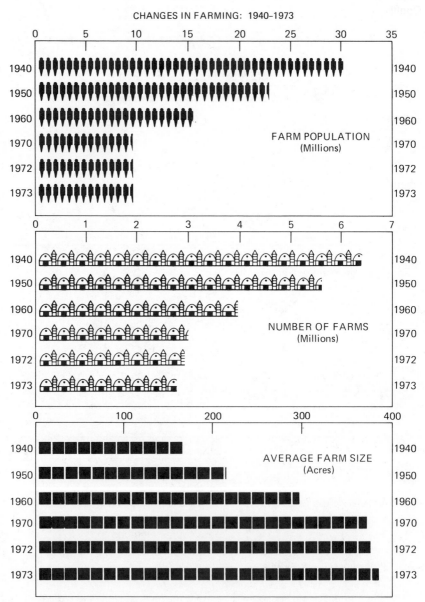

CHANGES IN FARMING: 1940–1973

FARM POPULATION
(Millions)

NUMBER OF FARMS
(Millions)

AVERAGE FARM SIZE
(Acres)

Source: Chart prepared by U.S. Bureau of the Census. Data from U.S. Department of Agriculture, Statistical Reporting and Economic Research Service.

284

1. Consider the *Changes in Farming* data. The pictograph suggests what type of correlation, positive, negative, or zero, between farm population and the number of farms?
2. What type of correlation exists between the number of farms and the average farm size?
3. What type of correlation exists between the farm population and the average farm size?
4. True or false: Correlation coefficients range between 0 and 1.
5. True or false: When the coefficient of correlation has a value of 0, this means that the variables are perfectly correlated.
6. In many respects the standard error of the estimate very closely resembles the standard deviation. Do you agree with this statement? Explain.
7. When $r = \pm 1.00$, what is the value of

$$\sqrt{\frac{\Sigma(Y - Y_p)^2}{n - 2}}$$

8. When $r = 0$, what is the value of

$$\sqrt{\frac{\Sigma(Y - Y_p)^2}{n - 2}}$$

9. The correlation coefficient r indicates the _____ of the relationship between two variables.
10. A value of $r = -0.93$ for the correlation coefficient indicates a _____ relationship between the variables.

Form *B*

1. The ages and systolic blood pressure of ten men after seeing a horror movie was as follows:

Age (yrs), X	Systolic Blood Pressure, Y
35	120
40	125
45	130
50	135
55	140
39	128
43	130
52	135
63	150
29	145

a. Draw a scatter diagram for the data.
b. Calculate the coefficient of correlation.
c. Find the least squares prediction equation.

2. What is the purpose of a scatter diagram?
3. Why is correlation analysis included in the study of linear regression? Can we study these two subjects independently?
4. A leading psychologist believes that IQ scores and the number of hours per week that a child watches television are related. The following data was collected for six students to verify this claim:

IQ Score, X	Number of Hours Spent Watching T.V., Y
110	10
105	12
115	16
121	6
100	14
90	18

Find the least squares prediction equation.
5. The following chart indicates the number of years of schooling beyond high school that an employee of a perfume company has and the employee's salary in dollars:

Number of Years of Schooling, X	0	1	2	3	4
Salary, Y	10,000	12,000	13,000	15,000	16,000

a. Draw a scatter diagram for the data and then compute the coefficient of correlation.
b. Does a higher education mean a higher salary?
6. The following chart indicates how the energy consumption by an individual in the United States changed over the years:

Year	Natural Gas Consumption Per Capita (in millions), X	Crude Petroleum Consumption Per Capita (in millions), Y
1960	249	103
1965	278	111
1970	330	145
1971	338	148
1972	346	158
1973	360	165
1974	369	175

Source: Energy Facts. The Library of Congress, November, 1974.

a. Compute the coefficient of correlation.
b. Find the least squares prediction equation.
7. Consider the following correlations. Are any of the correlations between IQ and achievement in specific school subjects or skills significant?

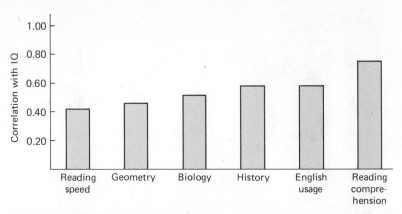

8. The following bar graph shows the scores achieved on a skills test by men when they were 20 years old and again when they were 60 years old. Is there a correlation between age and skills? Explain your answer.

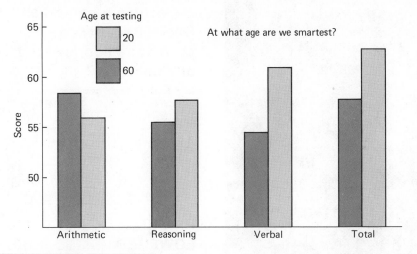

SUGGESTED READING

Draper, N. and Smith, H. *Applied Regression Analysis.* New York: Wiley, 1966.

Guilford, J. P. *Fundamental Statistics in Psychology and Education.* 3rd ed. New York: McGraw-Hill, 1956.

Jencks, C. *Inequality: A Reassessment of the Effect of Family and Schooling in America.* New York: Basic Books, 1972.

Johnson, N. L. and Leone, F. C. *Statistics and Experimental Design in Engineering and the Physical Sciences.* Belmont, California: Wadsworth, 1968.

Ostle, B. *Statistics in Research.* 2nd ed. Ames: Iowa State University Press, 1963.

Richmond, Samuel B. *Statistical Analysis.* 2nd ed. New York: Ronald Press, 1964.

Steel, Robert and Torrie, James. *Principles and Procedures in Statistics.* New York: McGraw-Hill, 1960.

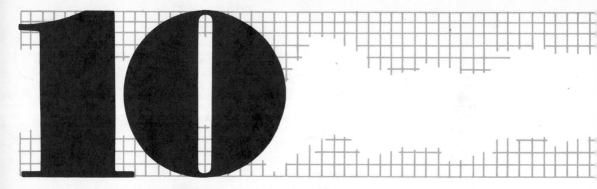

10

SAMPLING

Gary Settle/NYT Pictures

Should the $2,700 limit on income a person on Social Security can earn be raised, lowered, or eliminated?

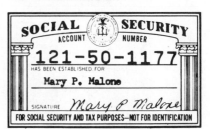

RAISED		**65%**
LOWERED		**4%**
ELIMINATED		**15%**
DON'T KNOW		**16%**

On April 6, 1976 both ABC and NBC television networks projected that Morris Udall would win the Democratic primary in Wisconsin. Their predictions were based upon samples from selected precincts, and did not take certain districts into consideration. When all the rural votes were counted Jimmy Carter came out on top. Many newspapers were so confident of their predictions that they printed, erroneously, the morning editions of their newspapers with the headline "CARTER UPSET BY UDALL".

In another random sample people were asked to express their opinion on a Social Security question. How are people randomly selected for such an opinion poll?

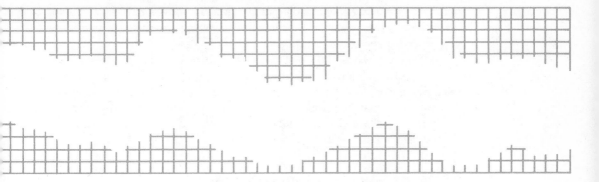

Preview

In this chapter we will discuss the following topics:

1. *Random Sampling*
We discuss what is meant by a random sample and how to obtain it.

2. *Table of Random Digits*
This is a table of numbers where each number that appears is obtained by a process that gives every digit an equally likely chance of coming up.

3. *Stratified Sampling*
This is a sampling procedure that is used when we want to obtain a sample with a specified number of people from different categories.

4. *Distribution of Sample Means*
When repeated samples are taken from a population, the frequency distribution of the values of the sample means is called the distribution of the sample means.

5. *Standard Error of the Mean*
This represents the standard deviation of the distribution of sample means.

6. *The Central Limit Theorem*
This tells us that the distribution of the sample means is basically a normal distribution. We discuss how we can use this theorem to make predictions about and calculate probabilities for the sample means.

Introduction

Suppose we are interested in determining whether the average automobile driver in the United States is in favor of some form of gasoline rationing. Must we ask each driver before making any statement concerning the desirability of gasoline rationing? Since there are so many drivers in the United States, it would require an enormous amount of work to interview each driver and gather all the data.

Do we actually need the complete population data? Can a properly selected sample give us enough information to make predictions about the entire population? In many cases, obtaining the complete population data may be quite costly or even impossible.

Similarly, suppose we are interested in purchasing an electric bulb. Must we burn all electric bulbs produced by a particular company in order to determine the bulb's average life? This is very impractical. Maybe we can estimate the average life of a bulb by testing only a sample of 100 bulbs.

As we noticed earlier (Definition 1.2, page 8), this is exactly what statistical inference involves. Samples are studied to obtain valuable information about a larger group called the **population.** Any part of the population is called a **sample.** The purpose of sampling is to select that part which truly represents the entire population.

Any sample provides only partial information about the population from which it is selected. Thus, it follows that any statement we make (based on a sample) concerning the population may be subject to error. One way of minimizing this error is to make sure that the sample is randomly selected. How is this to be done?

In this chapter we will discuss how to select a random sample and how to interpret different sample results.

10.1   SELECTING A RANDOM SAMPLE

The purpose of most statistical studies is to make generalizations from samples about the entire population. Yet, not all samples lend themselves to such generalizations. Thus we cannot generalize about the average income of a working person in the United States by sampling only lawyers and doctors. Similarly, we cannot make any generalizations about gasoline rationing by sampling only nondrivers.

Over the years many incorrect predictions have been made on the basis of nonrandom samples. For example, in 1936 the *Literary Digest* was interested in determining who would win the coming presidential election. It decided to poll the voters by mailing 10 million ballots. On the basis of the approximately 2 million ballots returned, it predicted that Alfred E. Landon would be elected. An October 31 headline read

Landon	1,293,669
Roosevelt	972,897
Final returns in the Digest's poll of ten million voters	

Source: The Literary Digest, 1936.

Actually Franklin Roosevelt carried 46 of the 48 states and many of them by a landslide. The ten million people to whom the *Digest* sent ballots were selected from telephone listings and from the list of its own subscribers. The year 1936 was a depression year and many people could not afford telephones or magazine subscriptions. Thus, the *Digest* did not select a random sample of the voters of the United States. The *Literary Digest* soon went out of business.

Again in 1948 the polls predicted that Dewey would win the presidential election. One newspaper even printed the morning edition of its newspaper with the headline "DEWEY WINS BY A LANDSLIDE." Of course, Truman won the election and laughed when presented with a copy of the newspaper predicting his defeat.

In both examples the reason for the incorrect prediction is that it was based on information obtained from poor samples. It is for this reason that statisticians insist that samples be randomly selected.

DEFINITION 10.1

A **random sample** is a sample selected from a
population in such a way that each member of the
population has an equal chance of being selected.

DEFINITION 10.2

Random sampling is a procedure to be followed which
results in a random sample.

It may seem that the selection of a random sample is an easy task. Unfortunately this is not the case. You may think that we can get a random sample of voters by opening a telephone book and selecting every tenth name. This will not give a random sample since many voters either do not have phones or else have unlisted numbers. Furthermore, many young voters and most women are not listed. These people, who are members of the voting population, do not have an equal chance of being selected.

To further illustrate the nature of random sampling, suppose that the administration of a large southern college, with an enrollment of 30,000 students, is considering revising its grading system. The administration is interested in replacing its present grading system with a pass–fail system. Since not all students agree with this proposed change, the administration has decided to poll 1000 students. How is this to be done? Polling a thousand students in the school cafeteria or in the student lounge will not result in a random sample since there may be many students who neither eat in the cafeteria nor go to the student lounge.

One way of obtaining a random sample is to write each student's name on a separate piece of paper and then put all the pieces in a large bowl where they can be thoroughly mixed. A paper is then selected from the bowl. After a name is read the paper is replaced and another paper is selected. This procedure is repeated until 1000 names are obtained. Any name that is repeated is used only once. In this manner a random sample of 1000 names can be obtained. Great care must be exercised to make sure that the bowl is thoroughly mixed after a piece of paper is selected. Otherwise, the papers on the bottom of the bowl do not have an equal chance of being selected and the sample will not be random.

Although using slips of paper in a bowl will result in a random sample if properly done, this fish bowl method becomes unmanageable as the number of people in the population increases. The job of numbering slips of paper can be completely avoided by using a **table of random numbers** or a **table of random digits.**

What are random numbers? They are the digits 0, 1, 2, ... , 9 arranged in a random fashion, that is, in such a way that all the digits appear with approximately the same frequency. Table VII in the Appendix is an example of a table of random numbers. How are such tables constructed? Today most tables of random numbers are constructed with the help of electronic computers. Yet a simple spinner such as the one shown in Figure 10.1 will also generate such a table of random numbers. After each spin we use the digit selected by the arrow. The resulting sequence of random digits could then be used to construct random number tables of 5 digits each, for example, as shown in Table VII. Although this method will generate random numbers, it is not practical. After a while some numbers will be favored over others as the spinner begins to wear out.

Thus, the best way to select a random sample is to use a table of random numbers obtained with the help of an electronic computer. Table VII in the Appendix is such a table. In this table the various digits are scattered at random and arranged in groups of five for greater legibility.

Let us now return to our example. As a first step in using this table, each student is assigned a number from 00001 to 30000. To obtain a sample of 1000 students we merely read down column 1 and select the first 1000 students whose numbers are listed. Thus the following students would be selected:

Figure 10.1

A spinner.

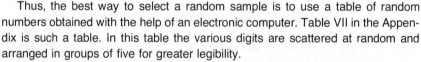

10,480   22,368   24,130   29,918   09,429 ...

Notice that we skip the number 42,167 since no student has this number. The same is true for the numbers 37,570, 77,921, 99,562. ... We disregard any numbers larger than 30,000 that are obtained from this table since there are no students associated with these numbers.

To further illustrate the proper use of this table, suppose that a hotel has 150 guests registered for a weekend. The management wishes to select a sample of 15 people at random to rate the quality of its service. It should proceed as follows. Assign each guest a different number from 1 to 150 as 001, 002, . . . , 150. Then select any column in the table of random numbers. Suppose the fourth column is selected. Three digit numbers are then read off the table by reading down the column. If necessary they continue on to another column or another page. Starting on top of column 4, they get

$$020, 853, 972, 616, 166, 427, 699, \ldots$$

Since the guests have numbers between 001 and 150 only, they ignore any number larger than 150. From column 4 they get

$$020, 079, 102, 034, 081, 099, 143, 073, 129$$

From column 5 they get

$$078, 061, 091, 133, 040, 023$$

Thus, the management should interview the guests whose numbers are 20, 79, 102, 34, 81, 99, 143, 73, 129, 78, 61, 91, 133, 40, and 23.

Comment   Whenever we speak of a random sample in this text we will assume that it has been selected in the manner just described.

EXERCISES

1. In one town each welfare recipient has a case number from 1 to 3000. Government officials have decided to verify that a welfare recipient is actually eligible for such aid by selecting a random sample of 20 recipients and thoroughly investigating each claim. Use columns 13 and 14 of Table VII in the Appendix to find which case numbers will be selected.
2. In one jail there are 987 prisoners, each of whom has been assigned a number from 1 to 987. A committee of 18 prisoners is to be selected from among them to discuss their grievances with the warden. Use column 5 of Table VII in the Appendix to find which prisoners should be selected.
3. A certain book has 496 pages. To check for typographical errors the publisher has decided to randomly select 25 pages and to thoroughly check them. Use column 1 of Table VII in the Appendix to find which pages will be checked.

4. Each member of the security force of a local department store carries a badge. These badges are numbered in order from 1 to 217. A random sample of 20 officers is to be selected for training in new antishoplifting techniques. Use columns 12 and 13 of Table VII in the Appendix to find which officers should be selected.

5. Each scientific calculator that is produced by the Adams Precision Company during a year carries a 4–digit serial number from 1 to 9000, in addition to the date of manufacture. During 1976 the company manufactured 5638 calculators and numbered them in order, 1, 2, . . . , 5638. Twenty of these calculators are to be selected randomly and inspected for quality and workmanship. Use column 9 of Table VII in the Appendix to find which calculators will be selected.

6. A state tax department has decided to randomly select 15 individual income tax returns for 1976 and to thoroughly audit them. Each return has been filed according to the taxpayer's Social Security number. Use the last 4 digits of the Social Security number and column 5 of Table VII in the Appendix to find which returns will be audited.

10.2   STRATIFIED SAMPLING

Although random sampling, as discussed in Section 10.1, is the most popular way of selecting a sample, there are times when **stratified samples** are preferred. To obtain a stratified sample, we divide the entire population into a number of groups or strata. The purpose of such stratification is to obtain groups of people that are more or less equal in some respect. We select a random sample, as discussed in Section 10.1, from each of these groups or stratum. We then combine these subsamples into one large sample.

Thus, in the example discussed in the beginning of Section 10.1, the administration of the college may first divide the entire student body into four groups: freshmen, sophomores, juniors, and seniors. Then they select a random sample from each of these groups, and finally they combine these four subsamples into one large sample. If the four subsamples are of the same size, they can obtain a more accurate poll of student opinion by stratified sampling. However, the cost of obtaining a stratified sample is higher than that of obtaining a random sample since the administration must spend money for dividing the student body into four groups.

The method of stratifying samples is especially useful in pre-election polls. Past experience indicates that different subpopulations often demonstrate particularly different voting preferences.

10.3   CHANCE VARIATION AMONG SAMPLES

Imagine that a cigarette manufacturer is interested in knowing the average tar content of a new brand of cigarettes that is about to be sold. The Food and Drug

Administration requires such information to be indicated alongside all advertisements that appear in magazines, newspapers, etc.

The manufacturer decides to send random samples of 100 cigarettes each to 20 different testing laboratories. With the information obtained from these samples, the manufacturer hopes to be able to estimate the mean or average milligram tar content of the cigarette.

Since we will be discussing both samples and populations, let us pause for a moment to indicate the notation that we will use to distinguish between samples and populations. See Table 10.1.

TABLE 10.1

NOTATION FOR SAMPLE AND POPULATION

TERM	SAMPLE	POPULATION
Mean	$\bar{X}$	μ
Standard deviation	s	σ
Number	N	n

From Chapter 3 we have the following formulas.

FORMULA 10.1

Mean

Sample	Population
$\bar{X} = \dfrac{\Sigma X}{N}$	$\mu = \dfrac{\Sigma x}{n}$

FORMULA 10.2

Standard deviation

Sample	Population
$s = \sqrt{\dfrac{\Sigma(X - \bar{X})^2}{N - 1}}$	$\sigma = \sqrt{\dfrac{\Sigma(x - \mu)^2}{n}}$

Comment Up to now we have been using μ and σ for both samples and populations. For simplicity we did not distinguish between them. In practice, Greek letters are used for population measurements and Roman italic letters are used for measurements obtained from samples. We shall follow this convention from this point on.

Let us now return to our example. Since each sample sent to a laboratory is randomly selected, it is reasonably safe to assume that there will be differences among the means of each sample. The 20 laboratories report the following average milligram content per cigarette:

14.8, 16.2, 14.8, 15.8, 15.3, 13.9, 16.9, 15.9, 14.3, 15.2
14.9, 16.2, 15.6, 15.5, 13.4, 15.1, 15.7, 14.8, 14.4, 15.3

These figures indicate that the sample means vary considerably from sample to sample.

The manufacturer decides to take the average of these 20 sample means and gets

$$\bar{X} \doteq \frac{\Sigma X}{N} = \frac{14.8 + 16.2 + \cdots + 15.3}{20} = \frac{304}{20} = 15.2$$

He now uses this overall average of the sample means, 15.2, as his estimate of the true population mean.

How reliable is this estimate? Although we cannot claim for certain that the population mean is 15.2, we can feel reasonably confident that 15.2 is not a bad estimate of the population mean since it is based on 20×100, or 2000, observations.

Thus we can obtain a fairly good estimate of the population mean by calculating the mean of the samples. If we let $\mu_{\bar{X}}$, read as mu sub X bar, represent the mean of the samples, then we say that $\mu_{\bar{X}}$ is a good estimate of μ.

What about the standard deviation? Let us calculate the standard deviation of the sample means. Recall that the formula for the standard deviation is

$$\sqrt{\frac{\Sigma(x - \mu)^2}{n}}$$

Since μ is unknown, we have to replace it with an estimate. The most obvious replacement is $\bar{X}$. To account for this replacement, we divide by $N - 1$ instead of by N. Thus the formula for the standard deviation is given by Formula 10.3.

FORMULA 10.3

The standard deviation of the sample means is

$$\sqrt{\frac{\Sigma(X - \bar{X})^2}{N - 1}}$$

EXAMPLE 1

Calculate the sample standard deviation for data of the average tar content of the 20 laboratories.

SOLUTION

We arrange the data as follows:

X	$X - \bar{X}$	$(X - \bar{X})^2$
14.8	$14.8 - 15.2 = -0.4$	0.16
16.2	$16.2 - 15.2 = 1$	1.00
14.8	$14.8 - 15.2 = -0.4$	0.16
15.8	$15.8 - 15.2 = 0.6$	0.36
15.3	$15.3 - 15.2 = 0.1$	0.01
13.9	$13.9 - 15.2 = -1.3$	1.69
16.9	$16.9 - 15.2 = 1.7$	2.89
15.9	$15.9 - 15.2 = 0.7$	0.49
14.3	$14.3 - 15.2 = -0.9$	0.81
15.2	$15.2 - 15.2 = 0$	0
14.9	$14.9 - 15.2 = -0.3$	0.09
16.2	$16.2 - 15.2 = 1$	1.00
15.6	$15.6 - 15.2 = 0.4$	0.16
15.5	$15.5 - 15.2 = 0.3$	0.09
13.4	$13.4 - 15.2 = -1.8$	3.24
15.1	$15.1 - 15.2 = -0.1$	0.01
15.7	$15.7 - 15.2 = 0.5$	0.25
14.8	$14.8 - 15.2 = -0.4$	0.16
14.4	$14.4 - 15.2 = -0.8$	0.64
15.3	$15.3 - 15.2 = 0.1$	0.01
$\Sigma X = 304$		$\Sigma(X - \bar{X})^2 = 13.22$

Using formula 10.3 we get

$$s = \sqrt{\frac{\Sigma(X - \bar{X})^2}{N - 1}}$$

$$= \sqrt{\frac{13.22}{20 - 1}} = \sqrt{\frac{13.22}{19}}$$

$$= \sqrt{0.696}$$

$$= 0.83$$

Thus the standard deviation of the samples is 0.83.

In practice the standard deviation is not calculated by using Formula 10.3 since the computations required are time consuming. Instead we can use a shortcut formula given as Formula 10.4. The advantage in using Formula 10.4 is that we do not have to calculate $\bar{X}$ and $X - \bar{X}$ and square $X - \bar{X}$. We only have to calculate ΣX and ΣX^2. These represent the sum of the X's and the sum of the squares of the X's respectively. Then we use Formula 10.4.

FORMULA 10.4

The standard deviation of the sample means is

$$\sqrt{\frac{N(\Sigma X^2) - (\Sigma X)^2}{N(N-1)}}$$

EXAMPLE 2

Using Formula 10.4, find the standard deviation of the sample means for the data of Example 1 in this section.

SOLUTION

We arrange the data as follows:

X	X^2
14.8	219.04
16.2	262.44
14.8	219.04
15.8	249.64
15.3	234.09
13.9	193.21
16.9	285.61
15.9	252.81
14.3	204.49
15.2	231.04
14.9	222.01
16.2	262.44
15.6	243.36
15.5	240.25
13.4	179.56
15.1	228.01
15.7	246.49
14.8	219.04
14.4	207.36
15.3	234.09
$\Sigma X = 304$	$\Sigma X^2 = 4634.02$

Using Formula 10.4 we get

$$s = \sqrt{\frac{N(\Sigma X^2) - (\Sigma X)^2}{N(N-1)}} = \sqrt{\frac{20(4634.02) - (304)^2}{20(19)}}$$

$$= \sqrt{\frac{92680.4 - 92416}{380}} = \sqrt{\frac{264.4}{380}} = \sqrt{0.696} = 0.83$$

Thus, the standard deviation of the samples is 0.83. This is the same result we obtained using Formula 10.3.

EXAMPLE 3

A large office building has six elevators, each with a capacity for 10 people. The elevator operator of each elevator has determined the average weight of the people in their elevators. The results follow:

Elevator	1	2	3	4	5	6
Average Weight (lbs.)	125	138	145	137	155	140

Find the overall average of these sample means. Also find the standard deviation of these samples by first using Formula 10.3 and then Formula 10.4.

SOLUTION

We arrange the data as follows:

X	$X - \bar{X}$	$(X - \bar{X})^2$	X^2
125	$125 - 140 = -15$	225	15,625
138	$138 - 140 = -2$	4	19,044
145	$145 - 140 = 5$	25	21,025
137	$137 - 140 = -3$	9	18,769
155	$155 - 140 = 15$	225	24,025
140	$140 - 140 = 0$	0	19,600
$\Sigma X = 840$		$\Sigma(X - \bar{X})^2 = 488$	$\Sigma X^2 = 118,088$

Then

$$\bar{X} = \frac{\Sigma X}{N} = \frac{840}{6} = 140$$

Using Formula 10.3 we get

$$s = \sqrt{\frac{\Sigma(X - \bar{X})^2}{N - 1}} = \sqrt{\frac{488}{5}} = \sqrt{97.6} = 9.88$$

Using Formula 10.4 we get

$$s = \sqrt{\frac{N(\Sigma X^2) - (\Sigma X)^2}{N(N - 1)}} = \sqrt{\frac{6(118,088) - (840)^2}{6(5)}}$$

$$= \sqrt{\frac{708528 - 705600}{30}} = \sqrt{97.6} = 9.88$$

Thus the mean of the samples is 140 and the standard deviation by Formula 10.3 or Formula 10.4 is 9.88.

EXERCISES

1. A taxi driver keeps accurate records on the average number of trips he makes to the airport each week. During the first six weeks of 1976 he made 22, 13, 19, 28, 11, and 9 trips. Find the mean and standard deviation of the average number of trips the taxi driver makes to the airport.

2. The president of the JFK Corporation claims that too many people are using the company Xerox machines to photocopy their personal papers. The average number of copies made by these machines in one week was 207, 386, 123, 294, and 305. Find the mean and standard deviation of the number of copies made.

3. Several student groups at Staten Island Community College have conducted surveys on the number of minutes that each student drives around the school looking for a parking space. These groups reported the following results in minutes: 22, 9, 12, 18, 9, 16, and 5. Find the mean and standard deviation of these samples.

4. Each day five hospitals report the average pollen count in the air. On Monday the five hospitals reported the following pollen counts:

Hospital	1	2	3	4	5
Average Pollen Count Reported	49	68	58	87	33

Find the mean and standard deviation of these samples.

5. The average number of consumer complaints the Attorney General's offices received last week concerning dishonest merchants was 41, 39, 68, 112, and 15. Find the mean and standard deviation of these samples.

6. A school principal has been experimenting with a new reading comprehension program. The average time required by the seven seventh-grade classes to understand a particular paragraph is 24, 13, 19, 28, 32, 27, and 25 minutes. Find the mean and standard deviation of these samples.

7. Several recent surveys indicated that the average amount of garbage collected daily in one city is 4.6, 3.9, 2.7, 5.6, and 3.2 tons. Find the mean and standard deviation of these samples.

8. A soda dispensing machine has not been functioning properly. It should dispense 7 ounces of soda per cup; however, this has not been happening. The owner samples 100 cups of soda for eight consecutive days and determines that the average amount of soda dispensed is not 7 ounces but varies as follows:

Day	1	2	3	4	5	6	7	8
Average Amount of Soda Dispensed (in ounces)	6.6	8.3	5.9	6.1	8.4	5.7	7.9	7.1

Find the mean and standard deviation of these samples.

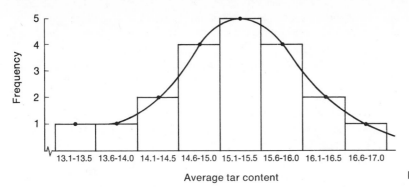

13.1-13.5 13.6-14.0 14.1-14.5 14.6-15.0 15.1-15.5 15.6-16.0 16.1-16.5 16.6-17.0

Average tar content

Figure 10.2

10.4 DISTRIBUTION OF SAMPLE MEANS

Let us refer to the example discussed at the beginning of Section 10.3. The manufacturer decides to draw the histogram for the average tar content he obtained from the 20 laboratories. This is shown in Figure 10.2. Notice that the value of $\bar{X}$ is actually a random variable since its value is different from sample to sample. In repeated samples different values of $\bar{X}$ were obtained. Yet they are all close to the 15.2 we obtained as the average of the sample means. Moreover, exactly 70% of the sample means are between 14.37 and 16.03 which represents 1 standard deviation away from the mean in either direction. Also, 90% of the sample means are between 13.54 and 16.86 which represents 2 standard deviations away from the mean in either direction. Thus Figure 10.2 actually represents the distribution of $\bar{X}$ since it tells us how the means of the samples vary from sample to sample. We refer to the distribution of $\bar{X}$ as the **distribution of sample means** or as **the sampling distribution of the mean.** Although the first terminology is much clearer, the second is more commonly used.

Strictly speaking, Figure 10.2 is not a complete distribution of $\bar{X}$ since it is based on only 20 samples. To obtain the complete distribution of sample means, we would have to take thousands of samples of 100 cigarettes each. Of course, in practice we do not take thousands of samples from the same population.

Comment Notice that the sample means form a normal distribution. We will have more to say about this in Section 10.5.

What can we say about this distribution? What is its mean? its standard deviation? How does this distribution compare with the distribution of *all* the cigarettes? To answer this question, the manufacturer decides to draw the histogram for the tar content of all the 2000 cigarettes. This is shown in Figure 10.3.

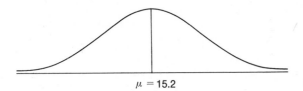

$\mu = 15.2$

Figure 10.3

Let us now compare these two distributions. Notice that both distributions are centered around the same number, 15.2. Thus it is reasonable to assume that $\mu_{\bar{x}} = \mu$. Also notice that the distribution of the sample means is not spread out as much as (that is, has a smaller standard deviation than) the distribution of the tar content of all the cigarettes. The reason for this should be obvious. When *all* the cigarettes are considered, several have a very high tar content and several have a very low tar content. These appear on the tail ends of the distribution of Figure 10.3. However, it is unlikely that an entire sample of 100 cigarettes will have a tar content of 18.5. Thus, the distribution of $\bar{X}$ has very little frequency at large distances from the mean.

We use the symbol $\sigma_{\bar{x}}$ to represent the standard deviation of the sampling distribution of the mean. We have the following formula for $\sigma_{\bar{x}}$:

FORMULA 10.5

The standard deviation of the sampling distribution of the mean is referred to as the **standard error of the mean** and is defined as

$$\sigma_{\bar{x}} = \frac{\sigma}{\sqrt{N}}$$

where σ is the standard deviation of the entire population and N is the number in each sample.

The standard error of the mean, $\sigma_{\bar{x}}$, plays a very important rule in statistics as will be illustrated in the remainder of this chapter.

Comment   It should be obvious from Formula 10.5 that the larger the sample size, the smaller will be the variation of the means. Thus, as we take larger and larger samples, we can expect the mean of the samples, $\mu_{\bar{x}}$, to be close to the mean of the population, μ. This is illustrated in Figure 10.4.

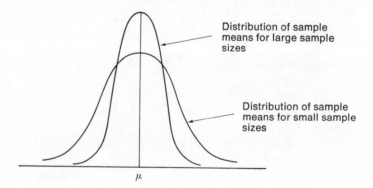

Distribution of sample means for large sample sizes

Distribution of sample means for small sample sizes

Figure 10.4

EXAMPLE 1

Using the data of Example 1 in Section 10.3 on page 296, find an estimate of the standard deviation of the population from which it was obtained.

SOLUTION

The standard deviation of the sample means was 0.83. There were 20 laboratories, each of which tested 100 cigarettes, so $N = 100$. Using Formula 10.5 we get

$$\sigma_{\bar{x}} = \frac{\sigma}{\sqrt{N}}$$

$$0.83 = \frac{\sigma}{\sqrt{100}}$$

$$0.83 = \frac{\sigma}{10} \quad \text{(We multiply both sides by 10)}$$

$$0.83(10) = \sigma$$
$$8.3 = \sigma$$

Thus, the standard deviation of the population is 8.3, whereas the standard deviation for the sample means is only 0.83.

EXAMPLE 2

Using the data of Example 3 in Section 10.3 on page 299, find an estimate of the standard deviation of the population from which it was obtained.

SOLUTION

The standard deviation of the sample means was 9.88. Each elevator holds 10 people so $N = 10$. Applying Formula 10.5 we get

$$\sigma_{\bar{x}} = \frac{\sigma}{\sqrt{N}}$$

$$9.88 = \frac{\sigma}{\sqrt{10}}$$

$$9.88 = \frac{\sigma}{3.16}$$

$$9.88(3.16) = \sigma$$
$$31.22 = \sigma$$

Thus, the standard deviation of the population is 31.22. Again this is considerably larger than the standard deviation of the sample means which was only 9.88.

303

Let us summarize our discussion up to this point. Using the distribution of the sample means of the laboratories and the distribution of the tar content of all the 2000 cigarettes, we conclude the following:

1. The mean of the distribution of sample means and the mean of the original population is the same.
2. The standard deviation of the distribution of the sample means is less than the standard deviation of the original population. The exact relationship is referred to as the standard error of the mean and is found by using Formula 10.5.
3. The distribution of the sample means is normally distributed.

Comment The last statement is so important that it is referred to as the **Central Limit Theorem.** Since much of the work of statistical inference is based on this theorem, we will discuss its importance, as well as its applications in detail, in the following sections.

10.5   THE CENTRAL LIMIT THEOREM

One of the most important theorems in probability is the **Central Limit Theorem.** This theorem, first established by De Moivre in 1733 (see the discussion on page 222), was named "The Central Limit Theorem of Probability" by G. Polya in 1920. The theorem may be summarized as follows:

THE CENTRAL LIMIT THEOREM

Let X be a random variable. The following three things will be true about the distribution of sample means.

1. The distribution of the sample means will be approximately normally distributed.
2. The mean of the sampling distribution will be equal to the mean of the population. Symbolically

$$\mu_{\bar{x}} = \mu$$

3. The standard deviation of the sampling distribution will be equal to the standard deviation of the population divided by the square root of the number of items in each sample. Symbolically,

$$\sigma_{\bar{x}} = \frac{\sigma}{\sqrt{N}}$$

Comment   If the sample size is large enough, the sampling distribution will be normal, even if the original distribution is not. Large enough usually means larger than 30 items in the sample.

Comment   The N referred to in the theorem refers to the size of each sample, not to the number of samples.

Since the Central Limit Theorem is so important, we will discuss its applications in the next section.

10.6   APPLICATIONS OF THE CENTRAL LIMIT THEOREM

In this section we use the Central Limit Theorem to predict the behavior of sample means. To apply the standardized normal distribution discussed in Chapter 8, we have to change Formula 8.1 somewhat. Recall that

$$z = \frac{X - \mu}{\sigma}$$

In advanced probability it is shown that when dealing with sample means this formula becomes that given as Formula 10.6.

FORMULA 10.6

$$z = \frac{\bar{X} - \mu}{\sigma/\sqrt{N}}$$

The following examples will illustrate how the Central Limit Theorem is applied.

EXAMPLE 1

The average height of all the workers in a hospital is known to be 65 inches with a standard deviation of 2.3 inches. If a sample of 36 people is selected at random, what is the probability that the average height of these 36 people will be between 64 and 65.5 inches?

SOLUTION

We use Formula 10.6. Here $\mu = 65$, $\sigma = 2.3$, and $N = 36$. Thus, $\bar{X} = 64$ corresponds to

$$z = \frac{64 - 65}{2.3/\sqrt{36}} = \frac{-1}{0.38} = -2.63$$

and $\bar{X} = 65.5$ corresponds to

$$z = \frac{65.5 - 65}{2.3/\sqrt{36}} = \frac{0.5}{0.38} = 1.32$$

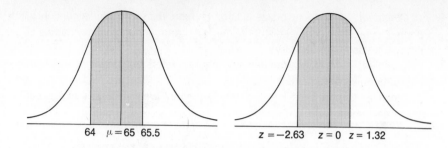

Figure 10.5

Thus, we are interested in the area of a standard normal distribution between $z = -2.63$ and $z = 1.32$. See Figure 10.5.

From Table V in the Appendix we find that the area between $z = 0$ and $z = -2.63$ is 0.4957 and that the area between $z = 0$ and $z = 1.32$ is 0.4066. Adding we get

$$0.4957 + 0.4066 = 0.9023$$

Thus, the probability that the average height of the sample of 36 people is between 64 and 65.5 inches is 0.9023.

EXAMPLE 2

The average amount of money that a depositor of the Second National City Bank has in an account is $5,000 with a standard deviation of $650. A random sample of 36 accounts is taken. What is the probability that the average amount of money that these 36 depositors have in their accounts is between $4800 and $5300?

SOLUTION

We use Formula 10.6. Here $\mu = 5000$, $\sigma = 650$, and $N = 36$. Thus, $\bar{X} = 4800$ corresponds to

$$z = \frac{4800 - 5000}{650/\sqrt{36}} = \frac{-200}{108.33} = -1.85$$

and $\bar{X} = 5300$ corresponds to

$$z = \frac{5300 - 5000}{650/\sqrt{36}} = \frac{300}{108.33} = 2.77$$

Thus we are interested in the area between $z = -1.85$ and $z = 2.77$. See Figure 10.6.

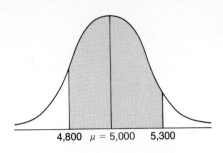
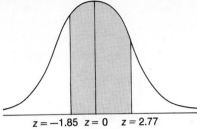

4,800 $\mu = 5,000$ 5,300 $z = -1.85$ $z = 0$ $z = 2.77$ **Figure 10.6**

From Table V in the Appendix we find that the area between $z = 0$ and $z = -1.85$ is 0.4678 and that the area between $z = 0$ and $z = 2.77$ is 0.4972. Adding these two we get

$$0.4678 + 0.4972 = 0.9650$$

Thus, the probability is 0.9650 that the average amount of money these depositors have in their accounts is between $4800 and $5300.

EXAMPLE 3

The average purchase of a customer in a large department store is $4.00 with a standard deviation of $0.85. If 49 customers are selected at random, what is the probability that their average purchases will be less than $3.70?

SOLUTION

We use Formula 10.6. Here $\mu = 4.00$, $\sigma = 0.85$, and $N = 49$. Thus, $\bar{X} = 3.70$ corresponds to

$$z = \frac{3.70 - 4.00}{0.85/\sqrt{49}} = \frac{-0.30}{0.12} = -2.50$$

Thus we are interested in the area to the left of $z = -2.5$. See Figure 10.7.

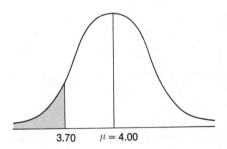
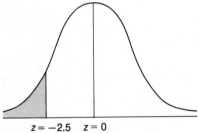

3.70 $\mu = 4.00$ $z = -2.5$ $z = 0$ **Figure 10.7**

From Table V in the Appendix we find that the area from $z = 0$ to $z = -2.5$ is 0.4938. Thus, the area to the left of $z = -2.5$ is $0.5000 - 0.4938$, or 0.0062. Therefore, the probability that the average purchase will be less than \$3.70 is 0.0062.

EXAMPLE 4

The Smith Trucking Company claims that the average weight of its delivery trucks when fully loaded is 6000 pounds with a standard deviation of 120 pounds. Thirty-six trucks are selected at random and their weights recorded. Within what limits will the weights of 90% of the trucks lie?

SOLUTION

Here $\mu = 6000$ and $\sigma = 120$. We are looking for two values within which the weights of 90% of the trucks will lie. From Table V we find that the area between $z = 0$ and $z = 1.65$ is approximately 0.45. Similarly, the area between $z = 0$ and $z = -1.65$ is approximately 0.45. See Figure 10.8. Using Formula 10.6 we have

$$z = \frac{\bar{X} - \mu}{\sigma/\sqrt{N}}$$

If $z = 1.65$, then

$$1.65 = \frac{\bar{X} - 6000}{120/\sqrt{36}}$$

$$= \frac{\bar{X} - 6000}{20}$$

$$1.65(20) = \bar{X} - 6000$$
$$33 + 6000 = \bar{X}$$
$$6033 = \bar{X}$$

If $z = -1.65$, then

$$-1.65 = \frac{\bar{X} - 6000}{120/\sqrt{36}}$$

$$= \frac{\bar{X} - 6000}{20}$$

$$-1.65(20) = \bar{X} - 6000$$
$$6000 - 33 = \bar{X}$$
$$5967 = \bar{X}$$

Thus, 90% of the trucks will weigh between 5967 and 6033 pounds.

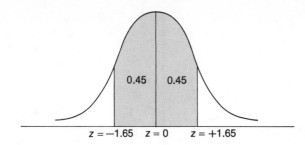

0.45   0.45

z = −1.65   z = 0   z = +1.65   **Figure 10.8**

EXERCISES

1. The average weight of a suitcase carried by a passenger boarding an airplane is 32 pounds with a standard deviation of 6 pounds. If a sample of 49 passengers is selected at random, what is the probability that the average weight will be less than 30 pounds?

2. The average height of a college student at a university is 68 inches with a standard deviation of 6.3 inches. If a sample of 50 students is selected, what is the probability that the average height of these students will be greater than 70 inches?

3. The average length of a ruler produced by the Meter Corporation is 12 inches with a standard deviation of 0.21 inches. If 3 dozen rulers are selected at random, what is the probability that their average length will be greater than 12.08 inches?

4. The Goldcup Baking Company claims that the average weight of its chocolate cupcakes is 5 ounces with a standard deviation of 0.15 ounces. The local consumer's complaint bureau selects a random sample of 25 cupcakes and records their weight. Within what limits should the weights of 95% of the cupcakes lie?

5. The average amount of money contributed to charity by a person in Misertown is $100 per year with a standard deviation of $38. A random sample of 36 people is selected. What is the probability that the average amount of money contributed by these people will be greater than $115?

6. Global Airlines claims that the average number of people who pay for in-flight movies, when the plane is fully loaded, is 42 with a standard deviation of 8. A sample of 36 fully loaded planes is taken. Find the probability that fewer than 38 people paid for the in-flight movies.

7. The average amount of time the tennis players of a racket club play on a given day is 120 minutes with a standard deviation of 26 minutes. A survey of 40 players is taken. What is the probability that they will play tennis for more than 110 minutes?

8. In a certain city, scholarship test scores for all high school seniors have a mean of 83 with a standard deviation of 9.1. One particular high school has 81 seniors. Find the probability that the average test score for this group will be at least 80.

9. The average length of a cigar produced by the Smokey Cigar Company is 5 inches with a standard deviation of 0.3 inches. A random sample of 30 cigars is selected. What is the probability that the average length of these cigars will be less than 4.8 inches?

SELF-STUDY GUIDE

In this chapter we discussed the nature of random sampling and how to go about selecting a random sample. The most convenient way of selecting a random sample is to use a table of random digits.

In some situations, as we pointed out, stratified samples are preferred.

When repeated random samples are taken from the same population, different sample means are obtained. The average of these sample means can be used as an estimate of the population mean. Of course, these sample means form a distribution. If the samples are large enough, the Central Limit Theorem tells us that they will be normally distributed. Furthermore, the mean of the sampling distribution is the same as the population mean. The standard deviation of the sampling distribution is less than the population standard deviation. The Central Limit Theorem led to many useful applications.

You should be able to identify each symbol in the following formulas, understand the relationships among the symbols expressed in each formula, understand the significance of each formula, and use the formulas in solving problems.

1. Population mean

$$\mu = \frac{\Sigma x}{n}$$

2. Sample mean

$$\bar{X} = \frac{\Sigma X}{N}$$

3. Population standard deviation

$$\sigma = \sqrt{\frac{\Sigma(x - \mu)^2}{n}}$$

4. Sample standard deviation

$$s = \sqrt{\frac{\Sigma(X - \bar{X})^2}{N - 1}}$$

5. Standard deviation of sample means

$$\sqrt{\frac{N(\Sigma X^2) - (\Sigma X)^2}{N(N - 1)}}$$

6. Standard error of the mean

$$\sigma_{\bar{X}} = \frac{\sigma}{\sqrt{N}}$$

7. Central Limit theorem

$$z = \frac{\bar{X} - \mu}{\sigma / \sqrt{N}}$$

You should now be able to demonstrate your knowledge of the following ideas presented in this chapter by giving definitions, descriptions, or specific examples. Page references are given for each term so that you can check your answer.

Population (page 290)
Sample (page 290)
Random sample (page 291)
Random sampling (page 291)
Table of random numbers (page 292)
Table of random digits (page 292)

Stratified sample (page 294)
Distribution of sample means (page 301)
Sampling distribution of the mean
    (page 301)
Standard error of the mean (page 302)
The Central Limit Theorem (page 304)

The tests of the following section will be more useful if you take them after you have studied the examples and solved the exercises given in this chapter.

MASTERY TESTS

Form *A*

1. Picking every tenth name that appears in a phone book will result in a random sample. Do you agree? Explain.   *no*
2. We can obtain a random sample as follows. Pick a number from a bowl containing all the numbers from 1 to 1000. Record the number. Throw the piece of paper away. Pick another number and follow the same procedure over and over again. Do you agree that we will get a random sample? Explain your answer.   *no*
3. True or false: The entire population must first be divided into categories to obtain a stratified sample.   *T*
4. True or false: The mean of the sampling distribution of X's equals the mean of the sample.   *T*

5. The standard deviation of the sampling distribution is *a.* greater than, *b.* less than, or *c.* equal to the standard deviation of the population.
6. True or false: The sample means will always be normally distributed. *approx*
7. True or false: In the Central Limit Theorem, *N* refers to the number of samples.
8. True or false: The Central Limit Theorem can be used to predict the behavior of sample means.
9. True or false: The purpose of stratified samples is to obtain groups of people that are approximately equal in some respect.
10. True or false: The distribution of the sample means will not be as spread out as the population distribution.

Form *B*

1. Suppose that the yellow pages of a local telephone directory alphabetically lists 100 auto repair shops. Also, suppose that your car is broken and that you wish to call 5 randomly selected auto repair shops to get an estimate for the cost of repairing the car. Use column 12 of Table VII to find which auto repair shops you should call. 06527   01915   07684   09250
2. A television manufacturer attaches a serial number from 1 to 2000 on each television set that he manufactures during a week. A random sample of 20 sets are to be selected and inspected for quality, workmanship, etc. Use columns 6 and 8 of Table VII to find which sets shall be inspected.
3. If on six consecutive days students waited an average of 5, 15, 12, 9, 6, and 19 minutes before seeing their guidance counselor calculate the mean and standard deviation of the average number of minutes that many students wait before seeing this guidance counselor.
4. A congresswoman polled her constituents on five consecutive days and determined that the average yearly salary of these people was $14,980, $9,850, $8,890, $14,850, and $12,800. Find the mean and standard deviation of these samples.
5. A television rating group has conducted eight random samples of viewer preference for evening movies. A certain network received ratings of 40, 38, 58, 46, 59, 35, 22, and 30 points. Find the mean and standard deviation of these samples.
6. Several recent surveys have indicated that the average life of a washing machine is 4.8, 3.8, 3.9, 5.6, 3.6, and 2.9 years. Find the mean and standard deviation of these samples.
7. Five police stations have each been testing 100 tires of a particular brand. The average life of the tires at each of these stations was 48,000, 36,000, 28,000, 33,000, and 22,000 miles. Find the mean and standard deviation of these samples.

SUGGESTED READING

Hicks, C. R. *Fundamental Concepts in the Design of Experiments.* New York: Holt, Rinehart and Winston, 1964.

Johnson, N. L. and Leone, F. C. *Statistics and Experimental Design in Engineering and the Physical Sciences.* New York: Wiley, 1964.

Mendenhall, W. L., Ott, L. and Scheaffer, R. L. *Elementary Survey Sampling.* Belmont, Calif: Wadsworth, 1968.

Mendenhall, W. L. *Introduction to Probability and Statistics.* 3rd ed. North Scituate, Mass.: Duxbury Press, 1971.

Slonim, M. *Sampling.* New York: Simon & Schuster, 1960.

"For Better Polls." *Business Week,* June 27, 1949.

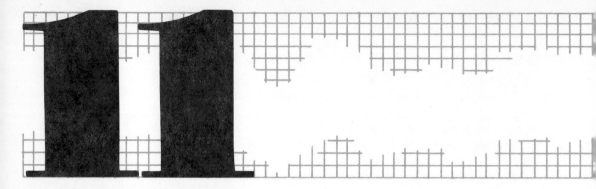

ESTIMATION

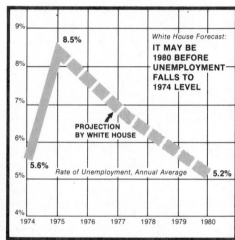

9%				White House Forecast:		
	8.5%			IT MAY BE 1980 BEFORE UNEMPLOYMENT FALLS TO 1974 LEVEL		
8%						
7%						
6%		PROJECTION BY WHITE HOUSE				
5%	5.6%	Rate of Unemployment, Annual Average				5.2%
4%						
	1974	1975	1976	1977	1978	1979 1980

U.S. News & World Report, February 16, 1976. Reprinted from *U.S. News & World Report,* Copyright © 1976 U.S. News & World Report, Inc.

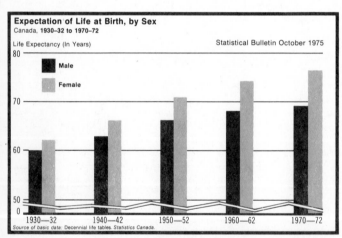

Expectation of Life at Birth, by Sex
Canada, **1930–32 to 1970–72**

Life Expectancy (In Years)   Statistical Bulletin October 1975

■ Male
▨ Female

Source of basic data: Decennial life tables. Statistics Canada.

Courtesy of The Metropolitan Life Insurance Company.

Very often we must use presently available information to make predictions about the future. The right graph gives a White House point estimate of the inflation rate for the next few years. Great care must be exercised in making such estimates since there are many factors that can offset such an estimate.

The left graph gives the estimate of the life expectancy of men and women. Such information is used by insurance companies in determining the premium to charge for life insurance.

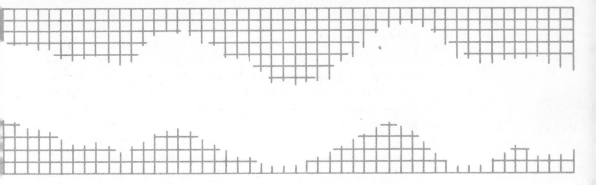

Preview

In this chapter we will discuss the following topics:

1. *Statistical Estimation*
Sample data can often be used to estimate certain unknown quantities. This use of samples is called statistical estimation.

2. *Population Parameters*
These are statistical descriptions of the population.

3. *Point and Interval Estimates*
Sample data can be used to obtain point estimates. These give estimates as a single number. As opposed to this, an interval estimate sets up an interval within which the parameter will lie.

4. *Degree of Confidence*
This tells us the probability that the interval will actually contain the quantity that we are trying to estimate.

5. *Estimating Population Means and Standard Deviations*
We apply the Central Limit Theorem to set up confidence intervals for the mean and standard deviation.

6. *Student's t–Distribution*
When the sample size is small, we use the *t*–distribution values to set up the confidence intervals.

7. *Estimating Population Proportions*
We apply the Central Limit Theorem to set up confidence intervals for the population proportions.

8. *Determining Sample Size*
We discuss how to determine the size of a sample.

Introduction

We have mentioned on several occasions that statistical inference is the process by which statisticians make predictions about a population on the basis of samples. Much information can be gained from a sample. As we mentioned in Chapter 10, the sample mean can be used as an estimate of the population mean. Also, we can obtain an estimate of the population standard deviation on the basis of samples. Thus, one use of sample data is to *estimate* certain unknown quantities of the population. This use of samples is referred to as **statistical estimation.**

On the other hand, sample data can also be used to either accept or reject specific claims about populations. To illustrate this use of sample data, suppose a manufacturer claims that the average milligram tar content per cigarette of a particular brand is 15 with a standard deviation of 0.5. Repeated samples are taken to test this claim. If these samples show that the average tar content per cigarette is 22, then the manufacturer's claim is incorrect. If these samples show that the average tar content is within "predictable limits," then his claim is accepted. Thus, sample data can also be used to either accept or reject specific claims about populations. This use of samples is referred to as **hypothesis testing.**

Statistical inference can be divided into two main categories: problems of estimation and tests of hypotheses. In this chapter we will discuss statistical estimation. In Chapter 12 we will analyze the nature of hypothesis testing.

11.1   POINT AND INTERVAL ESTIMATES

In most statistical problems we do not know certain population values such as the mean and the standard deviation. Somehow we want to use the information obtained from samples to estimate their values. These values, which really are statistical descriptions of the population, are often referred to as **population parameters.**

Suppose we are interested in determining the average life of an electric refrigerator under normal operating conditions. A sample of 50 refrigerators is taken and their lives are as follows:

6.9	7.6	5.7	3.6	7.7	6.6	7.2	7.3	10.6	5.9
8.2	5.7	7.6	8.7	7.9	8.8	7.0	8.1	7.3	6.8
5.7	11.1	8.5	8.9	7.6	5.6	9.0	9.2	6.8	8.3
6.1	9.7	9.8	7.4	6.8	7.3	8.3	9.9	7.5	7.8
7.7	7.4	9.1	7.3	5.5	8.1	6.7	8.8	7.6	5.3

The average life of these refrigerators is 7.6 years. Since this is the only information available to us, we would logically say that the mean life of *all* similar refrigerators is 7.6 years. This estimate of 7.6 years for the population mean is called a **point estimate** since this estimate is a single number. Of course, this estimate may be a poor estimate, but it is the best we can get under the circumstances.

Our confidence in this estimate would be improved considerably if the sample size were larger. Thus, we would have much greater confidence in an estimate that is based on 5000 refrigerators or 50,000 refrigerators than in one that is based only on 50 refrigerators.

One major disadvantage with a point estimate is that the estimate does not indicate the extent of the possible error. Furthermore, a point estimate does not specify how confident we can be that the estimate is close in value to the parameter that it is estimating. Yet, point estimates are quite popular in estimating population parameters.

Another type of estimate that is often used by statisticians, which overcomes the disadvantages mentioned in the previous paragraph, is **interval estimation.** In this method we first find a point estimate. Then we use this estimate to construct an interval within which we can be reasonably sure that the true parameter will lie. Thus, in our example a statistician may say that 95% of all refrigerators will have a mean life of between 5.3 and 9.9 years. An interval such as this is called a **confidence interval.** The lower and upper boundaries, 5.3 and 9.9 respectively, of the interval are called **confidence limits.** The probability that the interval will actually contain the quantity that we are trying to estimate is called the **degree of confidence.**

Generally speaking, as we increase the degree of certainty, namely, the degree of confidence, the confidence interval will become wider. Thus, if the length of an interval is very small (with a specific degree of confidence), then a fairly accurate estimate has been obtained.

When estimating the parameters of a population, statisticians use both point and interval estimates. In the next few sections we will indicate how point and interval estimates are obtained.

11.2   ESTIMATING THE POPULATION MEAN ON THE BASIS OF A LARGE SAMPLE

In Chapter 10 we indicated that the sample mean can be used as an estimate of the population mean. Moreover, the larger the sample size the better the estimate. Yet, as we pointed out in Section 11.1, there are some disadvantages with using point estimates.

The Central Limit Theorem (see page 304) says that the sample means will be normally distributed if the sample sizes are large enough. Generally speaking, statisticians say that a sample size is considered large if N is greater than 30. We can use the Central Limit Theorem to help us construct confidence intervals. This is done as follows.

Since the sample means are normally distributed, we can expect 95% of the $\bar{X}$'s to fall between

$$\bar{X} - 1.96\sigma_{\bar{X}} \quad \text{and} \quad \bar{X} + 1.96\sigma_{\bar{X}}$$

(Since from a normal distribution chart we note that 0.95 probability implies that $z = 1.96$ or -1.96.)

Recall (Formula 10.5, page 302) that

$$\sigma_{\bar{X}} = \frac{\sigma}{\sqrt{N}}$$

Thus, 95% of the $\bar{X}$'s are expected to fall between

$$\bar{X} - 1.96\frac{\sigma}{\sqrt{N}} \quad \text{and} \quad \bar{X} + 1.96\frac{\sigma}{\sqrt{N}}$$

This is shown in Figure 11.1.

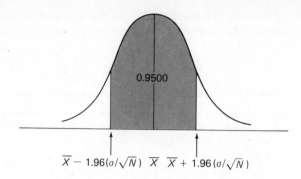

Figure 11.1

$$\overline{X} - 1.96(\sigma/\sqrt{N}) \quad \overline{X} \quad \overline{X} + 1.96(\sigma/\sqrt{N})$$

In order to determine the interval estimate of μ, we must first know the value of the population standard deviation, σ. Although this value is usually unknown, we can use the sample standard deviation as an approximation for σ. We then have the confidence interval for μ given in Formula 11.1.

FORMULA 11.1

Let $\overline{X}$ be a sample mean and s be the sample standard deviation. Then an interval is called a **95% confidence interval for** μ if the lower boundary of the confidence interval is $\overline{X} - 1.96 \dfrac{s}{\sqrt{N}}$ and if the upper boundary of the confidence interval is $\overline{X} + 1.96 \dfrac{s}{\sqrt{N}}$

Comment Formula 11.1 tells us how to find a 95% confidence interval for μ. This means that there is a 95% probability that the population mean will fall within this interval. We must still realize that 5% of the time the population mean will fall outside this interval. This is true because the sample means are normally distributed and 5% of the values of a random variable in a normal distribution will fall further away than 2 standard deviations from the mean (see page 227).

Comment Depending upon the nature of the problem, some statisticians will often prefer a 99% confidence interval for μ or a 90% confidence interval for μ. The boundaries for these intervals are

	Lower boundary	Upper boundary
99% Confidence interval	$\overline{X} - 2.58 \dfrac{s}{\sqrt{N}}$	$\overline{X} + 2.58 \dfrac{s}{\sqrt{N}}$
90% Confidence interval	$\overline{X} - 1.64 \dfrac{s}{\sqrt{N}}$	$\overline{X} + 1.64 \dfrac{s}{\sqrt{N}}$

Thus, as we reduce the size of the interval, we reduce our confidence that the true mean will fall within that interval.

The following examples will illustrate how we establish confidence intervals.

EXAMPLE 1

A coffee vending machine fills 100 cups of coffee before it has to be refilled. On Monday the mean number of ounces in a filled cup of coffee was 7.5. The population standard deviation is known to be 0.25 ounces. Find 95% and 99% confidence intervals for the mean number of ounces of coffee dispensed by this machine.

SOLUTION

We use Formula 11.1. Here $N = 100$, $\sigma = 0.25$, and $\bar{X} = 7.5$. To construct a 95% confidence interval for μ, we have

Lower boundary	Upper boundary
$= \bar{X} - 1.96 \dfrac{\sigma}{\sqrt{N}}$	$= \bar{X} + 1.96 \dfrac{\sigma}{\sqrt{N}}$
$= 7.5 - 1.96 \left(\dfrac{0.25}{\sqrt{100}} \right)$	$= 7.5 + 1.96 \left(\dfrac{0.25}{\sqrt{100}} \right)$
$= 7.5 - 0.05$	$= 7.5 + 0.05$
$= 7.45$	$= 7.55$

Thus, we conclude that the population mean will lie between 7.45 and 7.55 ounces with a probability of 0.95. To construct a 99% confidence interval for μ, we have

Lower boundary	Upper boundary
$= \bar{X} - 2.58 \dfrac{\sigma}{\sqrt{N}}$	$= \bar{X} + 2.58 \dfrac{\sigma}{\sqrt{N}}$
$= 7.5 - 2.58 \left(\dfrac{0.25}{\sqrt{100}} \right)$	$= 7.5 + 2.58 \left(\dfrac{0.25}{\sqrt{100}} \right)$
$= 7.5 - 0.06$	$= 7.5 + 0.06$
$= 7.44$	$= 7.56$

Thus, we conclude that the population mean will lie between 7.44 and 7.56 ounces with a probability of 0.99. In this example we did not have to use s as an estimate of σ since we were told that the population standard deviation was known to be 0.25.

EXAMPLE 2

A sample survey of 81 movie theaters showed that the average length of the main feature film was 90 minutes with a standard deviation of 20 minutes. Find
a. a 90% confidence interval for the mean of the population.
b. a 95% confidence interval for the mean of the population.

SOLUTION

We use Formula 11.1. Here $N = 81$, $s = 20$, and $\bar{X} = 90$.
a. To construct a 90% confidence interval for μ, we have

Lower boundary	Upper boundary
$= \bar{X} - 1.64 \dfrac{s}{\sqrt{N}}$	$= \bar{X} + 1.64 \dfrac{s}{\sqrt{N}}$
$= 90 - 1.64 \left(\dfrac{20}{\sqrt{81}} \right)$	$= 90 + 1.64 \left(\dfrac{20}{\sqrt{81}} \right)$
$= 90 - 3.64$	$= 90 + 3.64$
$= 86.36$	$= 93.64$

Thus, a 90% confidence interval for μ is 86.36 to 93.64 minutes.
b. To construct a 95% confidence interval for μ, we have

Lower boundary	Upper boundary
$= \bar{X} - 1.96 \dfrac{s}{\sqrt{N}}$	$= \bar{X} + 1.96 \dfrac{s}{\sqrt{N}}$
$= 90 - 1.96 \left(\dfrac{20}{\sqrt{81}} \right)$	$= 90 + 1.96 \left(\dfrac{20}{\sqrt{81}} \right)$
$= 90 - 4.36$	$= 90 + 4.36$
$= 85.64$	$= 94.36$

Thus, a 95% confidence interval for μ is 85.64 to 94.36 minutes.
    Notice that as we increase the size of the confidence interval, our confidence that this interval contains μ also increases.

EXAMPLE 3

The management of the Night-All Corporation recently conducted a survey of its 196 employees to determine the average number of hours that each employee sleeps at night. The company statistician submitted the following information to the management:

$$\Sigma X = 1479.8 \quad \text{and} \quad \Sigma (X - \bar{X})^2 = 1755$$

where X is the number of hours slept by each employee. Find a 95% confidence interval estimate for the average number of hours that each employee sleeps at night.

SOLUTION

In order to use Formula 11.1, we must first calculate $\bar{X}$ and s. Using the given information, we have

$$\bar{X} = \frac{\Sigma X}{N}$$

$$\bar{X} = \frac{1479.8}{196}$$

$$= 7.55$$

$$\text{and} \quad s = \sqrt{\frac{\Sigma(X - \bar{X})^2}{N - 1}}$$

$$= \sqrt{\frac{1755}{195}}$$

$$= \sqrt{9} = 3$$

Now we can use Formula 11.1 with $\bar{X} = 7.55$, $s = 3$, and $N = 196$. To find the 95% confidence for μ, we have

Lower boundary	Upper boundary
$= \bar{X} - 1.96 \dfrac{s}{\sqrt{N}}$	$= \bar{X} + 1.96 \dfrac{s}{\sqrt{N}}$
$= 7.55 - 1.96\left(\dfrac{3}{\sqrt{196}}\right)$	$= 7.55 + 1.96\left(\dfrac{3}{\sqrt{196}}\right)$
$= 7.55 - 0.42$	$= 7.55 + 0.42$
$= 7.13$	$= 7.97$

Thus, the management can conclude with a probability of 0.95 that the average number of hours that an employee sleeps at night is between 7.13 and 7.97 hours.

EXERCISES

1. A sample of 49 families in Death Valley found that a family of 4 spends an average of $42 a week for food. The standard deviation was 5.1. Find a 95% confidence interval for the average amount of money that a family of 4 in Death Valley spends for food.

2. According to 1975 Internal Revenue Service records, the average amount of money deducted for medical expenses by 100 randomly selected taxpayers whose adjusted gross income was in the $10,000–15,000 range was $346. The standard deviation was $31. Find a 90% confidence interval for medical expenses by taxpayers in this income bracket.

3. A survey of 36 drivers in Gotham City shows that these drivers paid an average of $48 within the past year to have their cars tuned. The standard deviation was $6.25. Find a 99% confidence interval for the cost of a tune-up in Gotham City.

4. According to the Bureau of Labor Statistics records, the average spendable weekly earnings of 75 manufacturing workers with 3 dependents during 1975 was $166.04. The standard deviation was $9.63. Find a 95% confidence interval for the average salary of these workers.

5. The average life of 50 stereo equipment components produced by the Groovy Corporation is 2 years with a standard deviation of 0.3 years. Find a 90% confidence interval for the average life of the stereo equipment component.

6. The average cost of a pack of cigarettes in 60 randomly selected stores in Holtsville was 55 cents with a standard deviation of 5 cents. Find a 90% confidence interval for the cost of a pack of cigarettes.

7. A sample of 49 check cashing businesses in New York City shows that the average check cashed is for $200 with a standard deviation of $12. Find the 99% confidence interval for the average amount of money involved when a check is cashed.

8. The average length of 49 tuna fish caught by Charlie is 55 inches with a standard deviation of 9 inches. Find a 95% confidence interval for the average length of a tuna fish caught by Charlie.

11.3 ESTIMATING THE POPULATION MEAN ON THE BASIS OF A SMALL SAMPLE

In Section 11.2 we indicated how to determine confidence intervals for μ when the sample size is larger than 30. Unfortunately, this is not always the case. Suppose a sample of 16 bulbs is randomly selected from a large shipment and has a mean life of 100 hours with a standard deviation of 5 hours. Using only the methods of Section 11.2, we cannot determine confidence intervals for the mean life of a bulb since the sample size is less than 30.

Fortunately, in such situations we can base confidence intervals for μ on a distribution which is in many respects similar to the normal distribution. This is the **Student's t–distribution.** This distribution was first studied by William S. Gosset, who was a statistician for Guinness, an Irish brewing company. Gosset was the first to recognize the need for developing methods for interpreting information obtained from small samples. Yet his company did not allow any of its employees to publish anything. So, Gosset secretly published his findings in 1907 under the name "Student." To this day this distribution is referred to as the Student's t–distribution.

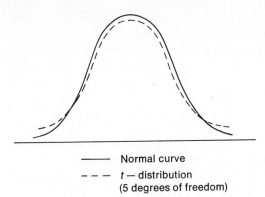

Normal curve

- - - t – distribution
(5 degrees of freedom)

Figure 11.2

Relationship between the normal
distribution and the t-distribution.

Figure 11.2 indicates the relationship between the normal distribution and the
t–distribution. Notice that the t–distribution is also symmetrical about zero, which is
its mean. However, the shape of the t–distribution depends upon a parameter
called the **number of degrees of freedom.** In our case the number of degrees of
freedom. Thus, the 1.96 of Formula 11.1 of Section 11.2 can be replaced by the
$d.f. = N - 1$.

Table VIII in the Appendix indicates the value of t for different degrees of
freedom. Thus, the 1.96 of Formula 11.1 of Section 11.2 can be replaced by the
$t_{0.025}$ value as listed in this table, depending upon the number of degrees of freedom. When using the $t_{0.025}$ value of Table VIII, 95% of the area under the curve of
the t–distribution will fall between $-t_{0.025}$ and $t_{0.025}$ as shown in Figure 11.3.

We then have the following formula.

FORMULA 11.2

Let $\bar{X}$ be a sample mean and let s be the sample
standard deviation. We have the following 95%
small-sample confidence interval for μ:

$$\text{Lower boundary: } \bar{X} - t_{0.025}\frac{s}{\sqrt{N}}$$

$$\text{Upper boundary: } \bar{X} + t_{0.025}\frac{s}{\sqrt{N}}$$

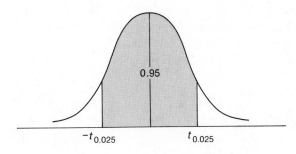

0.95

$-t_{0.025}$ $t_{0.025}$

Figure 11.3

Comment In addition to the $t_{0.025}$ values, Table VIII in the Appendix also lists many other values of t. Thus the $t_{0.005}$ values are used when we want a 99% confidence interval for μ, the $t_{0.050}$ values are used when we want a 90% confidence interval for μ, the $t_{0.10}$ values are used when we want an 80% confidence interval for μ, etc.

EXAMPLE 1

A survey of 16 taxi drivers found that the average tip they receive is 40 cents with a standard deviation of 8 cents. Find a 95% confidence interval for estimating the average amount of money that a taxi driver receives as a tip.

SOLUTION

We will use Formula 11.2. Here we have $N = 16$, $\bar{X} = 40$, and $s = 8$. We first find the number of degrees of freedom which is $N - 1$, or $16 - 1 = 15$. Then we find the appropriate value of t from Table VIII. The $t_{0.025}$ value with 15 degrees of freedom is 2.131. Finally we establish the confidence interval. We have

$$\text{Lower boundary} = \bar{X} - t_{0.025}\,\frac{s}{\sqrt{N}} \qquad \text{Upper boundary} = \bar{X} + t_{0.025}\,\frac{s}{\sqrt{N}}$$

$$= 40 - 2.131\left(\frac{8}{\sqrt{16}}\right) \qquad\qquad = 40 + 2.131\left(\frac{8}{\sqrt{16}}\right)$$

$$= 40 - 4.26 \qquad\qquad\qquad\qquad = 40 + 4.26$$

$$= 35.74 \qquad\qquad\qquad\qquad\qquad = 44.26$$

Thus, a 95% confidence interval for the average amount of money that a taxi driver will receive as a tip is 35.74 to 44.26 cents.

EXAMPLE 2

A survey of the hospital records of 25 randomly selected patients suffering from a particular disease indicated that the average length of stay in the hospital is 10 days. The standard deviation is estimated to be 2.1 days. Find a 99% confidence interval for estimating the mean length of stay in the hospital.

SOLUTION

We use Formula 11.2. Here $N = 25$, $\bar{X} = 10$, and $s = 2.1$. We first find the number of degrees of freedom which is $N - 1$, or $25 - 1 = 24$. Then we find the appropriate value of t from Table VIII

in the Appendix. The $t_{0.005}$ value with 24 degrees of freedom is
2.797. Finally we establish the confidence interval. We have

Estimating the
Population Mean on
the Basis of a Small
Sample

Lower boundary $= \bar{X} - t_{0.005} \dfrac{s}{\sqrt{N}}$ Upper boundary $= \bar{X} + t_{0.005} \dfrac{s}{\sqrt{N}}$

$$= 10 - 2.797\left(\frac{2.1}{\sqrt{25}}\right) \qquad\qquad = 10 + 2.797\left(\frac{2.1}{\sqrt{25}}\right)$$

$$= 10 - 1.17 \qquad\qquad\qquad\qquad = 10 + 1.17$$

$$= 8.83 \qquad\qquad\qquad\qquad\qquad = 11.17$$

Thus, a 99% confidence interval for the average length of stay in
the hospital is 8.83 to 11.17 days.

EXAMPLE 3

A music teacher asks six randomly selected students how many
hours a week each practices the electric guitar. The teacher
receives the following answers: 10, 12, 8, 9, 16, 5. Find a 90%
confidence interval for the average length of time that these
students practice.

SOLUTION

In order to use Formula 11.2 we must first calculate the sample
mean and the sample standard deviation. We have

$$\bar{X} = \frac{\Sigma X}{N} = \frac{10 + 12 + 8 + 9 + 16 + 5}{6} = \frac{60}{6} = 10$$

To calculate the standard deviation we arrange the data as follows:

X	$X - \bar{X}$	$(X - \bar{X})^2$
10	$10 - 10 = 0$	0
12	$12 - 10 = 2$	4
8	$8 - 10 = -2$	4
9	$9 - 10 = -1$	1
16	$16 - 10 = 6$	36
5	$5 - 10 = -5$	25
	Total $= 70$	

$$s = \sqrt{\frac{\Sigma(X - \bar{X})^2}{N - 1}} = \sqrt{\frac{70}{6 - 1}} = \sqrt{14} = 3.74$$

325

Now we find the appropriate value of t from Table VIII. The $t_{0.05}$ value with $N - 1$, or $6 - 1 = 5$, degrees of freedom is 2.015. Finally we establish the confidence interval. We have

Lower boundary $= \bar{X} - t_{0.05} \dfrac{s}{\sqrt{N}}$ ⠀⠀⠀ Upper boundary $= \bar{X} + t_{0.05} \dfrac{s}{\sqrt{N}}$

$= 10 - 2.015 \left(\dfrac{3.74}{\sqrt{6}}\right)$ ⠀⠀⠀⠀⠀⠀ $= 10 + 2.015 \left(\dfrac{3.74}{\sqrt{6}}\right)$

$= 10 - 3.08$ ⠀⠀⠀⠀⠀⠀⠀⠀⠀⠀⠀ $= 10 + 3.08$

$= 6.92$ ⠀⠀⠀⠀⠀⠀⠀⠀⠀⠀⠀⠀⠀⠀⠀ $= 13.08$

Thus, a 90% confidence interval for the average length of time that a student practices the electric guitar is 6.92 to 13.08 hours.

EXERCISES

1. Ten brands of beer were tested to see how long each kept its head and flavor after the can was opened. The average time was 12 minutes with a standard deviation of 3.1 minutes. Find a 95% confidence interval for the average length of time a can of beer retains its flavor.

2. Five brands of paint were tested to see how long it took them to dry. The average time was 8 hours with a standard deviation of 30 minutes. Find a 90% confidence interval for the average drying time of these paints.

3. Dr. Frankenstein is interested in knowing how the blood pressure of a man is affected by watching a horror movie. Sixteen men had their blood pressures measured before and after watching a horror movie. The changes in their blood pressure were as follows:

$$-4, +5, -1, -3, -9, +6, -3, -8, +5, -6, -11, 0, +3, -7, +2, +10$$

Find a 99% confidence interval for the mean change in the blood pressure.

4. The nutritional content of 5 leading cereals (on a particular nutritional scale) is as follows:

Kind of Cereal	Nutritional Content
Special Z	82
Soggies	63
Wheat Flakes	88
Defrosted Flakes	72
Sugar Flops	50

Find a 95% confidence interval for the mean nutritional content of these cereals.

5. The Stingy Corporation is conducting a survey to determine the amount of money given by each of its employees for charity. It received the following amounts from 6 employees: $14, $2, $5, $0, $6, $3. Find a 90% confidence interval for the average amount of money given to charity by an employee of the Stingy Corporation.

6. Seven students were randomly selected and asked to park a car in a very tight spot. The students needed 12, 17, 9, 13, 6, 15, and 5 minutes respectively to park the car. Find a 90% confidence interval for the average amount of time needed to park the car.

7. At a recent social gathering nine members of the college administration consumed an average of 3 pints of liquor with a standard deviation of 1.68 pints. Find a 95% confidence interval for the average amount of liquor consumed.

8. A study of the television viewing of 18 families revealed that they watched an average of 10 hours of television on the weekend. The standard deviation was 2.32. Find a 95% confidence interval for the mean amount of time spent watching television on the weekend.

9. A leading child psychologist claims that fathers are not spending enough time with their children. To support the claim, he selected 16 families in Dover and found that the father spent an average of 8 hours per week with his children. The standard deviation was 1.69 hours. Find a 90% confidence interval for the average time spent by a father in Dover with his children.

10. Medical records of 9 hospitals in a large city indicate that the hospitals treated an average of 28 patients per week for burns. The standard deviation was 1.83. Find a 95% confidence interval for the mean number of burn victims treated.

11.4 THE ESTIMATION OF THE STANDARD DEVIATION

Until now we have been discussing how to obtain point and interval estimates of the population mean μ. In this section we discuss point and interval estimates for the population standard deviation.

Suppose we took a sample of 100 cigarettes of a particular brand and determined that the sample mean tar content was 15.2 with a standard deviation of 1.1. If this procedure were to be repeated many times, each time we would obtain different estimates for the population mean and standard deviation. Nevertheless, the mean of these estimates will approach the true population mean as more and more samples are included. When this happens we say that the average of the sample means is an **unbiased estimator** of the population mean.

Recall that on page 295 two formulas were given for the standard deviation. Suppose we were to calculate the standard deviation of a sample by using the formula

$$\sqrt{\frac{\Sigma(X - \bar{X})^2}{N}}$$

Then we cannot use this value as an unbiased estimate of the population standard deviation. Specifically, the average of such estimates would always be too small, no matter how many samples are included. It is for this reason that we divide by $N - 1$ instead of N. When this is done, it can be shown that

$$\sqrt{\frac{\Sigma(X - \bar{X})^2}{N - 1}}$$

is an unbiased estimate of σ. This means that as more and more samples are included, the standard deviation of these samples will approach the true population standard deviation. Thus a point estimate of σ is

$$\sqrt{\frac{\Sigma(X - \bar{X})^2}{N - 1}}$$

When dealing with large sample sizes, mathematicians have developed the following 95% confidence interval for σ.

FORMULA 11.3

Let s be the sample standard deviation and let N be the number in the sample. We have the following 95% confidence interval for σ:

$$\text{Lower boundary} = \frac{s}{1 + \frac{1.96}{\sqrt{2N}}}$$

$$\text{Upper boundary} = \frac{s}{1 - \frac{1.96}{\sqrt{2N}}}$$

EXAMPLE 1

A sample of 50 cigarettes was taken. The sample mean tar content was 15.2 with a standard deviation of 1.1. Find a 95% confidence interval for the population standard deviation.

SOLUTION

We use Formula 11.3. Here $N = 50$ and $s = 1.1$. The lower boundary of the confidence interval is

$$\frac{s}{1 + \frac{1.96}{\sqrt{2N}}} = \frac{1.1}{1 + \frac{1.96}{\sqrt{2(50)}}}$$

$$= \frac{1.1}{1 + 0.196}$$

$$= 0.92$$

The upper boundary of the confidence interval is

$$\frac{s}{1 - \frac{1.96}{\sqrt{2N}}} = \frac{1.1}{1 - \frac{1.96}{\sqrt{2(50)}}}$$

$$= \frac{1.1}{1 - 0.196}$$

$$= 1.37$$

Thus, a 95% confidence interval for σ is 0.92 to 1.37.

11.5   DETERMINING THE SAMPLE SIZE

Until now we have been discussing how sample data can be used to estimate various parameters of a population. Selecting a sample usually involves an expenditure of money. The larger the sample size, the greater the cost. Therefore, one question that must be answered before selecting a sample is how large should the sample be.

Generally speaking the size of the sample is determined by the desired degree of accuracy. Most problems specify the maximum allowable error. Yet we must realize that no matter what the size of a sample, any estimate may exceed the maximum allowable error. To be more specific, suppose we are interested in estimating the mean life of a calculator on the basis of a sample. Suppose also that we want our estimate to be within 0.75 years of the true value of the mean. In this case the **maximum allowable error,** denoted as e, is 0.75 years. How large a sample must be taken? Of course, the larger the sample size, the smaller the chance that our estimate will not be within 0.75 years of the value. If we want to be sure with a probability of 0.95 that our estimate will be within 0.75 years of the true value, then the sample size must satisfy

$$1.96\sigma_{\bar{x}} = 0.75$$

More generally, if the maximum allowable error is e, then we must have

$$e = 1.96\sigma_{\bar{x}}$$

Since

$$\sigma_{\bar{x}} = \frac{\sigma}{\sqrt{N}}$$

we get

$$e = 1.96\frac{\sigma}{\sqrt{N}}$$

Solving this equation for N gives Formula 11.4.

FORMULA 11.4

Let σ be the population standard deviation, e the maximum allowable error, and N the size of the sample that is to be taken. Then a sample of size

$$N = \left(\frac{1.96\sigma}{e}\right)^2$$

will result in an estimate of μ which is less than the maximum allowable error 95% of the time.

EXAMPLE 1

Suppose we wish to estimate the average life of a calculator to within 0.75 years of the true value. Past experience indicates that the standard deviation is 2.6 years. How large a sample must be selected if we want our answer to be within 0.75 years 95% of the time?

SOLUTION

We use Formula 11.4. Here $\sigma = 2.6$ and $e = 0.75$.

$$\text{Then } N = \left(\frac{1.96\sigma}{e}\right)^2$$

$$= \left(\frac{1.96(2.6)}{0.75}\right)^2 = (6.79)^2 = 46.10$$

Thus, the sample should consist of 47 calculators. (Note: In determination of sample size, any decimal is always rounded to the next highest number.)

EXAMPLE 2

The administration of a university in California desires to estimate the average teaching experience, measured in years, of its 2000 faculty members. How large a sample should be taken in order to be 95% confident that the sample mean does not differ from the population mean by more than 1/2 year? Past experience indicates that the standard deviation is 2.6 years.

SOLUTION

We use Formula 11.4. Here $\sigma = 2.6$ and $e = 0.50$. Then

$$N = \left(\frac{1.96\sigma}{e}\right)^2$$

$$= \left(\frac{1.96(2.6)}{0.50}\right)^2$$

$$= (10.19)^2$$

$$= 103.84$$

Thus, the administration should select 104 teachers to determine the average teaching experience of a teacher.

EXERCISES

In each of Exercises 1–7 assume that it is required that the maximum allowable error not be exceeded with a 95% degree of confidence.

1. An auto manufacturer wishes to estimate the average time needed by a driver to change a tire. How large a sample should be taken if the manufacturer wishes the estimate to be within 0.25 hours of the true value? Assume that $\sigma = 1.6$.

2. A local conservation group wishes to estimate the average time its members ride their bicycles. How large a sample should be taken if the conservation group wishes that the estimate be within 1.36 hours of the true value? Assume that $\sigma = 2.37$.

3. A scientist is experimenting with a new type of drug and wishes to estimate the average recovery time of people treated with this drug. How large a sample should be taken if the scientist wishes the estimate to be within 2 hours of the true value? Assume that $\sigma = 7.39$.

4. A young homemaker wishes to estimate the weekly cost of feeding her family. She is going to take a sample of many homemakers in the city. How large a sample should be taken if she wishes her estimate to be within $2.25 of the true value? Assume that $\sigma = \$5.85$.

5. A druggist is interested in determining how much money women spend on facial cosmetics. How large a sample should be taken if the druggist wishes the estimate to be within $1.79 of the true value? Assume that $\sigma = \$3.76$.

6. A dermatologist wishes to determine the average number of minutes that a person spends in the shower. How large a sample must be taken if the dermatologist wishes the estimate to be within 2.34 minutes? Assume that $\sigma = 7.83$.

7. In measuring the reaction time to a sudden driving hazard, a State Motor Vehicle Bureau found that the standard deviation is 0.06 seconds. How large a sample should the Bureau take if it wishes to estimate the average reaction time of a driver with a maximum allowable error of 0.02 seconds?

8. Forty dentists were asked to indicate their income. The standard deviation of their income was computed and found to be $1962. Find a 95% confidence interval for the population standard deviation.

9. A sample of 200 accounts in a savings bank indicated that the average amount of money on deposit is $8896. The standard deviation is $821.19. Find a 95% confidence interval for the standard deviation of *all* the accounts.

10. A sample of 36 second graders showed that a second grader spends an average of 10.3 hours per week playing with friends. The standard deviation was 1.64 hours. Find a 95% confidence interval for the population standard deviation.

11.6 THE ESTIMATION OF PROPORTIONS

So far, we have discussed the estimation of the population mean and the population standard deviation. Very often statistical problems arise for which the data is available in **proportion** or **count form** rather than in measurement form. For example, suppose a doctor has developed a new technique for predicting the sex of an unborn child. He tests his new method on 1000 pregnant women and correctly predicts the sex of 900 of the children. Is his new technique reliable? The doctor has correctly predicted the sex of 900 unborn children. Thus the **sample proportion** is $\frac{900}{1000}$, or 0.90. What is the **true proportion** of unborn children whose sex he can correctly predict?

Since we will be discussing both sample proportions and true population proportions, we will use the following notation:

π true population proportion
p sample proportion

In our case the doctor estimates the true population proportion, π, to be 0.90. This estimate is based upon the sample proportion p. How reliable is this estimate of the true population proportion? Suppose the doctor now tests his technique on 1000 different pregnant women. For what proportion of these 1000 unborn children will he be able to correctly predict the sex?

If the true population proportion is π, repeated sample proportions will be normally distributed. The average of the sample proportions will be π. The standard deviation of these sample proportions will equal

$$\sqrt{\frac{\pi(1 - \pi)}{N}}$$

Thus, if the doctor's technique is reliable, he should find that the sample proportion will be normally distributed with a mean of 0.9. The standard deviation will be

$$\sqrt{\frac{(0.9)(1 - 0.9)}{1000}} = \sqrt{0.00009}, \text{ or } 0.0095$$

We can summarize these results as follows.

FORMULA 11.5

Suppose we have a large population, a proportion of which has some particular characteristic. We select random samples of size N and determine the proportion of people in each sample with this characteristic. Then the sample proportions will be normally distributed with mean π and standard deviation

$$\sqrt{\frac{\pi(1 - \pi)}{N}}$$

Since the sample proportions are normally distributed, we can apply the standardized normal charts as we did for the sample means. The following examples will illustrate how this is done.

EXAMPLE 1

From past experience it is known that 70% of all airplane tickets sold by Global Airways are round trip tickets. A random sample of 100 passengers is taken. What is the probability that at least 75% of these passengers have round trip tickets?

SOLUTION

We use Formula 11.5. Here $N = 100$, $\pi = 0.70$, and the sample proportion p is 0.75. Since the sample proportions are normally distributed, we are interested in the area to the right of $x = 0.75$ in a

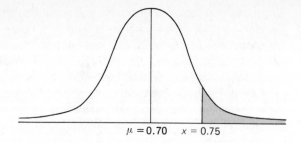

Figure 11.4
$\mu = 0.70$ $x = 0.75$

normal distribution whose mean is 0.70. See Figure 11.4. We first calculate the standard deviation of the sample proportions, denoted as σ_p:

$$\sigma_p = \sqrt{\frac{\pi(1 - \pi)}{N}}$$

$$= \sqrt{\frac{0.70(1 - 0.70)}{100}}$$

$$= \sqrt{0.0021}$$

$$= 0.046$$

Then $z = \dfrac{p - \pi}{\sigma_p}$

$$= \frac{0.75 - 0.70}{0.046}$$

(p replaces $\bar{X}$ and π replaces μ in Formula 10.6 on page 305).

$$= 1.09$$

From Table V the area between $z = 0$ and $z = 1.09$ is 0.3621. Therefore the area to the right of $z = 1.09$ is $0.5000 - 0.3621$, or 0.1379. Thus, the probability that at least 75% of the passengers have round trip tickets is 0.1379.

EXAMPLE 2

Fifty-four percent of all students in the Gamma Fraternity at Beta University have type O blood. Thirty-six fraternity members are selected at random. What is the probability that between 51% and 58% of these members have type O blood?

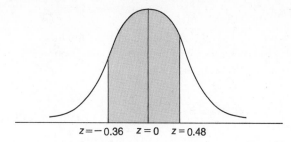

$z = -0.36$ $z = 0$ $z = 0.48$ **Figure 11.5**

SOLUTION

We will use Formula 11.5. Here $\pi = 0.54$ and $N = 36$. We first
calculate σ_p:

$$\sigma_p = \sqrt{\frac{\pi(1 - \pi)}{N}}$$

$$= \sqrt{\frac{0.54(1 - 0.54)}{36}}$$

$$= \sqrt{0.0069}$$

$$= 0.083$$

Now we can use Formula 11.5 with $\pi = 0.54$, $N = 36$, and
$\sigma_p = 0.083$:

$p = 0.51$ corresponds to $z = \dfrac{0.51 - 0.54}{0.083} = \dfrac{-0.03}{0.083} = -0.36$

$p = 0.58$ corresponds to $z = \dfrac{0.58 - 0.54}{0.083} = \dfrac{0.04}{0.083} = 0.48$

Thus, we are interested in the area between $z = -0.36$ and $z = 0.48$.
See Figure 11.5.

From Table V we find that the area between $z = 0$ and $z = -0.36$
is 0.1406 and the area between $z = 0$ and $z = 0.48$ is 0.1844. Adding
we get

$$0.1406 + 0.1844 = 0.3250$$

Thus, the probability that between 51% and 58% of these members
have type O blood is 0.3250.

In the same way that the sample mean is used to estimate the
population mean, we can use the sample proportions, p, as a point
estimate of the population proportion, π. However, this estimate
does not indicate the probability of its accuracy. Thus we set up
interval estimates for the true population proportion, π. We have
Formula 11.6.

FORMULA 11.6

Let p be a sample proportion and let π be the true population proportion. Then we have the following 95% confidence interval for π:

$$\text{Lower boundary} = p - 1.96 \sqrt{\frac{p(1-p)}{N}}$$

$$\text{Upper boundary} = p + 1.96 \sqrt{\frac{p(1-p)}{N}}$$

EXAMPLE 3

A union member reported that 80 out of 120 workers interviewed supported some form of work stoppage to further its demands for a shorter work week. Find a 95% confidence estimate of the true proportion of workers supporting the union's stand on a work stoppage.

SOLUTION

We use Formula 11.6. Here $p = 80/120$, or 0.67, and $N = 120$. To construct a 95% confidence interval for π, we have

Lower boundary

$$= p - 1.96 \sqrt{\frac{p(1-p)}{N}}$$
$$= 0.67 - 1.96 \sqrt{\frac{0.67(1-0.67)}{120}}$$
$$= 0.67 - 1.96 \sqrt{0.0018}$$
$$= 0.67 - 1.96 (0.043)$$
$$= 0.67 - 0.084$$
$$= 0.586$$

Upper boundary

$$= p + 1.96 \sqrt{\frac{p(1-p)}{N}}$$
$$= 0.67 + 1.96 \sqrt{\frac{0.67(1-0.67)}{120}}$$
$$= 0.67 + 1.96 \sqrt{0.0018}$$
$$= 0.67 + 1.96 (0.043)$$
$$= 0.67 + 0.084$$
$$= 0.754$$

Thus, the union official concluded with a probability of 0.95 that the true proportion of workers supporting the union claim is between 0.586 and 0.754. This is the 95% confidence interval.

EXAMPLE 4

A new pay telephone has been installed in the student dormitory. A dime was lost 45 of the first 300 times that it was used. Construct a 95% confidence interval for the true proportion of times that a user will lose a dime.

SOLUTION

We use Formula 11.6. Here $p = 45/300$, or 0.15, and $N = 300$.
To construct a 95% confidence interval for π, we have

Lower boundary

$$= p - 1.96 \sqrt{\frac{p(1-p)}{N}}$$

$$= 0.15 - 1.96 \sqrt{\frac{0.15(1-0.15)}{300}}$$

$$= 0.15 - 1.96 \sqrt{0.0004}$$

$$= 0.15 - 1.96(0.02)$$

$$= 0.15 - 0.039$$

$$= 0.111$$

Upper boundary

$$= p + 1.96 \sqrt{\frac{p(1-p)}{N}}$$

$$= 0.15 + 1.96 \sqrt{\frac{0.15(1-0.15)}{300}}$$

$$= 0.15 + 1.96 \sqrt{0.0004}$$

$$= 0.15 + 1.96(0.02)$$

$$= 0.15 + 0.039$$

$$= 0.189$$

Thus, a 95% confidence interval for the true proportion of times that
a user will lose a dime is 0.111 to 0.189.

EXERCISES

1. A random survey of 1000 people found that 400 of them were in favor of
   legalizing drugs. What is the probability that the proportion of people in favor of
   legalizing drugs is greater than 43%?
2. Should calculators be allowed in the classroom? In a recent survey of 500
   teachers 230 were in favor of allowing calculators in the classroom. What is
   the probability that the proportion of teachers favoring classroom use of calculators is less than 42%?
3. A local hospital claims that only 5% of the patients treated in the emergency
   room of the hospital are dissatisfied with the services provided. To test this
   claim, the hospital randomly selected 300 patients and asked them to rate the
   services provided. What is the probability that the proportion of patients dissatisfied with the service provided is between 2% and 8%?
4. A large bank in the western United States, which employs approximately
   50,000 people, claims that 46% of its top paying jobs are held by women. A
   local chapter of a women's liberation group is challenging this claim. A random
   sample of 100 of the company's top paying jobs is taken. What is the probability that the proportion of women having these top paying jobs is less than
   45%?
5. A random sample of 500 families in Layette indicated that 225 of them preferred female babysitters to male babysitters. Find a 95% confidence interval for
   the true proportion of all families living in Layette who prefer female babysitters.

6. One hundred eighty-six of the 220 students interviewed at a community college said they used public transportation to get to school. Find a 95% confidence interval for the true proportion of all students who use public transportation to get to school.

7. Three hundred twenty-eight of the 500 drivers interviewed said that they would never use retread tires. Find a 95% confidence interval for the true proportion of all drivers who would never use retread tires.

8. A bicycle manufacturer claims that 95% of all bikes on the road in a certain town are equipped with nighttime safety reflectors. A random sample of 75 bikes is selected. What is the probability that fewer than 91% of these bikes are not equipped with the reflectors?

9. A toy company claims that 7% of all its toy electric trains do not function properly. A random sample of 350 of these toy trains is taken. What is the probability that the proportion of electric trains that do not function properly is between 6% and 9%?

10. A State Consumer Fraud Bureau claims that 40% of all complaints to its offices are for incorrect bills received from the Watt Electric Supply Company. The company disagrees with this claim. A sample of 250 of the complaints received is taken. Find the probability that fewer than 90 of these complaints will be for incorrect bills.

SELF-STUDY GUIDE

In this chapter we indicated how sample data can be used to estimate the population mean and the population standard deviation. Both point and interval estimates were discussed. In most cases sample data is used to construct confidence intervals within which a given parameter is likely to lie with a specified probability. Sample data can also be used to make estimates about the true population proportion and to construct confidence intervals for the population proportion.

All estimates considered in this chapter were unbiased estimates. This means that the average of these estimates will approach the true population parameter that they are trying to estimate as more and more samples are included. For this reason we divide $\Sigma(X - \bar{X})^2$ by $N - 1$, not by N, in determining the sample standard deviation.

We also indicated how to determine the size of a sample to be used in gathering data. Depending upon the maximum allowable error, Formula 11.4 determines the sample size with a 95% degree of accuracy.

You should be able to identify each symbol in the following formulas, understand the relationships among the symbols expressed in each formula, understand the significance of each formula, and use the formulas in solving problems.

1.

Size of Sample	Parameter	Degree of Confidence	Lower Boundary	Upper Boundary
Large	Mean	90%	$\bar{X} - 1.64\dfrac{s}{\sqrt{N}}$	$\bar{X} + 1.64\dfrac{s}{\sqrt{N}}$
Large	Mean	95%	$\bar{X} - 1.96\dfrac{s}{\sqrt{N}}$	$\bar{X} + 1.96\dfrac{s}{\sqrt{N}}$
Large	Mean	99%	$\bar{X} - 2.58\dfrac{s}{\sqrt{N}}$	$\bar{X} + 2.58\dfrac{s}{\sqrt{N}}$
Small	Mean	95%	$\bar{X} - t_{0.025}\dfrac{s}{\sqrt{N}}$	$\bar{X} + t_{0.025}\dfrac{s}{\sqrt{N}}$
Large	Standard deviation	95%	$\dfrac{s}{1 + \dfrac{1.96}{\sqrt{2N}}}$	$\dfrac{s}{1 - \dfrac{1.96}{\sqrt{2N}}}$
Large	Proportion	95%	$p - 1.96\sqrt{\dfrac{p(1-p)}{N}}$	$p + 1.96\sqrt{\dfrac{p(1-p)}{N}}$

2. Size of sample

$$N = \left(\frac{1.96\sigma}{e}\right)^2$$

3. Mean of sampling proportions π
4. Standard deviation of sampling proportion

$$\sqrt{\frac{\pi(1-\pi)}{N}}$$

You should now be able to demonstrate your knowledge of the following ideas presented in this chapter by giving definitions, descriptions, or specific examples. Page references are given for each term so that you can check your answer.

Statistical estimation (page 315)
Hypothesis testing (page 316)
Population parameter (page 316)
Point estimate (page 316)
Interval estimate (page 317)
Confidence interval (page 317)
Confidence limits (page 317)
Degree of confidence (page 317)

Student's t–distribution (page 322)
Number of degrees of freedom (page 323)
Unbiased estimator (page 328)
Maximum allowable error (page 329)
Sample proportion (page 332)
True population proportion (page 332)

The tests of the following section will be more useful if you take them after you have studied the examples and solved the exercises in this chapter.

MASTERY TESTS

Form *A*

1. The lower and upper boundaries of a confidence interval are called _____.
2. True or false: As we increase the degree of confidence, that is, degree of certainty, the confidence interval will become narrower.
3. The *t*–distribution is symmetrical about the mean. The value of the mean is *a.* 0 *b.* −1 *c.* 1 *d.* 3 *e.* not enough information given
4. The shape of the *t*–distribution depends upon a parameter called *a.* the mean *b.* the median *c.* the standard deviation *d.* the degrees of freedom *e.* none of these
5. In a *t*–distribution 95% of the area under the curve will fall between *a.* $-t_{0.025}$ and $t_{0.025}$ *b.* $-t_{0.05}$ and $t_{0.05}$ *c.* $-t_{0.01}$ and $t_{0.01}$ *d.* $-t_{0.10}$ and $t_{0.10}$ *e.* none of these
6. True or false: $\bar{X}$ is an unbiased estimator of μ.
7. We can obtain an unbiased estimate of the population standard deviation by dividing $\Sigma(X - \bar{X})^2$ by *a.* N *b.* N − 1 *c.* N − 2 *d.* N + 2 *e.* none of these
8. True or false: The sample proportions of a large population will be normally distributed.
9. When standardized normal charts are used in analyzing proportions, the formula $z = \dfrac{x - \mu}{\sigma}$ becomes $z = $ _____.
10. The student's *t*–distribution is used when we wish to estimate *a.* the population mean *b.* the population standard deviation *c.* the sample size *d.* the population proportion *e.* none of these

Form *B*

1. Several students are in the student lounge discussing the amount of reading necessary to pass their history course. Five students indicate that the number of pages assigned by their teachers is 419, 378, 520, 480, and 363 respectively. Find a 95% confidence interval for the average number of pages assigned by a history teacher.
2. A random sample of 500 families in Smogsville indicated that 285 of them owned color television sets. Find a 95% confidence interval for the true proportion of all families living in Smogsville who own color television sets.
3. A sample of 36 depositors of the Bank of America found that the amount of money on deposit is $6000 with a standard deviation of $490. Find a 95% confidence interval for the average amount of money on deposit in this bank.
4. A senator claims that 60% of the voters are in favor of lowering the liquor tax. A random survey of 2000 voters from her community is taken. What is the probability that the proportion of people favoring this proposal is greater than 58%?
5. Seven returning passengers are randomly selected and asked to indicate how

many pictures or slides each snapped while vacationing in Europe. Their answers are 72, 90, 55, 12, 68, 54, and 69. Find a 95% confidence interval for the average number of pictures taken by these passengers while vacationing in Europe.

6. Eight members of a health club are randomly selected and asked to indicate the number of pounds that each lost in a month since joining the club. The results are 5, 15, 25, 0, 12, 4, 2, and 9 pounds. Find a 95% confidence interval for the average weight loss of a club member.

7. A science teacher has been keeping records on how long it takes a student to complete a particular experiment. For a group of 100 students the average time was 18 minutes with a standard deviation of 2.8 minutes. Find a 99% confidence interval for the mean time needed by a student to complete the experiment.

8. A television repairman wishes to estimate the average life of a picture tube. How large a sample should he take if he wishes his estimate to be within 0.28 years of the true value? Assume that $\sigma = 1.9$.

9. A manufacturer claims that 6% of the clothing produced by his company is labeled incorrectly with the wrong size. To test this claim a consumer's group takes a random sample of 600 dresses produced by this company. What is the probability that the proportion of incorrectly labeled dresses is between 3% and 9%?

10. A teacher wishes to estimate the average amount of time spent by a student on homework assignments. How large a sample should the teacher select if the teacher wishes the estimate to be within 0.78 hours of the true value? Assume that $\sigma = 1.88$ hours.

11. An airline company claims that 95% of all its planes arrive on time. A random sample of 88 plane arrivals is taken. What is the probability that fewer than 93% of the planes arrive on time?

12. The management of a ski resort wishes to know the average number of days a guest stays. It randomly selects 400 guests who registered previously and determines that the average length of stay is 3.7 days with a standard deviation of 1.6 days. Find a 95% confidence interval for the mean stay of all guests.

SUGGESTED READING

Adler, H. and Roessler, E. B. *Introduction to Probability and Statistics*. 5th ed. San Francisco, Calif: Freeman, 1972.

Freund, John. *Statistics, A First Course*. Englewood Cliffs, N.J.: Prentice-Hall, 1970.

Gallup Opinion Index. Princeton, New Jersey: American Institute of Public Opinion, January 1971.

Hogg, Robert and Craig, Allen. *Introduction to Mathematical Statistics*. 3rd ed. New York: MacMillan, 1970.

Mood, A. M. *Introduction to the Theory of Statistics*. New York: McGraw-Hill, 1967.

12

HYPOTHESIS
TESTING

Table courtesy of The
Metropolitan Life Insurance
Company.

U.S. News & World Report, February 9,
1976. Reprinted from *U.S. News &
World Report,* Copyright © 1976 U.S.
News & World Report, Inc.

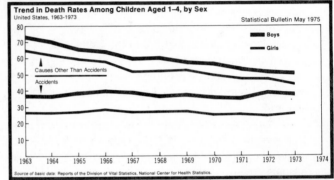

Trend in Death Rates Among Children Aged 1–4, by Sex
United States, 1963-1973 · Statistical Bulletin May 1975

Source of basic data: Reports of the Division of Vital Statistics, National Center for Health Statistics.

FEWEST HIGHWAY DEATHS IN A DOZEN YEARS

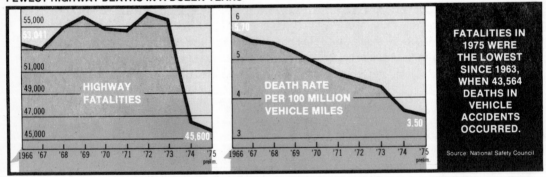

FATALITIES IN
1975 WERE
THE LOWEST
SINCE 1963,
WHEN 43,564
DEATHS IN
VEHICLE
ACCIDENTS
OCCURRED.

Source: National Safety Council

The top clipping indicates the trend in death rates, by sex, among children aged 1–4 years. Would you say that the average number of deaths as a result of accidents is significantly higher for boys than for girls? What about child deaths not resulting from accidents. Is the observed difference significant?

The bottom clipping indicates that the number of highway fatalities is steadily decreasing. The death rate per 100 million vehicle miles is also decreasing. In 1975 it was 3.50. One midwestern state had a death rate of 2.89 per 100 million vehicle miles. Is this state significantly below the national average?

Preview

In this chapter we will discuss the following topics:

1. *Hypothesis Testing*
We analyze how sample data can be used to reject or accept a claim about some aspect of a probability distribution. The claim to be tested is called the null hypothesis.

2. *Test Statistic*
This is a number that we compute to determine when to reject a given claim, that is, the null hypothesis.

3. *Critical Rejection Regions*
We reject a null hypothesis when the test statistic value falls within this region. How these regions are set up depends upon the specifications given within the problem.

4. *Two Types of Error*
We discuss the two errors that can be made when we use sample data to accept or reject a null hypothesis. We may incorrectly reject a true hypothesis or we may incorrectly accept a false hypothesis. In both cases an error is made.

5. *Level of Significance*
Even if we do not reject a claim, this does not mean that the sample data supports the claim. Sample data cannot prove a claim. However,
sample data can disprove a claim with a certain degree of probability.

6. *Tests Concerning Means, Differences between Means and Proportions*
We discuss the test statistics used when we wish to use sample data to determine whether observed differences between means and proportions are significant.

7. *Applications*
We apply the hypothesis testing procedures to wide-ranging problems.

Introduction

Many television commercials contain unusual performance claims. For example, consider the following commercials:

1. Four out of 5 dentists recommend Brand X sugarless gum for their patients who chew gum.

2. A particular car lasts an average of 11 years on American roads.

3. Detergent X produces the cleanest wash.

4. Brand X paper towels are stronger and more absorbent.

How much confidence can one have in such claims? Can they be verified statistically? Fortunately, in many cases the answer is yes. Samples are taken and claims are tested. We can then make a decision on whether to accept or to reject a claim on the basis of sample information. This process is called **hypothesis testing.** Perhaps this is one of the most important uses of samples.

As we indicated in Chapter 11, hypothesis testing is an important branch of statistical inference. Sample data provides us with estimates of population parameters. These estimates are in turn used in arriving at a decision to either accept or reject an hypothesis. By an **hypothesis** we shall mean an assumption about one or more of the population parameters that will either be accepted or rejected on the basis of the information obtained from a sample. In this chapter we will discuss methods for determining whether to accept or reject any hypothesis on the basis of sample data.

12.1 TESTING AGAINST AN ALTERNATE HYPOTHESIS

Suppose several players are in a gambling casino rolling a die. A bystander notices that in the first 120 rolls of the die, a 6 showed only 8 times. Is this reasonable or is the die loaded? The management claims that this unusual occurrence is to be attributed purely to chance and that the die is an honest die. The bystander claims otherwise.

If the die is an honest die, then Formula 7.5 (see page 213) for a binomial distribution tells us that the average number of 6s occurring in 120 rolls of the die is 20:

$$\mu = np$$
$$= 120(1/6)$$
$$= 20$$

If the die is loaded, then $\mu \neq 20$. (The symbol $\neq$ means "is not equal to.") Since the die is either an honest die or a loaded die, we must choose between the hypothesis $\mu = 20$ and the hypothesis $\mu \neq 20$. Thus, sample data will be used to either accept or reject the hypothesis $\mu = 20$. Such an hypothesis is called a **null hypothesis** and is denoted by H_0. Any hypothesis which differs from the null hypothesis is called an **alternate hypothesis** and is denoted as $H_1, H_2, \ldots$, etc. In our example

$$\text{null hypothesis, } H_0\colon \ \mu = 20$$
$$\text{alternate hypothesis, } H_1\colon \ \mu \neq 20$$

Notice that by formulating the alternate hypothesis as $\mu \neq 20$, we are indicating that we wish to perform a **two-sided, or two-tailed, test.** This means that if the die is not honest, then it may be loaded in favor of obtaining 6s more often than is expected or less often than is expected.

In our example the bystander strongly suspects that the die is loaded against obtaining 6s. Thus, his alternate hypothesis would be $\mu < 20$. (The symbol $<$ stands for "is less than".) Similarly if the bystander suspected that the die was loaded in favor of obtaining 6s more often than is expected, the alternate hypothesis would be $\mu > 20$. (The symbol $>$ stands for "is greater than.") In each of these cases the null hypothesis remains the same, H_0: $\mu = 20$. Such alternate hypotheses indicate that we wish to perform a **one-sided, or one-tailed, test.**

A decision as to whether to accept or reject the null hypothesis will be made on the basis of sample data. How is such a decision to be made? We must realize that even if we know for sure that the die is honest, it is very unlikely that we would get exactly twenty 6s in the 120 rolls of the die. Moreover, if we were to roll the die 120 times on many different occasions, we would find that the number of 6s appearing is around 20, sometimes more and sometimes less. It is therefore obvious that we must set up some interval which we will call the **acceptance region.** If the number of 6s appearing in 120 rolls of the die is within this acceptance region, then we will accept the null hypothesis. If the number of 6s obtained is outside this region, then we will reject the null hypothesis that the die is an honest die. The possibilities are shown in Figures 12.1, 12.2, and 12.3. The value that separates the rejection region from the acceptance region is called the **critical value.**

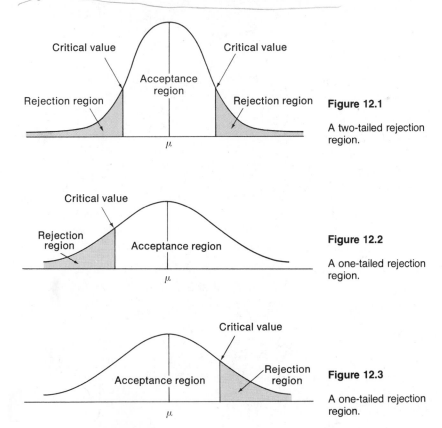

Figure 12.1

A two-tailed rejection region.

Figure 12.2

A one-tailed rejection region.

Figure 12.3

A one-tailed rejection region.

The type of symbol in the alternate hypothesis tells us what type of rejection region to use as shown in the following table:

If the symbol in the alternate hypothesis is	$<$	$\neq$	$>$
then the rejection region consists of	one region on the left side	two regions, one on each side	one region on the right side

Suppose we decide to accept the null hypothesis if the number of 6s obtained is between 15 and 25. Since in our case only eight 6s were obtained, we would reject the null hypothesis. In this case the acceptance region is 15 to 25. The two-tailed rejection region corresponds to the two tails, less than 15 and more than 25, as shown in Figure 12.4. When we reject a null hypothesis, we are claiming that the value of the population parameter, that is, the average number of 6s, is some value other than the one specified in the null hypothesis. Also, when the sample data indicates that we should reject a null hypothesis, we say that the observed difference is **significant.**

Our discussions in the last few paragraphs lead us to the following definitions.

DEFINITION 12.1

The **null hypothesis,** denoted by H_0, is the statistical hypothesis being tested.

DEFINITION 12.2

The **alternate hypothesis** denoted by $H_1, H_2 \ldots$, is the hypothesis that will be accepted when the null hypothesis is rejected.

DEFINITION 12.3

A **one-sided,** or **one-tailed, test** is a statistical test which has the rejection region located in the left tail *or* the right tail of the distribution.

DEFINITION 12.4

A **two-sided,** or **two-tailed, test** is a statistical test which has the rejection region located in both tails of the distribution.

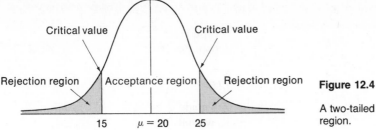

<figure>
Critical value

Critical value

Rejection region   Acceptance region   Rejection region

15     $\mu = 20$ 25
</figure>

Figure 12.4

A two-tailed rejection region.

Let us summarize the steps that must be followed in hypothesis testing:

1. State the null hypothesis which indicates the value of the population parameter to be tested.

2. State the alternate hypothesis which indicates the belief that the population parameter has a value other than the one specified in the null hypothesis.

3. Set up rejection and acceptance regions for the null hypothesis. The **region of rejection** is called the **critical region.** The remaining region is called the **acceptance region.**

4. Compute the value of the test statistic. A **test statistic** is a calculated number that is used to decide whether to reject or accept the null hypothesis. The formula for computing the value of the test statistic depends upon the parameter we are testing.

5. Reject the null hypothesis if the test statistic value falls within the rejection, that is, the critical, region. Otherwise accept the null hypothesis.

In this chapter we will be discussing various tests which enable us to decide whether to reject or accept a null hypothesis. Such tests are called **statistical tests of hypotheses** or **statistical tests of significance.**

12.2 TWO TYPES OF ERRORS

Since any decision to either accept or reject a null hypothesis is to be made on the basis of information obtained from sample data, there is a chance that we will be making an error. There are two possible errors that we can make. We may reject a null hypothesis when we really should accept it. Thus, returning to the die problem of Section 12.1 we may reject the claim that the die is honest even though it actually is honest. Alternately we may accept a null hypothesis when we should reject it. Thus, we may say that the die is honest when it really is a loaded die.

These two errors are referred to as a **Type-I** and a **Type-II error** respectively. In either case we have made a wrong decision. We define these formally as follows:

DEFINITION 12.5

A **Type-I error** is made when a true null hypothesis is rejected, that is, we reject a null hypothesis when we should accept it.

DEFINITION 12.6

A **Type-II error** is made when a false null hypothesis is accepted, that is, we accept a null hypothesis when we should reject it.

In the following box we indicate how these two errors are made:

	and we claim that	
	H_0 is true	H_0 is false
If H_0 is true	correct decision (no error)	Type-I error
H_0 is false	Type-II error	Correct decision (no error)

When deciding upon whether to accept or reject a null hypothesis, we always wish to minimize the probability of making a Type-I error or a Type-II error. Unfortunately the relationship between the probabilities of the two types of errors is of such a nature that if we reduce the probability of making one type of error, we usually increase the probability of making the other type. In most applied problems one type of error is more serious than the other. In such situations careful attention is given to the more serious error.

How much of a risk should a statistician take in rejecting a true hypothesis, that is, in making a Type-I error? Generally speaking, statisticians use the limits of 0.05 and 0.01. Each of these limits is called a **level of significance** or **significance level.** We have the following definition:

DEFINITION 12.7

The **significance level** of a test is the probability that the test statistic falls within the rejection region.

The **0.05 level of significance** is used when the statistician wishes that the risk of rejecting a true null hypothesis not exceed 0.05. The **0.01 level of significance** is used when the statistician wishes that the risk of rejecting a true null hypothesis not exceed 0.01.

In this book we will usually assume that we wish to correctly accept the null hypothesis 95% of the time and to incorrectly reject it only 5% of the time. Thus, the maximum probability of a Type-I error that we are willing to accept, that is, the significance level, will be 0.05. *The probability of making a Type-I error is denoted by the Greek letter α (pronounced alpha).* Therefore, the probability of making a correct decision is $1 - \alpha$.

As we indicated on page 345, when dealing with one-tailed tests, the critical region lies to the left or to the right of the mean. This is shown in Figure 12.5. When dealing with a two-tailed test, one-half of the critical region is to the left of the mean and one-half is to the right. The probability of making a Type-I error is evenly divided between these two tails, as shown in Figure 12.6.

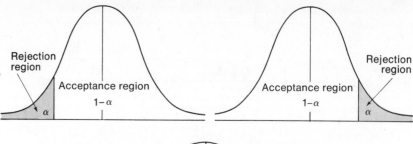

Figure 12.5

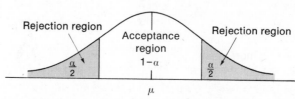

Figure 12.6

Comment   If the test statistic falls within the acceptance region, we do not reject the null hypothesis. When a null hypothesis is not rejected, this does not mean that what the null hypothesis claims is guaranteed to be true. It simply means that on the basis of the information obtained from the sample data there is not enough evidence to reject the null hypothesis.

12.3   TESTS CONCERNING MEANS

In this section we will discuss methods for determining whether we should accept or reject a null hypothesis about the mean of a population. We will illustrate the procedure with several examples.

Suppose a manufacturer claims that each family size bag of pretzels he sells weighs 12 ounces on the average, with a standard deviation of 0.8 ounces. A consumer's group decides to test this claim by accurately weighing 49 randomly selected bags of pretzels. If the mean weight of the sample is considerably different from the population mean, the manufacturer's claim will definitely be rejected. Thus, if the mean weight is 30 ounces or 5 ounces, the manufacturer's claim will be rejected. Only when the sample mean is close to the claimed population mean do we need statistical procedures to determine when to reject or accept a null hypothesis.

Let us assume that the sample mean of the 49 randomly selected bags of pretzels is 11.8 ounces. Since the sample mean, 11.8, is not the same as the population mean, 12, we wish to test the manufacturer's claim at the 5% level of significance.

The population parameter being tested in this case is the mean weight, μ, and the questioned value is 12 ounces. Thus,

$$\text{null hypothesis, } H_0: \mu = 12$$
$$\text{alternate hypothesis, } H_1: \mu \neq 12$$

The alternate hypothesis of not equal suggests a two-tailed rejection region. Therefore, the α of 0.05 is split equally between the two tails, as shown in Figure 12.7.

We now look in Table V to find which z-value has 0.4750 of the area between $z = 0$ and this z-value. From Table V we find that the z-value is 1.96. We label this on the diagram in Figure 12.7 and get the acceptance-rejection diagram shown in Figure 12.8.

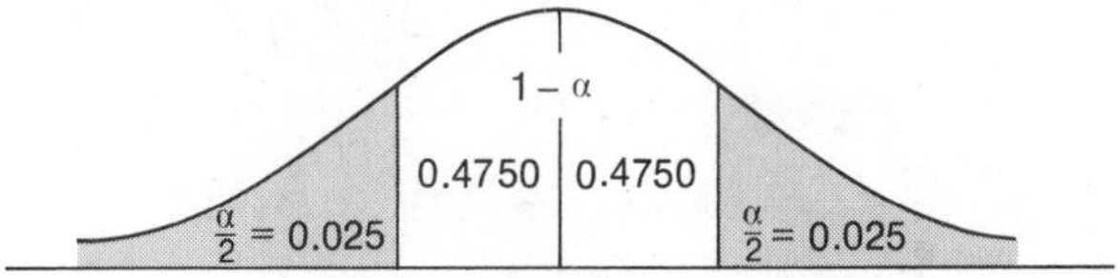

Figure 12.7

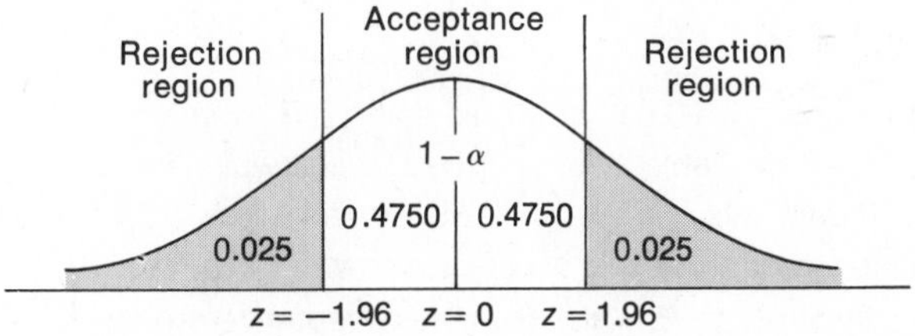

Figure 12.8

Since the Central Limit Theorem tells us that the sample means are normally distributed, we use

$$z = \frac{\bar{X} - \mu}{\sigma/\sqrt{N}}$$

as the test statistic and reject the null hypothesis if the value of the test statistic falls in the rejection region. In this test statistics, $\bar{X}$ is the sample mean and μ is the population mean as claimed in the null hypothesis. In our case we have

$$\begin{aligned} z &= \frac{\bar{X} - \mu}{\sigma/\sqrt{N}} \\ &= \frac{11.8 - 12}{0.8/\sqrt{49}} \\ &= \frac{-0.2}{0.8/7} \\ &= -1.75 \end{aligned}$$

Since this calculated value of z falls within the acceptance region, our decision is that we cannot reject H_0. The difference between the sample mean and the assumed value of the population mean is due purely to chance. We say that the difference is *not statistically significant.*

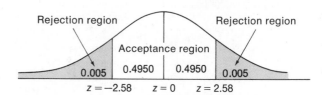

Figure 12.9

Figure 12.10

If the level of significance had been 0.01, we would split the 0.01 into two equal tails as shown in Figure 12.9. From Table V we find the z–value which has 0.4950 of the area between $z = 0$ and this z–value is 2.58. Thus, we reject the null hypothesis if the test statistic falls in the critical region shown in Figure 12.10.

Since we obtained a z–value of -1.75 which is in the acceptance region, we do not reject the null hypothesis.

Let us summarize the testing procedure outlined in the previous paragraphs.

1. First convert the sample mean into standard units using

$$z = \frac{\overline{X} - \mu}{\sigma/\sqrt{N}}$$

2. Then reject the null hypothesis about the population mean if z is less than -1.96 or greater than 1.96 when using a 5% level of significance.

3. If using a 1% level of significance, reject the null hypothesis if z is less than -2.58 or greater than 2.58.

We further illustrate the procedure with several examples.

EXAMPLE 1

A light bulb company claims that the 60 watt light bulb it sells has an average life of 1000 hours with a standard deviation of 75 hours. Sixty-four new bulbs were allowed to burn out to test this claim. The average lifetime of these bulbs was found to be 975 hours. Does this indicate that the average life of a bulb is not 1000 hours? Use a 5% level of significance.

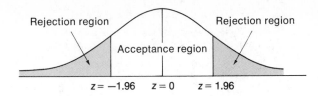

| z = −1.96 | z = 0 | z = 1.96 | **Figure 12.11** |

SOLUTION

In this case the population parameter being tested is μ, the average life of a bulb, and the value questioned is 1000. Since we are testing whether the average life of a bulb is or is not 1000 hours, we have

$$H_0: \mu = 1000$$
$$H_1: \mu \neq 1000$$

We are given that $\overline{X} = 975$, $\sigma = 75$, $\mu = 1000$, and $N = 64$. We first calculate the value of the test statistic, z. We have

$$z = \frac{\overline{X} - \mu}{\sigma/\sqrt{N}}$$

$$= \frac{975 - 1000}{75/\sqrt{64}}$$

$$= -2.67$$

We use the two-tailed rejection region shown in Figure 12.11. The value of $z = -2.67$ falls in the rejection region. Thus, we reject the null hypothesis that the average life of a bulb is 1000 hours.

Perhaps you are wondering why we used $\mu \neq 1000$ as the null hypothesis and not $\mu < 1000$. After all, who cares if the average life of a bulb is more than 1000 hours. The answer is that the manufacturer cares. When a manufacturer claims that the average life of a bulb is 1000 hours, he is concerned when bulbs last more or less than 1000 hours. If the mean life is less than 1000 hours, the manufacturer will lose business and consumer confidence. If the mean life is more than 1000 hours, the company will lose money.

EXAMPLE 2

A bank teller at a savings bank claims that the average amount of money on deposit in a savings account at this bank is $4800 with a standard deviation of $460. A random sample of 36 accounts is taken to test this claim. The average of these accounts is found to be $5000. Does this sample indicate that the average amount of money on deposit is not $4800? Use a 5% level of significance.

SOLUTION

In this case the population parameter being tested is μ, the average amount of money on deposit in a savings account. The value questioned is $4800. Since we are testing whether the average amount of money on deposit is $4800 or not, we have

$$H_0: \mu = 4800$$
$$H_1: \mu \neq 4800$$

We are given that $\bar{X} = 5000$, $\mu = 4800$, $\sigma = 460$, and $N = 36$ so that

$$z = \frac{\bar{X} - \mu}{\sigma/\sqrt{N}}$$

$$= \frac{5000 - 4800}{460/\sqrt{36}}$$

$$= 2.61$$

We use the same two-tailed rejection region as shown in Figure 12.11. The value of $z = 2.61$ falls in the rejection region. Thus, we reject the null hypothesis that the average amount of money on deposit is $4800.

EXAMPLE 3

The average score of a sixth-grader in a certain school district on the 1–2–3 math aptitude exam is 75 with a standard deviation of 8.1. A random sample of 100 students in one school was taken. The mean score of these 100 students was 71. Does this indicate that the students of this school are significantly less skilled in their mathematical abilities? Use a 5% level of significance.

SOLUTION

In this case the population parameter being tested is μ, the mean score on the math aptitude exam. The value questioned is $\mu = 75$. However, unlike the previous examples the sample data suggests

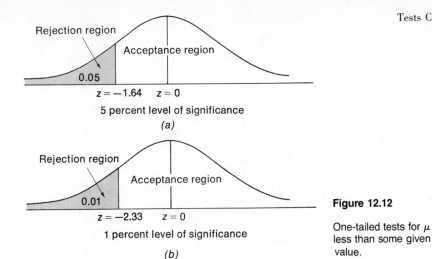

Rejection region

Acceptance region

0.05

$z = -1.64$ $z = 0$

5 percent level of significance

(a)

Rejection region

Acceptance region

0.01

$z = -2.33$ $z = 0$

1 percent level of significance

(b)

Figure 12.12

One-tailed tests for μ
less than some given
value.

that for this particular school the mean score is really less than 75. Thus, it is reasonable to set up a one-sided, or a one-tailed, test with the alternate hypothesis being that the population mean for this school is less than 75. We have

$$H_0: \mu = 75$$
$$H_1: \mu < 75$$

When dealing with one-tail (left-side) alternate hypotheses, we have the rejection regions illustrated in Figure 12.12. These values, like those in the two-tailed tests, are obtained from Table V. You should verify these results.

Now we can calculate the test statistic, z. Here $\bar{X} = 71$, $\mu = 75$, $\sigma = 8.1$, and $N = 100$. We have

$$z = \frac{\bar{X} - \mu}{\sigma/\sqrt{N}}$$

$$= \frac{71 - 75}{8.1/\sqrt{100}}$$

$$= -4.94$$

Since the z-value of -4.94 is in the rejection region, we conclude that the students of this school are significantly less skilled in their mathematical abilities.

EXAMPLE 4

The We-Haul Trucking Corp. claims that the average hourly salary of its mechanics is $9.25 with a standard deviation of $1.55. A random sample of 81 mechanics showed that the average hourly salary of these mechanics was only $8.95. Does this indicate that the average hourly salary of a mechanic is less than $9.25? Use a 1% level of significance.

SOLUTION

In this case the population parameter being tested is μ, the mean hourly salary. The questioned value is 9.25. The sample data suggests that $\mu < 9.25$. Thus, we will use a one-tailed test. We have

$$H_0: \mu = 9.25$$
$$H_1: \mu < 9.25$$

Here we are given that $\bar{X} = 8.95$, $\mu = 9.25$, $\sigma = 1.55$, and $N = 81$, so that

$$z = \frac{\bar{X} - \mu}{\sigma/\sqrt{N}}$$

$$= \frac{8.95 - 9.25}{1.55/\sqrt{81}}$$

$$= -1.74$$

We use the same one-tailed rejection region as shown in Figure 12.12. The value of $z = -1.74$ falls in the acceptance region. Thus, the sample data does not provide us with sufficient justification to reject the null hypothesis. We will therefore accept the null hypothesis that $\mu = \$9.25$.

EXAMPLE 5

A trash company claims that the average weight of any of its fully loaded garbage trucks is 11,000 pounds with a standard deviation of 800 pounds. A highway department inspector decides to check on this claim. He randomly checks 36 trucks and finds that the average weight of these trucks is 12,500 pounds. Does this indicate that the average weight of a garbage truck is more than 11,000 pounds? Use a 5% level of significance.

SOLUTION

In this case the population parameter being tested is μ, the average weight of a garbage truck. The value questioned is 11,000 pounds. The sample data suggests that the mean weight is really more than 11,000 pounds. Thus, the alternate hypothesis will be that the population mean is more than 11,000 pounds. We have

$$H_0: \mu = 11,000$$
$$H_1: \mu > 11,000$$

When dealing with a one-tail (right-side) alternate hypothesis, we have the rejection regions shown in Figure 12.13. Here we are given that $\bar{X} = 12,500$, $\mu = 11,000$, $\sigma = 800$, and $N = 36$, so that

$$z = \frac{\bar{X} - \mu}{\sigma/\sqrt{N}}$$

$$= \frac{12,500 - 11,000}{800/\sqrt{36}}$$

$$= 11.25$$

Since the z–value of 11.25 falls within the rejection region, we conclude that the average weight of a garbage truck is not 11,000 pounds. We reject the null hypothesis.

5 percent level of significance

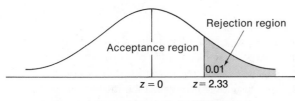

1 percent level of significance

Figure 12.13

One-tailed tests for μ greater than some given value.

EXAMPLE 6

An insurance company advertises that it takes 21 days on the
average to process an auto accident claim. The standard deviation
is 8 days. To check on the truth of this advertisement, a group of
investigators randomly selects 35 people who recently filed claims.
They find that it took the company an average of 24 days to
process these claims. Does this indicate that it takes the insurance
company more than 21 days on the average to process a claim?
Use a 1% level of significance.

SOLUTION

In this case the population parameter being tested is μ, the average
number of days needed to process a claim. The questioned value
is $\mu = 21$. The sample data suggests that $\mu > 21$. Thus, we will use
a one-tailed test. We have

$$H_0: \ \mu = 21$$
$$H_1: \ \mu > 21$$

Here we are given that $\bar{X} = 24$, $\mu = 21$, $\sigma = 8$, and $N = 35$, so that

$$z = \frac{\bar{X} - \mu}{\sigma/\sqrt{N}}$$

$$= \frac{24 - 21}{8/\sqrt{35}} = 2.22$$

Since the z-value of 2.22 falls within the acceptance region, we
cannot reject the null hypothesis. Thus, we cannot conclude that it
takes the insurance company more than 21 days to process a
claim.

In all the examples discussed so far, the sizes of the samples were large
enough to justify the use of the normal distribution. When the sample size is small,
we use the t-distribution instead of the normal distribution. The following examples
will illustrate how the t-distribution is used in hypothesis testing.

EXAMPLE 7

A manufacturer claims that each can of mixed nuts he sells contains
an average of 10 cashew nuts. A sample of 15 cans of these mixed
nuts has an average of 8 cashew nuts with a standard deviation of
3. Does this indicate that we should reject the manufacturer's claim?
Use a 5% level of significance.

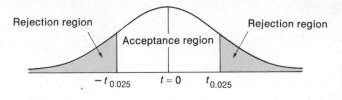

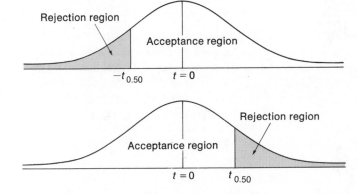

Figure 12.14

Two-tailed small-sample rejection region (five percent level of significance).

Figure 12.15

One-tailed small sample rejection region (5% level of significance).

SOLUTION

Since the sample size is only 15, the test statistic becomes

$$t = \frac{\bar{X} - \mu}{s/\sqrt{N}}$$

instead of z. Depending upon the number of degrees of freedom, we have the acceptance-rejection regions shown in Figures 12.14 and 12.15. In each case the value of t is obtained from Table VIII in the Appendix. It depends upon the number of degrees of freedom, which is $N - 1$.

Let us now return to our problem. The population parameter being tested is μ, the average number of cashew nuts in a can of mixed nuts. The questioned value is 10. We have

$$H_0: \mu = 10$$
$$H_1: \mu \neq 10$$

We will use a two-tailed rejection region as shown in Figure 12.14. Here we are given that $\bar{X} = 8$, $\mu = 10$, $s = 3$, and $N = 15$, so that

$$t = \frac{\bar{X} - \mu}{s/\sqrt{N}} = \frac{8 - 10}{3/\sqrt{15}} = -2.58$$

From Table VIII we find that the $t_{0.025}$ value for 15 − 1, or 14, degrees of freedom is 2.145, which means that we reject the null hypothesis if the test statistic is less than −2.145 or greater than 2.145. Since $t = -2.58$ falls within the critical region, we reject the manufacturer's claim that each can of mixed nuts contains an average of 10 cashew nuts.

EXAMPLE 8

A new weight reducing pill is being sold in a midwestern town. The manufacturer claims that any overweight person who takes this pill as directed will lose 15 pounds within a month. To test this claim, a doctor gives this pill to 6 overweight people and finds that they lose an average of only 12 pounds with a standard deviation of 4 pounds. Can we accept the manufacturer's claim? Use a 5% level of significance.

SOLUTION

In this case the population parameter being tested is μ, the average number of pounds lost when using this pill. The questioned value is $\mu = 15$. We have

$$H_0: \mu = 15$$
$$H_1: \mu < 15$$

Here we are given that $\bar{X} = 12$, $\mu = 15$, $s = 4$, and $N = 6$, so that

$$t = \frac{\bar{X} - \mu}{s/\sqrt{N}} = \frac{12 - 15}{4/\sqrt{6}} = -1.84$$

From Table VIII we find the $t_{0.05}$ value for 6 − 1, or 5, degrees of freedom is 2.015, which means that we reject the null hypothesis if the test statistic is less than −2.015. Since $t = -1.84$ does not fall within the rejection region, we cannot reject the manufacturer's claim that the average number of pounds lost when using this pill is 15.

EXERCISES

1. According to the manufacturer, the average life of a movie projector bulb is 25 hours with a standard deviation of 1.69 hours. A survey of 50 bulbs showed that the average life of these bulbs is 23 hours. Should we reject the manufacturer's claim? Use a 5% level of significance.

2. City officials claim that the typical fireman retires after an average of 23.2 years on the job. The standard deviation is 3.8 years. A survey of 45 retired firemen indicated that these firemen retired after 26.3 years. Should we reject the city officials' claim? Use a 1% level of significance.

3. College officials state that the entire registration process takes an average of 56 minutes with a standard deviation of 14.9 minutes. Forty students are randomly selected and it is found that for these students the registration process took 60 minutes. Does this indicate that the average time needed for registration is more than 56 minutes? Use a 5% level of significance.

4. Sally is stranded on a highway awaiting the arrival of an AAA truck to repair her car. The local AAA club says that it responds to emergency calls in an average of 30 minutes with a standard deviation of 7.2 minutes. After speaking to 30 friends who were in similar situations, Sally finds that the AAA responded to calls in an average of 35 minutes. Does this indicate that the average response time is more than 30 minutes? Use a 5% level of significance.

5. A sociologist claims that the average age at which a woman first gives birth in Fayetteville is 26 years with a standard deviation of 6.2 years. The records of 100 births at a maternity hospital are examined. It is found that the average age at which a woman has her first child is only 24.7 years. Should we reject the sociologist's claim? Use a 1% level of significance.

6. A drug manufacturer claims that 90% of all women who use the X-9 birth control pill as directed will not become pregnant. A sample of 100 women is taken and it is found that 83 of the women did not become pregnant. Does this indicate that the drug manufacturer's claim about the effectiveness of the birth control pill should be rejected? Use a 5% level of significance.

7. The game warden of a large park claims that the average length of a trout fish is 19 inches with a standard deviation of 3 inches. The average length of 58 trout fish caught by several fishermen is found to be only 16 inches. Does this indicate that the average length is less than 19 inches? Use a 5% level of significance.

8. Bank officials claim that the average new car loan is $4000 with a standard deviation of $600. To check on this claim, a new loan officer took a sample of 36 car loans and found that the average car loan is $3800. Does this indicate that the average car loan is less than $4000? Use a 1% level of significance.

9. The manufacturer of a certain foreign car sold in the United States claims that it will average 36 miles per gallon of gasoline. To test this claim, a consumer's group randomly selects ten cars and drives them under normal driving conditions. These cars average 29 miles to the gallon with a standard deviation of 8.3 miles. Should we accept the manufacturer's claim? Use a 5% level of significance.

10. A claim is made in a college newspaper that the average age of a college student is 26.2 years with a standard deviation of 7.86 years. A statistics teacher believes that the average age is considerably lower than 26.2 years. To support her belief, the teacher asks each of the 44 students in her class to indicate their age on a piece of paper. She finds that the average age is 24.8 years. Should we reject the newspaper claim? Use a 5% level of significance.

11. The average life expectancy of a cat is 13 years with a standard deviation of 6 years. A veterinarian claims to have developed a drug which increases the life expectancy of a cat by 8 years. To test this claim, another veterinarian injects 12 cats with this drug at birth. The average life of these cats was 15.4 years. Is this drug effective? Use a 5% level of significance.

12.4   TESTS CONCERNING DIFFERENCES BETWEEN MEANS

There are many instances in which we must decide whether the observed difference between two sample means is due purely to chance or whether the population means from which these samples were selected are really different. For example, suppose a teacher gave an IQ test to 50 girls and 50 boys and obtained the following test scores:

	Boys	Girls
Mean	78	81
Standard deviation	7	9

Is the observed difference between the scores significant? Are the girls smarter?

In problems of this sort the null hypothesis is that there is no difference between the means. Since we will be discussing more than one sample, we use the following notation. We let $\bar{X}_1$, s_1, and N_1 be the mean, standard deviation, and sample size respectively of one of the samples, and $\bar{X}_2$, s_2, and N_2 be the mean, standard deviation, and sample size respectively of the second sample. Decisions as to whether to reject or accept the null hypotheses are then based upon the test statistic z, where

$$ z = \frac{\bar{X}_1 - \bar{X}_2}{\sqrt{\dfrac{s_1^2}{N_1} + \dfrac{s_2^2}{N_2}}} $$

Depending upon whether the alternate hypothesis is $\mu_1 - \mu_2 \neq 0$, $\mu_1 < \mu_2$, or $\mu_1 > \mu_2$, we have a two-sided test or a one-sided test as indicated in Section 12.3. The following examples will illustrate how this test statistic is used.

EXAMPLE 1

Consider the example discussed at the beginning of this section. Is the observed difference between the two IQ scores significant? Use a 5% level of significance.

SOLUTION

Let $\bar{X}_1$, s_1, and N_1 represent the boy's mean score, standard deviation, and sample size and let $\bar{X}_2$, s_2, and N_2 be the corresponding girls' scores. Then the problem is whether the observed difference between the sample means is significant or not. Thus

$$ H_0: \mu_1 = \mu_2 $$
$$ H_1: \mu_1 \neq \mu_2 $$

Here we are given that $\bar{X}_1 = 78$, $s_1 = 7$, $N_1 = 50$, $\bar{X}_2 = 81$, $s_2 = 9$, and $N_2 = 50$, so that

$$ z = \frac{\bar{X}_1 - \bar{X}_2}{\sqrt{\dfrac{s_1^2}{N_1} + \dfrac{s_2^2}{N_2}}} = \frac{78 - 81}{\sqrt{\dfrac{7^2}{50} + \dfrac{9^2}{50}}} = \frac{-3}{\sqrt{\dfrac{49}{50} + \dfrac{81}{50}}} $$

$$ = \frac{-3}{\sqrt{2.60}} = -1.86 $$

We use the two-tail rejection region of Figure 12.11 (see page 353). The value of $z = -1.86$ falls in the acceptance region. Thus, we conclude that the sample data does not support the claim that there is a difference between the boys' IQ scores and the girls' IQ scores.

EXAMPLE 2

An executive who has two secretaries, Jean and Mark, is interested in knowing whether there is any difference in their typing abilities. Mark typed a 40 page report and made an average of 2.6 errors per page. The standard deviation was 0.6. Jean typed a 20 page report and made an average of 2.3 errors per page. The standard deviation was 0.8. Is there any difference between their performances? Use a 5% level of significance.

SOLUTION

Let $\bar{X}_1$, s_1, and N_1 represent Mark's scores and let $\bar{X}_2$, s_2, and N_2 represent Jean's scores. Then the problem is whether the observed difference between the sample means is significant or not. Thus

$$H_0: \mu_1 = \mu_2$$
$$H_1: \mu_1 \neq \mu_2$$

Here we are given that $\bar{X}_1 = 2.6$, $s_1 = 0.6$, $N_1 = 40$, $\bar{X}_2 = 2.3$, $s_2 = 0.8$, and $N_2 = 20$, so that

$$z = \frac{\bar{X}_1 - \bar{X}_2}{\sqrt{\dfrac{s_1^{\ 2}}{N_1} + \dfrac{s_2^{\ 2}}{N_2}}}$$

$$= \frac{2.6 - 2.3}{\sqrt{\dfrac{0.6^2}{40} + \dfrac{0.8^2}{20}}}$$

$$= \frac{2.3}{\sqrt{0.009 + .032}}$$

$$= \frac{2.3}{\sqrt{0.041}} = \frac{2.3}{0.202}$$

$$= 11.39$$

We use the two-tail rejection region of Figure 12.11 (see page 353). The value of $z = 11.39$ falls in the rejection region. Thus, there is a difference between the typing abilities of the two secretaries.

EXAMPLE 3

There are many advertisements on television about toothpastes. One such advertisement claims that children who use Smile toothpaste have fewer cavities than children who use any other brand. To test this claim, a consumer's group selected 100 children and divided them into 2 groups of 50 each. The children of group I were told to brush daily with only Smile toothpaste. The children of group II were told to brush daily with Vanish toothpaste. The experiment lasted one year. The following number of cavities were obtained:

Smile:	$\bar{X}_1 = 2.31$	$s_1 = 0.6$
Vanish:	$\bar{X}_2 = 2.68$	$s_2 = 0.4$

Is Smile more effective than Vanish in preventing cavities? Use a 5% level of significance.

SOLUTION

In this case the question is whether Smile is better than Vanish. This means that people who use Smile toothpaste will have fewer cavities than those who use Vanish. Thus,

$$H_0: \mu_1 = \mu_2$$
$$H_1: \mu_1 < \mu_2$$

Here we are given that $\bar{X}_1 = 2.31$, $s_1 = 0.6$, $N_1 = 50$, $\bar{X}_2 = 2.68$, $s_2 = 0.4$, and $N_2 = 50$, so that

$$z = \frac{\bar{X}_1 - \bar{X}_2}{\sqrt{\dfrac{s_1^2}{N_1} + \dfrac{s_2^2}{N_2}}}$$

$$= \frac{2.31 - 2.68}{\sqrt{\dfrac{(0.6)^2}{50} + \dfrac{(0.4)^2}{50}}} = \frac{-0.37}{\sqrt{0.0104}} = \frac{-0.37}{0.102}$$

$$= -3.63$$

We use the one-tail rejection region of Figure 12.13 (see page 357). The value of $z = -3.63$ falls in the rejection region. Thus, we reject the null hypothesis. The sample data would seem to support the manufacturer's claim. Actually, further studies are needed before making any definite decision about the effectiveness of Smile in preventing cavities.

EXAMPLE 4

The local chapter of a Women's Liberation group claims that a female college graduate earns less than a male college graduate. A survey of 40 men and 30 women indicated the following results:

	Average Starting Salary	Standard Deviation
Women	$9000	$600
Men	$9700	$900

Do these figures support the claim that women earn less? Use a 1% level of significance.

SOLUTION

Let $\bar{X}_1$, s_1, and N_1 represent the women's scores, and let $\bar{X}_2$, s_2, and N_2 represent the men's scores. The problem is whether the observed difference between the sample means is significant or not. Thus

$$H_0: \mu_1 = \mu_2$$
$$H_1: \mu_1 < \mu_2$$

Here we are given that $\bar{X}_1 = 9000$, $s_1 = 600$, $N_1 = 30$, $\bar{X}_2 = 9700$, $s_2 = 900$, and $N_2 = 40$, so that

$$z = \frac{\bar{X}_1 - \bar{X}_2}{\sqrt{\dfrac{s_1^2}{N_1} + \dfrac{s_2^2}{N_2}}}$$

$$= \frac{9000 - 9700}{\sqrt{\dfrac{(600)^2}{30} + \dfrac{(900)^2}{40}}} = \frac{-700}{\sqrt{12,000 + 20,250}}$$

$$= \frac{-700}{\sqrt{32,250}} = \frac{-700}{179.58}$$

$$= -3.9$$

We use the one-tail rejection region of Figure 12.12 (see page 355). The value of $z = -3.9$ falls in the rejection region. Thus, we reject the null hypothesis. There is a difference between the starting salary of men and women.

EXERCISES

In each of the following, use a 5% level of significance.

1. Forty-five glasses of Blue beer maintained flavor for an average of 35 minutes with a standard deviation of 6 minutes, while 53 glasses of Red beer maintained flavor for an average of 39 minutes with a standard deviation of 7.5 minutes. Is there a significant difference between the average time that the two beers maintain flavor?

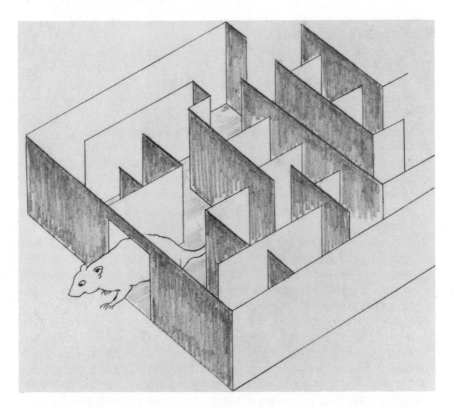

2. Dr. Mendel injected two groups of rats with two different drugs to determine how the drug affects the speed with which the rats run a maze. The 50 rats treated with drug A needed an average of 12 minutes to run the maze. The standard deviation was 2.3 minutes. The 40 rats treated with drug B needed an average of 15 minutes to run the maze. The standard deviation was 3.6. Is there a significant difference between the effects of the two drugs on the average time it takes the rats to run the maze?

3. A consumer's group claims that many supermarkets charge higher prices for an identical item in slum neighborhoods than in middle class neighborhoods. To investigate this claim a citizen purchases one container of milk from each of 38 different stores located in poorer neighborhoods and one container of

milk from each of 43 different stores located in middle class neighborhoods. The following results were obtained:

	Poorer Neighborhoods	Middle Class Neighborhoods
Average Price (in cents)	47	45.5
Sample Standard Deviation	6	4
Sample Size	38	43

Is there a significant difference between the average price of a container of milk in these neighborhoods?

4. A reading teacher is experimenting with two different techniques for teaching reading to second grade students. In one class the 32 students needed an average of 30 minutes to learn to read a paragraph. The standard deviation was 3.7 minutes. In a second class the 31 students needed an average of 26 minutes to learn to read a paragraph. The standard deviation was 4.8 minutes. Is there a significant difference between these reading techniques?

5. Sixty fences painted with Brand X rust inhibitor lasted an average of 7 years before repainting was necessary. The standard deviation was 1.96 years. Seventy-five fences painted with Brand Y rust inhibitor lasted an average of 8 years before repainting was necessary. The standard deviation was 2.32. Is there any significant difference in the durability of the two brands of rust inhibitor?

6. The average salary of the 36 assistant professors at a university is $13,800 with a standard deviation of 1400. The average salary of the 40 assistant professors at another university is $14,400 with a standard deviation of $900. Is there a significant difference between the average salary of assistant professors at these universities?

7. A State Motor Vehicle Department conducted tests on 200 randomly selected drivers in the state to determine the average time needed to react to a particular driving hazard. It divided the 200 drivers into 2 groups of 100 drivers each. Group A needed an average of 7 seconds to react to the hazard. The standard deviation was 0.08 seconds. Group B needed 8.5 seconds to react to the hazard. The standard deviation was 0.02 seconds. Is there any significant difference in the average reaction time of these two groups of drivers?

8. Mr. Blank, the company payroll supervisor, claims that the 80 workers on the thirteenth floor take a longer coffee break than the 50 workers on the eighth floor. To support his claim he gathered the following information:

	Average Time Spent on Coffee Break (in minutes)	Standard Deviation
Workers on 13th Floor	18	4
Workers on 8th Floor	16	7

Is Mr. Blank's claim justified?

9. In a recent school bowling tournament the 50 members of the girls' team scored an average of 225 points with a standard deviation of 13 points. The 46 members of the boys' team scored an average of 230 points with a standard deviation of 27 points. Is there any significant difference between average scores of the two teams?

10. Thirty-two bouncers at Pete's Bar and Grill bounced an average of 332 patrons during the month of January. The standard deviation was 12. The 36 bouncers at Bill's Bar and Grill bounced an average of 281 patrons during the month of January and the standard deviation was 28. Is there any significant difference between the average number of patrons bounced by these two bars?

12.5   TESTS CONCERNING PROPORTIONS

Suppose a congressman claims that 60% of the voters in his district are in favor of lowering the drinking age to 16 years. If a random sample of 400 voters showed that 221 of them favored the proposal, can we reject the congressman's claim? Questions of this type occur quite often and are usually answered on the basis of observed proportions. We assume that we can use the binomial distribution and that the probability of success is the same from trial to trial. We can therefore apply Formulas 7.5 and 7.6 (see page 213) which give us the mean and standard deviation of a binomial distribution. Thus,

$$\text{mean} \quad \mu = np$$
$$\text{standard deviation } \sigma = \sqrt{np(1 - p)}$$

The null hypothesis in such tests assumes that the observed proportion, p, is the same as the population proportion, π. Depending upon the situation, we have the following alternate hypotheses:

Null hypothesis $H_0: p = \pi$
Alternate hypothesis $G_1: p \neq \pi$ [which means a two-tailed test]
                        $H_2: p > \pi$ [which means a one-tailed (right-side) test]
                        $H_3: p < \pi$ [which means a one-tailed (left-side) test]

The test statistic is z, where

$$z = \frac{p - \pi}{\sqrt{\dfrac{\pi(1 - \pi)}{N}}}$$

We reject the null hypothesis if the test statistic falls in the critical, or rejection, region.

369

The following examples illustrate how we test proportions.

EXAMPLE 1

The Dean of Students at a college claims that only 12% of the students are on some form of drugs. To test this claim, a students' group takes a sample of 80 students. They find that 14 of these students are on drugs. Is the Dean's claim acceptable? Use a 5% level of significance.

SOLUTION

In this case the population parameter being tested is π, the true proportion of students on drugs. The questioned value is 0.12. Since we are testing whether the true proportion is 0.12 or not, we have

$$H_0: \pi = 0.12$$
$$H_1: \pi \neq 0.12$$

We are told that 14 of the 80 sampled students are on drugs, so that $p = 14/80 = 0.175$. Thus

$$z = \frac{p - \pi}{\sqrt{\dfrac{\pi(1 - \pi)}{N}}} = \frac{0.175 - 0.12}{\sqrt{\dfrac{(0.12)(1 - 0.12)}{80}}}$$

$$= \frac{0.055}{\sqrt{0.00132}} = \frac{0.055}{0.036}$$

$$= 1.53$$

We use the two-tail rejection region of Figure 12.11 (see page 353). The value of $z = 1.53$ falls in the acceptance region. Thus we cannot reject the null hypothesis and the Dean's claim that the true proportion of students on drugs is 12%.

EXAMPLE 2

In a recent press conference a congressman claimed that 55% of the American people supported the President's foreign policy. To test this claim a newspaper editor selected a random sample of 1000 people and 490 of them said that they supported the President. Is the Congressman's claim justified? Use a 1% level of significance.

SOLUTION

In this case the population parameter being tested is π, the true proportion of Americans who support the President. The questioned value is 0.55. Since we are testing whether the true proportion is 0.55 or not, we have

$$H_0:\ \pi = 0.55$$
$$H_1:\ \pi \neq 0.55$$

We are told that 490 of the 1000 people interviewed supported the President, so that

$$p = \frac{490}{1000} = 0.49$$

Thus $\quad z = \dfrac{p - \pi}{\sqrt{\dfrac{\pi(1 - \pi)}{N}}} \quad = \dfrac{0.49 - 0.55}{\sqrt{\dfrac{(0.55)(1 - 0.55)}{1000}}}$

$$= \frac{-0.06}{\sqrt{0.0002475}} = \frac{-0.06}{0.016}$$

$$= -3.75$$

Since the level of significance is 1%, we use the two-tailed rejection of Figure 12.10 (see page 352). The value of $z = -3.75$ falls in the rejection region. Thus, we reject the null hypothesis and the congressman's claim that 55% of the American people support the President's foreign policy.

EXAMPLE 3

A latest government survey indicates that 22% of the people in Camelot are illegally receiving some form of public assistance. The Mayor of Camelot believes that the figures are exaggerated. To test this claim, he carefully examines 75 cases and finds that 11 of these people are illegally receiving aid. Does this sample support the government's claim? Use a 5% level of significance.

SOLUTION

In this case the population parameter being tested is π, the true proportion of people illegally receiving financial aid. The questioned value is 0.22. Since we are testing whether the true proportion is 0.22 or lower, we have

$$H_0:\ \pi = 0.22$$
$$H_1:\ \pi < 0.22$$

371

We are told that 11 of the 75 cases examined are illegally receiving aid, so that

$$p = \frac{11}{75} = 0.15$$

Thus

$$z = \frac{p - \pi}{\sqrt{\dfrac{\pi(1 - \pi)}{N}}} = \frac{0.15 - 0.22}{\sqrt{\dfrac{(0.22)(1 - 0.22)}{75}}}$$

$$= -1.46$$

We use the one-tail rejection region of Figure 12.12 (see page 355). The value of $z = -1.46$ falls within the acceptance region. Thus, we cannot reject the null hypothesis that 22% of the people in Camelot are illegally receiving financial aid.

EXAMPLE 4

Government officials claim that approximately 29% of the residents of a state are opposed to building a nuclear plant to generate electricity. Local conservation groups claim that the true percentage is much higher. To test the government's claim an independent testing group selects a random sample of 81 state residents and finds that 38 of the people are opposed to the nuclear plant. Can we conclude that the government's claim is inaccurate? Use a 5% level of significance.

SOLUTION

In this case the population parameter being tested is π, the true proportion of state residents who are opposed to building the nuclear plant. The questioned value is 0.29. Since we are testing whether the true proportion is 0.29 or higher, we have

$$H_0: \pi = 0.29$$
$$H_1: \pi > 0.29$$

We are told that 38 of the 81 residents opposed the nuclear plant so that

$$p = \frac{38}{81} = 0.47$$

Thus

$$z = \frac{p - \pi}{\sqrt{\dfrac{\pi(1 - \pi)}{N}}} = \frac{0.47 - 0.29}{\sqrt{\dfrac{(0.29)(1 - 0.29)}{81}}}$$

$$= \frac{0.18}{\sqrt{0.002542}} = \frac{0.18}{0.0504} = 3.57$$

We use the one-tail rejection region of Figure 12.13 (see page 357). The value of $z = 3.57$ falls in the rejection region. Thus, we reject the null hypothesis that the true proportion of state residents opposed to the nuclear plant is 0.29.

EXERCISES

In each of the following exercises use a 5% level of significance.

1. A local planned parenthood group claims that approximately 19% of all women in the city are in favor of legalizing abortion. An abortion clinic claims that the true percentage is much higher. To test the parenthood group's claim, an independent group randomly selects 75 women. Sixteen of these favor legalizing abortion. Can we accept the planned parenthood group's claim?

2. A construction company claims that 55% of its workers are from minority groups. In a random sample of 80 of the company's workers it is found that 41 of the workers are from minority groups. Can we accept the claim that 55% of the company's workers are from minority groups?

3. Officials in New Amsterdam claim that 95% of all packages mailed in the city have correct postage on them. To verify this claim, a postal inspector randomly selects 150 packages and finds that 9 of them do not have the correct postage. Can we accept the official's claim that 95% of all packages mailed in the city have the correct postage on them?

4. A congressman claims that 60% of all voters in his district are in favor of lowering the drinking age to 16 years. If a random sample of 400 voters shows that 221 of them favor the proposal, can we reject the congressman's claim?

5. An airline company claims that approximately 16% of its stewardesses are married. The union claims that the true percentage is much higher. To test the company's claim, an independent group takes a random sample of 70 stewardesses. Thirteen of them are married. Can we conclude that the airline company's claim is inaccurate?

6. A large, northeastern U.S. city is in financial trouble and is having difficulty repaying its bondholders and creditors. The mayor claims that 60% of the citizens believe that the city should default on its payments to the banks. A citizen's watchdog commission believes that this figure is exaggerated. To test the mayor's claim, an independent group takes a random sample of 200

citizens. Only 110 of them favor default. Can we conclude that the mayor's claim is inaccurate?

7. A State Motor Vehicle Bureau claims that 20% of all cars on the state highways are equipped with defective air-pollution control devices. To test this claim, an independent group randomly selects 400 cars and checks their air-pollution control devices. Ninety-five of the cars are not properly equipped with an air-pollution control device. Can we conclude that the Motor Vehicle Bureau's claim is accurate?

8. An airline finds that approximately 5% of all people booking reservations cancel. In one month 260 of 5000 reservations were canceled. Does this support the airline's claim?

9. A housing manager claims that 18% of all families in his housing complex have a dog or a cat. In a random sample of 40 of the families living in the housing complex, it is found that 9 of these families have a dog or a cat. Is the housing manager's claim justified?

10. The ABC Driving School claims that 70% of all its students pass the road test on their first try. In a random sample of 38 students from this school it is found that 25 of the students passed their road test the first time. Can we accept the driving school's claim?

11. Dr. Peters claims to have developed a new ointment which is 88% effective in restoring hair to people who are bald. To test this claim, a consumer's group asks 49 bald people to use the ointment. Thirty-five have their hair restored. Can we accept the doctor's claim?

SELF-STUDY GUIDE

In this chapter we discussed hypothesis testing, which is a very important branch of statistical inference. Hypothesis testing is the process by which a decision is made to either reject or accept a null hypothesis about one of the parameters of the distribution. The decision to accept or reject a null hypothesis is based upon information obtained from sample data. Since any decision is subject to error, we discussed Type-I and Type-II errors.

We reject a null hypothesis when the test statistic falls in the critical, or rejection, region. The critical region is determined by two things: 1. whether we wish to perform a one-tail or two-tailed test and 2. the level of significance, 1% or 5%.

If a null hypothesis is not rejected, we cannot say that the sample data proves that what the null hypothesis says is necessarily true. It merely does not reject it. Some statisticians prefer to say that in this situation they *reserve judgement* rather than accept the null hypothesis.

We discussed methods for testing means, proportions, and differences between means. All the tests studied are summarized in Table 12.1. You should be able to identify each symbol in the formulas, understand the relationships among the symbols expressed in each formula, understand the significance of each formula, and use the formulas in solving problems.

TABLE 12.1

VARIOUS TESTS FOR ACCEPTING OR REJECTING A NULL HYPOTHESIS

POPULATION PARAMETER TESTED	SAMPLE SIZE	TYPE OF TEST	SIGNIFICANCE	TEST STATISTIC	REJECT NULL HYPOTHESIS IF
Mean	Large	Two-tailed	0.05	$z = \dfrac{\bar{X} - \mu}{\sigma/\sqrt{N}}$	$z < -1.96$ or $z > 1.96$
Mean	Large	Two-tailed	0.01		$z < -2.58$ or $z > 2.58$
Mean	Large	One-tailed	0.05		$z < -1.64$ or $z > 1.64$
Mean	Large	One-tailed	0.01		$z < -2.33$ or $z > 2.33$
Mean	Small	Two-tailed	0.05	$t = \dfrac{\bar{X} - \mu}{s/\sqrt{N}}$	$t < -t_{0.025}$ or $t > t_{0.025}$
Mean	Small	One-tailed	0.05		$t < -t_{0.05}$ or $t > t_{0.05}$
Difference between means	Large	Two-tailed	0.05 or 0.01	$z = \dfrac{\bar{X}_1 - \bar{X}_2}{\sqrt{\dfrac{s_1^2}{N_1} + \dfrac{s_2^2}{N_2}}}$	Same as above
Difference between means	Large	One-tailed	0.05 or 0.01		Same as above
Proportion	Large	Two-tailed	0.05 or 0.01	$z = \dfrac{p - \pi}{\sqrt{\dfrac{\pi(1 - \pi)}{N}}}$	Same as above
Proportion	Large	One-tailed	0.05 or 0.01		Same as above

You should now be able to demonstrate your knowledge of the following ideas presented in this chapter by giving definitions, descriptions, or specific examples. Page references are given for each term so that you can check your answer.

Hypothesis testing (page 344)
Hypothesis (page 344)
Null hypothesis (page 344)
Alternate hypothesis (page 344)
Two-sided, or two-tailed, test
  (page 344)
One-sided, or one-tailed, test
  (page 345)
Acceptance region (page 345)
Critical Value (page 345)
Critical region (page 347)
Rejection region (page 347)

Test statistic (page 347)
Statistic tests of hypotheses
  (page 347)
Statistical tests of significance
  (page 347)
Type-I error (page 348)
Type-II error (page 348)
Level of significance (page 349)
5% significance level (page 349)
1% significance level (page 349)
Alpha, α (page 349)
Statistically significant (page 351)

The tests of the following section will be more useful if you take them after you have studied the examples and solved the exercises in this chapter.

MASTERY TESTS

Form *A*

1. True or false: The significance level of a test is the probability that the test statistic will fall in the acceptance region.
2. Alpha, α, is the probability of making *a.* a Type-I error *b.* a Type-II error *c.* no error *d.* none of these
3. When we accept a null hypothesis instead of rejecting it, we are committing *a.* a Type-I error *b.* a Type-II error *c.* no error *d.* none of these
4. A number that is used to decide whether to reject or accept a null hypothesis is called a _____.
5. True or false: When a null hypothesis is accepted, this means that what the null hypothesis claims is guaranteed to be true.
6. The region of rejection of a null hypothesis is called the _____ region.
7. A statistical test which has the rejection region located in the left or right tail of the distribution is called a _____ test.
8. True or false: A hypothesis is an assumption about one or more of the population parameters that will be accepted or rejected on the basis of sample data.
9. The value that separates the rejection region from the acceptance region is called the _____.
10. If the alternate hypothesis consists of a $<$ symbol, then the rejection region consists of *a.* one region on the right side *b.* one region on the left side *c.* two regions, one on each side *d.* none of these.

1. Forty-five identical cars were driven an average of 19.8 miles on one gallon of Brand A gasoline. The standard deviation was 4.8 miles. These cars were then driven an average of 17.2 miles on one gallon of Brand B gasoline. The standard deviation was 5.9 miles. Is there a significant difference between the average number of miles driven with a gallon of each of these brands of gasoline? Use a 5% level of significance.

2. A large company claims that the average yearly salary of its workers is $9800. The union believes that the average salary is much less. In a random sample of 200 employees the union finds that the average salary is $9400 with a standard deviation of $1600. Should we accept the company's claim? Use a 5% level of significance.

3. The We-Move-It Company claims that a typical homeowner lives in his house an average of 7 years before selling it. The standard deviation is 2.98 years. To check this claim, a real estate firm interviews 64 homeowners. The results show that these homeowners sold their houses after an average of 7.8 years. Should we reject the moving company's claim? Use a 5% level of significance.

4. It is estimated that 25% of all students at a state college work full-time. In a random survey of 80 college students it is found that 24 students work full-time. Can we accept the estimate that 25% of all students at this state college work full-time? Use a 5% level of significance.

5. In Nova County a sample of 180 families with four children indicated that each family spends an average of $47.88 per week for food. The standard deviation was $8.43. A similar survey of 160 families in Vega County indicated that each family spends an average of $43.98 per week for food. The standard deviation was $9.88. Is the difference between the average weekly expenditure for food significant? Use a 1% level of significance.

6. A claim is made in the student newspaper that the average weight of a coed in the school's dance classes is 118 pounds with a standard deviation of 6.9 pounds. A reporter believes that the average weight is much less than 118 pounds. In a sample of 36 girls he finds that the average weight is 113 pounds. Should we reject the newspaper's claim? Use a 1% level of significance.

7. According to fire department officials in a certain city, 57% of all fire alarms are false. To check this, a city official examines the department records. If 81 of 124 fire alarms turn out to be false, can we accept the fire department claim? Use a 5% level of significance.

8. The mean lifetime of 60 transistor batteries produced by Company A is 41 hours with a standard deviation of 6.9 hours. The average lifetime of 50 similar batteries produced by Company B is 45 hours with a standard deviation of 8.8 hours. Is the difference in the lifetimes significant?

9. A farmer claims that by injecting his turkeys with a certain chemical the weight of a turkey will increase considerably. To test this claim, another farmer divided 200 turkeys into 2 groups of 100. Group A received the chemical while Group B did not. The following results were obtained over a 6 month period.

	Group A	Group B
Average Improvement (lbs)	6.2	5.9
Sample Standard Deviation (lbs)	3.3	2.8

Is the claim of the first farmer justified?

10. Fifty families who previously owned a Brand X washing machine indicated that their machines lasted an average of 5.8 years. The standard deviation was 0.98 years. Forty-five families who previously owned a Brand Y washing machine indicated that their machines lasted an average of 4.9 years. The standard deviation was 1.6 years. Is the difference between the average life of these two brands of washing machines significant?

11. The average age of the 80 girls in the Alpha Sorority is 19.9 years with a standard deviation of 0.9 years. The average age of the 60 girls in the Beta Sorority is 20.4 years with a standard deviation of 0.8 years. Is the difference between the average age of girls in these two sororities significant? Use a 5% level of significance.

12. A tax inspector claims that 11% of the merchants in the state charge the wrong sales tax. The Merchant's Association challenges this claim. In a random sample of 78 merchants it is found that 9 of the merchants charged the wrong sales tax. Can we accept the tax inspector's claim? Use a 1% level of significance.

13. A commuter estimates that in the last two years his train is late on the average 18 minutes each day. The standard deviation is 5 minutes. In the last 30 working days his train was late an average of only 12 minutes. Does this indicate an improvement of service? Use a 5% level of significance.

14. An English teacher claims that 70% of all students can pass the school proficiency exam. In a random sample of 68 seniors it is found that 54 seniors passed the exam. Can we accept the teacher's claim? Use a 1% level of significance.

15. The average hourly salary of a bank teller in one city is $2.85 with a standard deviation of $0.56. One large bank, which employs 85 tellers, pays an hourly rate of $2.70. Can this bank be accused of paying below the average hourly rate? Use a 5% level of significance.

16. A sociologist believes that women from suburban areas marry at a later age than women from urban areas. Do the following sample results support this claim? Use a 5% level of significance.

	Urban Women	Suburban Women
Average Age at which Women First Married	19.1	20.3
Standard Deviation	2.6	1.9
Number in Survey	38	45

17. According to the Red Cross, 45% of all the people in a southern city have type O blood. A random survey of 90 people found that 48 of them have type O blood. Can we accept the Red Cross claim? Use a 1% level of significance.

18. Court officials claim that any criminal case is disposed of in an average of 50 days with a standard deviation of 6 days. The local civil liberties organization claims that the average is considerably higher. An average of 56 days was required to dispose of 38 criminal cases. Can we accept the court officials' claim? Use a 5% level of significance.

19. The Chamber of Commerce claims that 10% of all businesses are polluting the atmosphere. Environmentalists claim that the percentage is higher. A random survey of 88 businesses shows that 12 of them are polluting the atmosphere. Can we accept the claim by the Chamber of Commerce? Use a 1% level of significance.

20. Forty typists in the east wing of an office building type an average of 33 letters per hour with a standard deviation of 7.6 letters. Fifty typists in the west wing type an average of 28 letters per hour with a standard deviation of 4.6 letters. Is the difference between the average number of letters typed by the workers in both wings significant? Use a 5% level of significance.

SUGGESTED READING

Adler, H. and Roessler, E. B. *Introduction to Probability and Statistics.* 5th ed. San Francisco, Calif.: Freeman & Co., 1972.

Freund, J. *Statistics, A First Course.* Englewood-Cliffs, N.J.: Prentice-Hall, 1970.

Hogg, Robert and Craig, Allen. *Introduction to Mathematical Statistics.* 3rd ed. New York: MacMillan, 1970.

Johnson, N. L. and Leone, F. C. *Statistics and Experimental Design in Engineering and the Physical Sciences.* New York: Wiley, 1964.

Lehmann, E. L. *Testing Statistical Hypotheses.* New York: Wiley, 1959.

Mendenhall, William. *Introduction to Probability and Statistics.* 3rd ed. North Scituate, Mass.: Duxbury Press, 1971.

Raiffa, H. and Schlaifer, R. *Applied Statistical Decision Theory.* Boston: Harvard Graduate School of Business Administration, 1961.

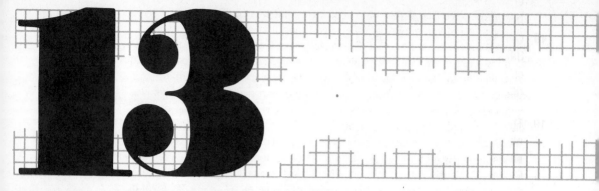

13

THE CHI-SQUARE DISTRIBUTION

By permission of the American Alliance for Health, Physical Education and Recreation, 1201 16th Street, N.W. Washington, D.C. 20036.

FITNESS TESTS FOR YOUR CHILDREN

Young people who measure their ratings on these physical–fitness tables can see where they stand in comparison with a sample of 7,600 boys and girls in public schools. The nationwide testing program was funded by the U.S. Office of Education.

STANDING LONG JUMP

BOYS

Age:	10	11	12	13	14	15	16	17
	ft. in.	ft. in.	ft. in.	ft. in.	ft. in.	ft. in.	ft. in.	ft. in.
Excellent	5-10	6-0	6-3	6-10	7-2	7-7	7-11	8-2
Good	5-6	5-9	6-0	6-5	6-10	7-3	7-6	7-10
Satisfactory	5-1	5-5	5-7	6-0	6-4	6-10	7-2	7-5
Poor	4-9	5-0	5-3	5-6	5-11	6-5	6-9	7-0

GIRLS

Age:	10	11	12	13	14	15	16	17
	ft. in.	ft. in.	ft. in.	ft. in.	ft. in.	ft. in.	ft. in.	ft. in.
Excellent	5-8	5-9	6-0	6-2	6-5	6-3	6-3	6-6
Good	5-2	5-5	5-8	5-10	6-0	6-0	5-11	6-2
Satisfactory	4-10	5-1	5-2	5-5	5-7	5-6	5-6	5-9
Poor	4-6	4-8	4-10	5-1	5-2	5-2	5-1	5-3

50-YARD DASH
(scores in seconds and tenths)

BOYS

Age:	10	11	12	13	14	15	16	17
Excellent	7.5	7.2	7.0	6.7	6.4	6.2	6.2	6.0
Good	7.8	7.5	7.3	7.0	6.6	6.4	6.4	6.3
Satisfactory	8.0	7.9	7.6	7.3	7.0	6.8	6.6	6.5
Poor	8.6	8.3	8.0	7.6	7.4	7.0	6.8	6.8

GIRLS

Age:	10	11	12	13	14	15	16	17
Excellent	7.5	7.5	7.2	7.0	7.0	7.0	7.1	7.0
Good	8.0	7.8	7.5	7.3	7.2	7.2	7.4	7.3
Satisfactory	8.4	8.1	8.0	7.7	7.6	7.6	7.7	7.6
Poor	8.9	8.5	8.3	8.1	8.0	8.0	8.0	8.0

SIT-UPS
(number performed in 60 seconds)

BOYS

Age:	10	11	12	13	14	15	16	17
Excellent	44	45	48	50	52	52	52	51
Good	40	41	43	47	48	49	49	47
Satisfactory	35	37	38	41	43	44	43	42
Poor	29	31	33	35	38	40	40	39

GIRLS

Age:	10	11	12	13	14	15	16	17
Excellent	40	40	41	43	42	40	41	
Good	35	36	37	38	39	38	36	38
Satisfactory	30	31	32	32	33	33	32	32
Poor	24	26	27	27	29	29	27	28

PULL-UPS FOR BOYS
(number of pull-ups)

Age:	10	11	12	13	14	15	16	17
Excellent	7	6	7	9	10	12	12	13
Good	4	5	5	6	8	10	10	11
Satisfactory	2	3	3	4	5	7	8	8
Poor	1	1	1	2	3	5	6	6

FLEXED-ARM HANG FOR GIRLS
(scores in seconds)

Age:	10	11	12	13	14	15	16	17
Excellent	29	30	27	25	29	28	24	28
Good	21	21	21	20	23	21	17	19
Satisfactory	12	13	12	11	13	12	10	10
Poor	6	7	6	6	7	7	5	6

Source: Youth Fitness Test Manual by American Alliance for Health, Physical Education, and Recreation.

The clipping gives the performance scores of boys and girls in different fitness tests according to age and height. By looking at the data, can we conclude that boys are more physically fit than girls or is the sex of the child independent of performance on these different tests? Such information is very valuable because it enables an individual to measure one's performance against that of others.

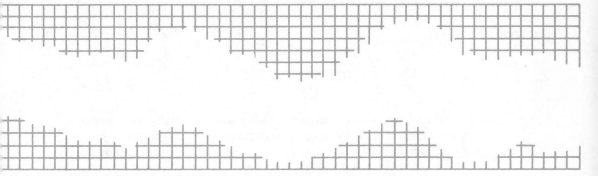

Preview

In this chapter we will discuss the following topics:

1. *Chi-Square Tests*
These tests provide the basis for testing whether more than two population proportions can be considered equal.

2. *Contingency Tables*
These are tabular arrangements of data into a two-way classification. The chi-square test statistic tells us whether the two ways of classifying the data are independent.

3. *Expected Frequencies*
These are the numbers, that is, frequencies, that should appear in each of the boxes of a contingency table.

4. *Goodness of Fit*
The chi-square test statistic will be applied to determine whether an observed frequency distribution is in agreement with the expected mathematical distribution.

Introduction

In Chapter 12 we discussed methods for testing whether the observed difference between two sample means is significant. In this chapter we will analyze whether differences among two or more sample proportions are significant or whether they are due purely to chance. For example, suppose a college professor distributes a faculty-evaluation form to the 150 students of his Psychology 26 class. Of the many questions on the evaluation form, the following are two examples:

1. What is your grade point average? Assume A = 4 points, B = 3 points, C = 2 points, D = 1 point, and F = 0 points.
   a. 3.0 to 4.0   *b.* 2.0 to 2.99   *c.* below 2.0
2. Would you be willing to take another course with this teacher?
   a. Yes   *b.* No

The results for these two questions are summarized as follows:

		GRADE POINT AVERAGE		
		3.0–4.0	2.0–2.99	Below 2.0
WOULD YOU	Yes	28	36	11
TAKE THIS				
TEACHER AGAIN?	No	22	44	9
	Total	50	80	20

The teacher may be interested in knowing whether these ratings are influenced by the student's grade point average. In the 3.0–4.0 category the proportion of students who said that they would take this teacher for another course is 28/50. In the 2.0–3.0 category the proportion is 36/80, and in the below 2.0 category the proportion is 11/20. Is it true that students with a higher grade point average tend to rate the teacher differently than students with a lower grade point average?

Questions of this type occur quite often. In this chapter we will study the chi-square distribution which is of great help in studying differences between proportions.

13.1   THE CHI-SQUARE DISTRIBUTION

To illustrate the method that is used when analyzing several sample proportions, let us return to the example discussed in the introduction. Let p_1 be the true proportion of students in the 3.0–4.0 category who will take another course with this teacher. Similarly let p_2 and p_3 represent the proportion of students in the 2.0–3.0 and below 2.0 categories who will take another course with this teacher. The null hypothesis that we wish to test is

$$H_0: p_1 = p_2 = p_3$$

This means that a student's grade point average does not affect the student's decision to take this teacher again. The alternate hypothesis is that $p_1, p_2,$ and p_3 are different. This means that a student's grade point average does affect the student's decision.

If the null hypothesis is true, the observed difference between the proportions in each of the grade point average categories is due purely to chance. Under this assumption we combine all the samples into one and consider it as one large sample. We then obtain the following estimate of the true proportion of *all* students in the school who are willing to take another course with this teacher. We get

$$\frac{28 + 36 + 11}{50 + 80 + 20} = \frac{75}{150} = 0.5$$

Thus our estimate of the true proportion of students who are willing to take another course with this teacher is 0.5.

There are 50 students in the 3.0–4.0 category. We would therefore expect 50(0.5), or 25, of these students to indicate yes they would take another course with this teacher, and we would expect 50(0.5), or 25, no's. Similarly in the 2.0–3.0 category there are 80 students so we would expect 80(0.5), or 40, yes answers and 40 no answers. Also in the below 2.0 category there are 20 students so that we would expect 20(0.5), or 10, yes answers and 10 no answers. The numbers that should appear are called **expected frequencies.** In the following table we have indicated these numbers in parentheses below the ones that were actually observed. We call the numbers that were actually observed **observed frequencies.**

GRADE POINT AVERAGE

	3.0–4.0	2.0–3.0	Below 2.0
Yes	28 (25)	36 (40)	11 (10)
No	22 (25)	44 (40)	9 (10)
Total	50	80	20

Notice that the expected frequencies and the observed frequencies are not the same. If the null hypothesis that $p_1 = p_2 = p_3$ is true, the observed frequencies should be fairly close to the expected frequencies. Since this rarely will happen, we need some way of determining when these differences are significant.

When is the difference between the observed frequencies and the expected frequencies significant? To answer this question we calculate a test statistic called the **chi-square statistic.**

FORMULA 13.1

Let E represent the expected frequency and let O represent the observed frequency. Then the **chi-square test statistic,** denoted as χ^2, is defined as

$$\chi^2 = \Sigma \frac{(O - E)^2}{E}$$

Comment In using Formula 13.1 we must calculate the square of the difference for each box, that is, cell, of the table. Then we add these squares together. Returning to our example we have

$$\chi^2 = \frac{(28 - 25)^2}{25} + \frac{(36 - 40)^2}{40} + \frac{(11 - 10)^2}{10} + \frac{(22 - 25)^2}{25} + \frac{(44 - 40)^2}{40} + \frac{(9 - 10)^2}{10}$$

$$= \quad 0.36 \quad + \quad 0.40 \quad + \quad 0.10 \quad + \quad 0.36 \quad + \quad 0.40 \quad + \quad 0.10$$

$$= 1.72$$

The value of the χ^2 statistic is 1.72.

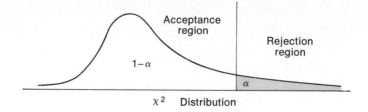

Figure 13.1 χ^2 Distribution

It should be obvious from Formula 13.1 that the value of χ^2 will be 0 when there is perfect agreement between the observed frequencies and the expected frequencies since in this case $O - E = 0$. Generally speaking, if the value of χ^2 is small, the observed frequencies and the expected frequencies will be pretty close to each other. On the other hand, if the value of χ^2 is large, this indicates that there is a considerable difference between the observed frequency and the expected frequency.

To determine when the value of the χ^2 statistic is significant, we use the **chi-square distribution.** This is pictured in Figure 13.1. We reject the null hypothesis when the value of the chi-square statistic falls in the rejection region of Figure 13.1. Table IX in the Appendix gives us the critical values, that is, dividing line, depending upon the number of degrees of freedom. Thus, $\chi^2_{0.05}$ represents the dividing line which cuts off 5% of the right tail of the distribution. *The number of degrees of freedom is always 1 less than the number of sample proportions that we are testing.*

In our example we are comparing 3 proportions so that the number of degrees of freedom is $3 - 1$, or 2. Now we look at Table IX in the Appendix to find the χ^2 value that corresponds to 2 degrees of freedom. We have $\chi^2_{0.05} = 5.991$. Since the test statistic value which we obtained, $\chi^2 = 1.72$, is much less than the table value of 5.991, we do not reject the null hypothesis. The difference between what was expected and what actually happened is due purely to chance.

Comment   Although we will usually use the 5% level of significance, Table IX in the Appendix also gives us the χ^2 values for the 1% level of significance. We use these values when we wish to find the dividing line which cuts off 1% of the right tail of the distribution.

Let us further illustrate the χ^2 test with several examples.

EXAMPLE 1

There are 10,000 students at a college. Twenty-seven hundred are freshmen, 2300 are sophomores, 3000 are juniors, and 2000 are seniors. Recently a new president was appointed. Two-thousand students attended the reception party for the president. The attendance breakdown is shown in the following table:

	Freshmen	Sophomores	Juniors	Seniors
Attended	300	700	650	350
Did Not Attend	2400	1600	2350	1650
Total	2700	2300	3000	2000

Test the null hypothesis that the proportion of freshmen, sophomores, juniors, and seniors that attended the reception is the same. Use a 5% level of significance.

SOLUTION

In order to compute the χ^2 test statistic, we must first compute the expected frequency for each box, that is, cell. To do this we obtain an estimate of the true proportion of students who attended the reception. We have

$$\frac{300 + 700 + 650 + 350}{2700 + 2300 + 3000 + 2000} = \frac{2000}{10,000} = 0.20$$

Out of 2700 freshmen we would expect 2700(0.20), or 540, to attend and 2700 − 540, or 2160, not to attend. Similarly, out of 2300 sophomores we would expect 2300(0.20), or 460, to attend and 2300 − 460, or 1840, not to attend. Also out of 3000 juniors we would expect 3000(0.20), or 600, to attend and 3000 − 600, or 2400, not to attend. Finally, out of 2000 seniors, we would expect 2000(0.20), or 400, to attend and 2000 − 400, or 1600, not to attend. We have indicated these expected frequencies just below the observed values in the following chart:

	Freshmen	Sophomores	Juniors	Seniors
Attended	300 (540)	700 (460)	650 (600)	350 (400)
Did Not Attend	2400 (2160)	1600 (1840)	2350 (2400)	1650 (1600)
Total	2700	2300	3000	2000

385

Now we calculate the value of the χ^2 statistic. We have

$$\chi^2 = \Sigma\frac{(O-E)^2}{E}$$

$$= \frac{(300-540)^2}{540} + \frac{(700-460)^2}{460} + \frac{(650-600)^2}{600} + \frac{(350-400)^2}{400}$$

$$+ \frac{(2400-2160)^2}{2160} + \frac{(1600-1840)^2}{1840} + \frac{(2350-2400)^2}{2400}$$

$$+ \frac{(1650-1600)^2}{1600}$$

$$= 106.67 + 125.22 + 4.17 + 6.25 + 26.67 + 31.30 + 1.04 + 1.56$$
$$= 302.88$$

There are 4 proportions that we are testing so that there are $4 - 1$, or 3, degrees of freedom. From Table IX in the Appendix we find the $\chi^2_{0.05}$ value with 3 degrees of freedom is 7.815. The value of the χ^2 test statistic ($\chi^2 = 302.88$) is definitely greater than 7.815. Hence we reject the null hypothesis. The proportions of freshmen, sophomores, juniors, and seniors that attended the reception are not the same.

EXAMPLE 2

A survey of the marital status of the students in 3 classes was taken. The following table indicates the results of the survey.

	Class 1	Class 2	Class 3	
Married	11	17	8	36
Single	29	33	22	84
Total	40	50	30	120

Test the null hypothesis that the proportion of students that are married in each of these classes is the same. Use a 5% level of significance.

$$X = \frac{36}{30}\;\;\frac{}{120}$$

SOLUTION

We must first compute the expected frequency for each cell. To do this we obtain an estimate of the true proportion of students who are married. We have

$$\frac{11+17+8}{40+50+30} = \frac{36}{120} = 0.3$$

$X = \frac{36}{40}\;\frac{}{120}$

$X = 12$

$X = \frac{36}{50}\;\frac{}{120}$

Thus, the estimate of the true proportion is 0.3. Out of the 40 students in class 1 we would expect 40(0.3), or 12, students to be married and 40 − 12, or 28, not to be married. In class 2 we would expect 50(0.3), or 15, students to be married and 50 − 15, or 35, students not to be married. In class 3 we would expect 30(0.3), or 9, students to be married and 30 − 9, or 21, not to be married. We have indicated these expected frequencies in parentheses in the following table:

	Class 1	Class 2	Class 3
Married	11	17	8
	(12)	(15)	(9)
Single	29	33	22
	(28)	(35)	(21)
Total	40	50	30

Now we calculate the value of the χ^2 statistic. We have

$$\chi^2 = \Sigma \frac{(O - E)^2}{E}$$

$$= \frac{(11 - 12)^2}{12} + \frac{(17 - 15)^2}{15} + \frac{(8 - 9)^2}{9} + \frac{(29 - 28)^2}{28}$$

$$+ \frac{(33 - 35)^2}{35} + \frac{(22 - 21)^2}{21}$$

$$= 0.08 + 0.27 + 0.11 + 0.04 + 0.11 + 0.05$$

$$= 0.66$$

There are 3 proportions that we are testing so that there are 3 − 1, or 2, degrees of freedom. From Table IX in the Appendix we find the $\chi^2_{0.05}$ value with 2 degrees of freedom is 5.991. Since the value of the test statistic, 0.66, is less than 5.991, we do not reject the null hypothesis.

Comment Experience has shown us that the χ^2 test can only be used when the expected frequency in each cell is at least 5. If the expected frequency of a cell is not larger than 5, this cell must be combined with other cells until the expected frequency is at least 5. We will not, however, concern ourselves with this situation.

EXERCISES

1. A survey of 500 randomly selected women in Los Angeles showed that 201 of them smoke. A similar survey of 400 women in San Francisco showed that 180 of them smoke. Using a 1% level of significance, test the null hypothesis that there is no difference between the corresponding proportions of women who smoke.

2. In a survey of 800 families in Dallas a sociologist found that 112 of these families attend church or synagogue on a regular basis. A similar survey of 600 families in Houston found that 62 of these families attend church or synagogue on a regular basis. Using a 5% level of significance, test the null hypothesis that there is no difference between the corresponding proportions of families who attend church or synagogue.

3. A survey of 300 randomly selected women in Chicago found that 182 were using birth control pills as a contraceptive. A similar survey of 180 women in New Orleans found that 81 of them were using birth control pills. Using a 5% level of significance, test the null hypothesis that there is no difference between the proportion of women using birth control pills in both cities.

4. Of the 400 seniors interviewed, 354 said they thought Professor Gonzales should be promoted to Associate Professor for the coming school year. Two-hundred thirteen of the 250 juniors interviewed and 132 of the 150 sophomores interviewed expressed similar views. Freshmen were not included in the survey since they were not familiar with Professor Gonzales. Using a 5% level of significance, test the null hypothesis that the proportion of students who think Professor Gonzales should be promoted is the same for all three classes of students.

5. Hilda is interested in registering for a statistics course with a teacher who is a very easy grader. There are 4 teachers scheduled to teach the course next semester. Hilda checks the bulletin boards and determines that each of the teachers passed the following number of students last semester:

	Professor Chi	Professor Lichtenfeld	Professor Holder	Professor Cocker
Number of Students Passed	48	40	35	43
Number of Students Failed	8	5	4	7

Using a 5% level of significance, test the null hypothesis that all four statistics teachers passed the same proportion of students.

6. Seven hundred females were asked if they find a certain actor attractive. The following table indicates their answers according to their age group:

	AGE (IN YEARS)		
	Under 25	25–35	Over 35
Yes	168	100	97
No	142	90	103

Using a 5% level of significance, test the null hypothesis that there is no difference between the proportion of females in the different age groups who think the actor to be attractive.

7. A newspaper reporter randomly selected 400 people from three sections of the United States to determine whether they are in favor of the proposed Women's Equal Rights Amendment to the Constitution. Their answers vary as follows:

$-65; \frac{73}{3}$ from 110

Geographic Location of Person	In Favor of Proposed Amendment	Opposed to Amendment	
East	68 _65.73_	42 _44.27_	110
Midwest	29 _35.85_	31 _24.15_	60
South	142 _137.42_	88 _92.58_	230
	239	161	400

Using a 5% level of significance, test the null hypothesis that the proportion of people in favor of the proposed amendment is the same for all three geographic locations.

8. A survey of 1000 people was taken to determine how many of them are in favor of passing new gun control legislation. The following table indicates the results of the survey according to the age of the person interviewed:

		AGE (IN YEARS)	
		Under 30	Over 30
IN FAVOR OF NEW GUN CONTROL LEGISLATION?	Yes	353	336
	No	147	164

Using a 1% level of significance, test the null hypothesis that there is no difference between the proportion of people in either age group who are in favor of new gun control legislation.

9. An advertiser is interested in knowing whether a person's income determines whether or not they read the Daily Times. The following table indicates the results of a sample of 600 randomly selected people:

		YEARLY INCOME		
		Below $15,000	$15,000–$25,000	Above $25,000
DO YOU READ THE DAILY TIMES?	Yes	79	104	130
	No	121	96	70

Using a 5% level of significance, test the null hypothesis that the proportion of people who read the *Daily Times* is the same for all three income groups.

10. Three-hundred politicians were polled to find out their view on the proposed sales tax increase. The results of the survey are as follows:

		Democrats	Republicans
SHOULD THE SALES TAX BE INCREASED?	Yes	38	48
	No	112	102

Using a 5% level of significance, test the null hypothesis that the proportion of Democrats and Republicans in favor of increasing taxes is the same.

11. Eleven hundred hamburgers sold by three hamburger chain stores were tested to determine if the percentage of fat content in each was within legal limits. The following table indicates the number of hamburgers that were within the legal limits for each chain store:

		CHAIN STORE		
		A	B	C
WITHIN LEGAL LIMIT?	Yes	354	213	132
	No	146	137	118

Using a 5% level of significance, test the null hypothesis that the proportion of hamburgers that have an illegal fat content is the same for all three chain stores.

13.2 CONTINGENCY TABLES

A very useful application of the χ^2 test discussed in Section 13.1 occurs in connection with contingency tables. Contingency tables are used when we wish to determine whether two variables of classification are related or dependent one upon the other. For example, consider the following chart which indicates the eye color and hair color of 100 randomly selected girls:

	Brown Eyes	Blue Eyes
Light Hair	10	33
Dark Hair	44	13

Is eye color independent of hair color or is there a significant relationship between hair color and eye color?

Contingency tables are especially useful in the social sciences where data is collected and often classified into two main groups. We might be interested in determining whether a relationship exists between these two ways of classifying the data or whether they are independent. We have the following definition:

DEFINITION 13.1

A **contingency table** is an arrangement of data into a two-way classification. One of the classifications is entered in rows and the other in columns.

When dealing with contingency tables, remember that the null hypothesis always assumes that the two ways of classifying the data are independent. We use the χ^2 test statistic as discussed in Section 13.1. The only difference is that we compute the expected frequency for each cell in a slightly different way.

FORMULA 13.2

The expected frequency of any cell in a contingency table is found by multiplying the total of the row with the total of the column to which the cell belongs. The product is then divided by the total sample size.

The following examples will illustrate how we apply the χ^2 test to contingency tables.

EXAMPLE 1

Let us consider the contingency table given at the beginning of this section. Is eye color independent of hair color?

SOLUTION

In order to compute the χ^2 test statistic, we must first compute the row total, column total, and the total sample size. We have

$$
\begin{aligned}
\textit{Row Totals:} \quad & 10 + 33 = 43 \\
& 44 + 13 = 57 \\
\textit{Column Totals:} \quad & 10 + 44 = 54 \\
& 33 + 13 = 46 \\
\textit{Total Sample Size:} \quad & 10 + 33 + 44 + 13 = 100
\end{aligned}
$$

We indicate these values in the following table:

	Brown Eyes	Blue Eyes	Row Total
Light Hair	10 (23.22)	33 (19.78)	43
Dark Hair	44 (30.78)	13 (26.22)	57
Column Total	54	46	

The expected value for the cell in the first row first column is obtained by multiplying the first row total with the first column total and then dividing the product by the total sample size. We get

$$\frac{54 \times 43}{100} = 23.22$$

For the first row second column we have

$$\frac{46 \times 43}{100} = 19.78$$

For the second row first column we have

$$\frac{54 \times 57}{100} = 30.78$$

For the second row second column we have

$$\frac{46 \times 57}{100} = 26.22$$

These values are entered in parentheses in the appropriate cell. We now use Formula 13.1 of the last section and calculate the χ^2 statistic. We have

$$\chi^2 = \Sigma \frac{(O - E)^2}{E}$$

$$= \frac{(10 - 23.22)^2}{23.22} + \frac{(33 - 19.78)^2}{19.78} + \frac{(44 - 30.78)^2}{30.78} + \frac{(13 - 26.22)^2}{26.22}$$

$$= 7.53 + 8.84 + 5.68 + 6.67$$
$$= 28.72$$

The χ^2 test statistic has a value of 28.72. If the contingency table has r rows and c columns, then the number of degrees of freedom is (r − 1)(c − 1). In this example there are 2 rows and 2 columns, so there are $(2-1) \cdot (2-1)$, or $1 \cdot 1$, which is 1 degree of freedom. From Table IX in the Appendix we find that the $\chi^2_{0.05}$ value with 1 degree of freedom is 3.841. Since we obtained a value of 28.72, we reject the null hypothesis and conclude that hair color and eye color are *not* independent.

EXAMPLE 2

A sociologist is interested in determining whether the occurrence of different types of crimes varies from city to city. An analysis of 1100 reported crimes produced the following results:

| | **TYPE OF CRIME** | | | | |
	Rape	Auto Theft	Robbery and Burglary	Other	Total
City A	76 (61.35)	112 (146.34)	87 (72.66)	102 (96.65)	377
City B	64 (68.83)	184 (164.20)	77 (81.52)	98 (108.44)	423
City C	39 (48.82)	131 (116.45)	48 (57.82)	82 (76.91)	300
Total	179	427	212	282	1100

Does this data indicate that the occurrence of a type of crime is dependent upon the location of the city? Use a 5% level of significance.

SOLUTION

We first calculate the expected frequency for each cell. We have

First Row:
$$\frac{(179)(377)}{1100} = 61.35 \qquad \frac{(427)(377)}{1100} = 146.34$$

$$\frac{(212)(377)}{1100} = 72.66 \qquad \frac{(282)(377)}{1100} = 96.65$$

Second Row:
$$\frac{(179)(423)}{1100} = 68.83 \qquad \frac{(427)(423)}{1100} = 164.20$$

$$\frac{(212)(423)}{1100} = 81.52 \qquad \frac{(282)(423)}{1100} = 108.44$$

Third Row:
$$\frac{(179)(300)}{1100} = 48.82 \qquad \frac{(427)(300)}{1100} = 116.45$$

$$\frac{(212)(300)}{1100} = 57.82 \qquad \frac{(282)(300)}{1100} = 76.91$$

Now we calculate the χ^2 test statistic getting

$$\chi^2 = \Sigma \frac{(O - E)^2}{E}$$

$$= \frac{(76 - 61.35)^2}{61.35} + \frac{(112 - 146.34)^2}{146.34} + \cdots + \frac{(82 - 76.91)^2}{76.91}$$

$$= 24.49$$

There are 3 rows and 4 columns so that there are 6 degrees of freedom since

$$(3 - 1)(4 - 1) = 2 \cdot 3 = 6$$

From Table IX in the Appendix we find that the $\chi^2{}_{0.05}$ value with 6 degrees of freedom is 12.592. Since we obtained a χ^2 value of 24.49, we reject the null hypothesis and conclude that the type of crime and the location of the city are not independent.

EXERCISES

1. The following table shows the educational level of the 240 faculty members of Stevensville University. Using a 5% level of significance, test the null hypothesis that there is no relationship between the sex of the faculty member and the degree earned.

	HIGHEST DEGREE EARNED		
	Bachelor's	Master's	Doctorate
Male	7	98	29
Female	13	82	11

2. The following chart indicates the height and hair color of the 350 girls in the dormitory:

		HAIR COLOR		
		Brown	Blonde	Red
	Tall	38	79	16
HEIGHT	Average	59	91	8
	Short	48	10	1

Using a 5% level of significance, test the null hypothesis that hair color is not influenced by height.

3. One hundred fifty students in Professor Jones's three statistics classes were asked to evaluate his performance as a teacher. The results are summarized below. Using a 1% level of significance, test the null hypothesis that these ratings were not influenced by the particular class that a student is attending.

	Easy Tests Explains Well	Easy Tests Explains Poorly	Difficult Tests Explains Well	Difficult Tests Explains Poorly
Class A	21	11	16	8
Class B	14	9	24	10
Class C	7	2	12	16

4. A survey was taken among the faculty, students, and administration of a college to determine their reaction to having students placed on the collegewide faculty review committee. This committee will determine which teachers are to be fired. The results of this survey are as follows:

	Faculty	Students	Administration
Opposed	148	62	47
No Opinion	39	108	18
In Favor	43	339	35

Using a 1% level of significance, test the null hypothesis that the attitude regarding this proposal is independent of the status of the individual.

5. An environmental group took a survey of 2000 people in different age groups to determine which environmental problems concerned them most. The following are the results:

	Water Pollution	Air Pollution	Over-Population	Energy Conservation
Under 25 yrs	192	128	224	76
25–40 yrs	422	146	54	166
Over 40 yrs	66	366	38	122

Using a 1% level of significance, test the null hypothesis that the environmental problem which most concerns an individual is independent of the person's age.

395

6. A study recently undertaken to determine whether a man's occupation influences the whiskey he prefers is summarized in the following table:

| | | TYPE OF WHISKEY PREFERRED | | |
		Scotch	Bourbon	Rye
OCCUPATION	Blue Collar Worker	69	78	106
	White Collar Worker	75	83	109

Using a 1% level of significance, test the null hypothesis that occupations and whiskey preferences are independent.

7. Suppose a social scientist believes there is a relationship between the conduct and the intelligence of school children. The following is data collected for one large school:

| | | INTELLIGENCE | | |
		Low	Average	High
CONDUCT	Satisfactory	68	96	138
	Unsatisfactory	49	35	78

Using a 5% level of significance, test the null hypothesis that the conduct and intelligence of children of this school are independent.

8. In the recent elections the voters of a town were asked whether they favored building a new courtroom. The following results were obtained:

| | COST | |
	Above 10 Million Dollars	Below 10 Million Dollars
Opposed	4372	3324
In Favor	1268	2756

Using a 1% level of significance, test the null hypothesis that a decision regarding this proposal is independent of its cost.

9. The number of magazine subscriptions ordered by different families according to their income in one town is shown in the following table:

		Newsweek	Time	Sports Illustrated
	Above $20,000	58	65	49
INCOME	$10,000–$20,000	69	76	69
	Below $10,000	55	71	66

Using a 5% level of significance, test the null hypothesis that for this town the magazine ordered is independent of the family's income.

10. Five hundred guests at the Rocky Beach Hotel were asked which sport they prefer. Their answers are summarized as follows:

	Tennis	Golf	Swimming	Horseback Riding
Men	58	144	29	28
Women	42	62	82	55

Using a 5% level of significance, test the null hypothesis that the sport preferred is not affected by the sex of the guest.

For each of the following you may wish to use a calculator.

*11. The leading causes of death in the United States during 1973 are shown in the following table. Using a 5% level of significance, test the null hypothesis that the cause of death is not affected by a person's age.

		AGE		
		Under 65	65 and Over	Total
	Disease of Heart and Blood Vessels	278,842	783,318	1,062,160
	Cancer	146,627	204,667	351,294
CAUSE OF DEATH	Pneumonia and Influenza	30,784	31,815	62,599
	Accidents	101,517	14,780	116,297
	Diabetes	12,108	26,117	38,225
	All Other Causes	202,317	140,111	342,428

Source: National Center for Health Services. U.S. Public Health Service HEW and The American Heart Association.

*12. The following chart gives the cause of accidental death in the United States during 1973:

	ACCIDENTAL CAUSE OF DEATH							
	Motor Vehicle	Falls	Drowning	Fires Burns	Inges-tion of Food Objects	Fire-arms	Poison (Solid and Liquid)	Poison by Gas
Male	39,941	8307	7426	3895	1780	2269	2354	1217
Female	15,570	8119	1299	2608	1233	349	1329	435

Source: National Safety Council, 1973.

Using a 5% level of significance, test the null hypothesis that the cause of death is independent of the sex of the person.

13.3   GOODNESS OF FIT

In addition to the applications mentioned in the previous sections, the chi-square test statistic can also be used to determine whether an observed frequency distribution is in agreement with the expected mathematical distribution. For example, when a die is rolled, we assume that we have a binomial distribution and that the probability of any one face coming up is 1/6. Thus, if a die is rolled 120 times, we would expect each face to come up approximately 20 times since $\mu = np$ and $\mu = 120(1/6) = 20$.

Suppose we actually rolled a die 120 times and obtained the results shown in Table 13.1.

TABLE 13.1

EXPECTED AND OBSERVED FREQUENCIES WHEN A DIE WAS TOSSED 120 TIMES

Die Shows	Expected Frequency	Observed Frequency
1	20	18
2	20	21
3	20	17
4	20	21
5	20	19
6	20	24
	Total = 120	Total = 120

In this table we have also indicated the expected frequencies. Are these observed frequencies reasonable? Do we actually have an honest die?

To check whether the differences between the observed frequencies and the expected frequencies are due purely to chance or are significant, we use the chi-square test statistic of Formula 13.1. We reject the null hypothesis that the observed differences are not significant only when the test statistic falls in the rejection region.

In our case the value of the test statistic is

$$\chi^2 = \Sigma \frac{(O - E)^2}{E}$$

$$= \frac{(18 - 20)^2}{20} + \frac{(21 - 20)^2}{20} + \frac{(17 - 20)^2}{20} + \frac{(21 - 20)^2}{20} + \frac{(19 - 20)^2}{20} + \frac{(24 - 20)^2}{20}$$

$$= 1.60$$

There are $6 - 1$, or 5, degrees of freedom. From Table IX in the Appendix we find that the $\chi^2{}_{0.05}$ value with 5 degrees of freedom is 11.070. Since the test statistic has a value of only 1.60, which is considerably less than 11.070, we do not reject the null hypothesis. Any differences between the observed frequencies and the expected frequencies are due purely to chance.

The following examples will further illustrate how the chi-square test statistic can be used to test **goodness of fit,** that is, to determine whether the observed frequencies fit with what was expected.

EXAMPLE 1

The number of phone calls received by a local chapter of Alcoholics Anonymous is indicated below:

	M	T	W	T	F
Number of Calls Received	173	153	146	182	193

Using a 5% level of significance, test the null hypothesis that the number of calls received is independent of the day of the week.

SOLUTION

We first calculate the number of expected calls. If the number of calls received is independent of the day of the week, we would expect to receive

$$\frac{173 + 153 + 146 + 182 + 193}{5} = 169.4$$

calls per day. We can then set up the following:

	M	T	W	T	F
Observed Number of Calls	173	153	146	182	193
Expected Number of Calls	169.4	169.4	169.4	169.4	169.4

Now we calculate the value of the chi-square test statistic. We have

$$\chi^2 = \Sigma \frac{(O - E)^2}{E}$$

$$= \frac{(173 - 169.4)^2}{169.4} + \frac{(153 - 169.4)^2}{169.4} + \frac{(146 - 169.4)^2}{169.4}$$

$$+ \frac{(182 - 169.4)^2}{169.4} + \frac{(193 - 169.4)^2}{169.4}$$

$$= 0.0765 + 1.5877 + 3.2323 + 0.9372 + 3.2878$$

$$= 9.1215$$

There are $5 - 1$, or 4, degrees of freedom. From Table IX in the Appendix we find that the $\chi^2_{0.05}$ value with 4 degrees of freedom is 9.488. Since the test statistic has a value of 9.1215, which is less than 9.488, we do not reject the null hypothesis and the claim that the number of calls received is independent of the day of the week.

EXAMPLE 2

A scientist has been experimenting with rats. As a result of certain injections, he claims that when two black rats are mated, the offspring will be black, white, and gray in the proportion 5:4:3. This means that the probability of a black offspring is 5/12, the probability of a white offspring is 4/12, and the probability of a gray rat is 3/12. Many rats were mated after being injected with the chemical. Of 180 newborn rats 71 were black, 69 were white, and 40 were gray. Can we accept the scientist's claim that the true proportion is 5:4:3? Use a 5% level of significance.

SOLUTION

We first calculate the expected frequencies. Out of 180 rats we would expect 180(5/12), or 75, of them to be black. Similarly, out of 180 rats we would expect 180(4/12), or 60, of them to be white, and we would expect 180(3/12), or 45, of them to be gray. We now set up the following table:

Color of Rat	Expected Frequency	Observed Frequency
Black	75	71
White	60	69
Gray	45	40

The value of the χ^2 test statistic is

$$\chi^2 = \Sigma\frac{(O - E)^2}{E}$$

$$= \frac{(71 - 75)^2}{75} + \frac{(69 - 60)^2}{60} + \frac{(40 - 45)^2}{45}$$

$$= 2.12$$

There are $3 - 1$, or 2, degrees of freedom. From Table IX in the Appendix we find that the $\chi^2_{0.05}$ value with 2 degrees of freedom is 5.991. Since the test statistic has a value of 2.12, which is less than 5.991, we do not reject the null hypothesis and the scientist's claim.

1. A gynecologist delivered 30 babies in the fall, 26 babies in the winter, 41 babies in the spring, and 19 babies in the summer. Using a 5% level of significance, test the null hypothesis that the number of babies delivered is independent of the season of the year.

2. A principal claims that third grade girls in her school do not care about the color of their dresses. In a sample of 475 girls the following were the colors of dresses worn by these girls:

$$
\begin{array}{ll}
81 & Red \\
102 & Green \\
88 & Blue \\
96 & Tan \\
108 & Brown
\end{array}
$$

Using a 5% level of significance, test the null hypothesis that third grade girls do not care about the color of their dresses.

3. A hunter claims that deer, rabbits, pheasant, and fox are present in the northern part of a forest in the ratio of 5:4:3:2. During 1976, 280 animals were killed in this area. Eighty deer, 95 rabbits, 55 pheasants, and 50 fox were killed. Using a 5% level of significance, test the null hypothesis that the true proportion is 5:4:3:2.

4. The number of new claims received by the State Unemployment Insurance Commission during the week of June 7–11 is as follows:

	M June 7	T June 8	W June 9	T June 10	F June 11
Number of New Claims Received	276	212	198	246	253

Using a 5% level of significance, test the null hypothesis that the number of new claims received is independent of the day of the week.

5. A die was tossed 600 times and the following faces showed:

Face Showing	1	2	3	4	5	6
Observed Frequency	89	106	118	79	93	115

Using a 5% level of significance, test the null hypothesis that the die is an honest die.

6. A manufacturer of mixed nuts claims that each can of mixed nuts he sells contains peanuts, cashews, pecans, and butternuts in the proportion 4:3:2:1. A sample of 300 nuts contains 100 peanuts, 99 cashew nuts, 48 pecans, and 53 butternuts. Using a 5% level of significance, test the null hypothesis that the true proportion is 4:3:2:1.

SELF-STUDY GUIDE

In this chapter we discussed the chi-square distribution and how it can be used to test hypotheses that differences between expected frequencies and observed frequencies are due purely to chance.

The chi-square test statistic can also be used to analyze whether the two factors of a contingency table are independent or not. This is very useful, especially in the social sciences, where the data is often grouped according to two factors.

Finally we applied the chi-square test statistic to test whether observed frequency distributions are in agreement with expected mathematical frequencies.

When using the χ^2 test statistic, we must take great care in determining the number of degrees of freedom. Also, as we pointed out in a comment on page 387, each expected cell frequency must be at least 5 for the χ^2 test statistic to be applied.

You should be able to identify each symbol in the following formulas, understand the relationships among the symbols expressed in each formula, understand the significance of each formula, and use the formulas in solving problems.

1. $$\chi^2 = \Sigma \frac{(O - E)^2}{E} \quad \text{where} \quad O = \text{observed frequency}$$

    $$\text{and} \quad E = \text{expected frequency}$$

2. The expected frequency for any cell of a contingency table is

$$\frac{(\text{total of row to which cell belongs}) \cdot (\text{total of column to which cell belongs})}{\text{total sample size}}$$

3. The number of degrees of freedom for a contingency table is

$(r - 1)(c - 1)$ where c = number of columns and r = number of rows

You should now be able to demonstrate your knowledge of the following ideas presented in this chapter by giving definitions, descriptions, or specific examples. Page references are given for each term so that you can check your answer.

Expected frequency (page 383)   The chi-square distribution (page 384)
Observed frequency (page 383)   Contingency table (page 391)
Chi-square test statistic (page 383)   Goodness of fit (page 399)

The tests of the following section will be more useful if you take them after you have studied the examples and solved the exercises in this chapter.

Form *A*

1. When there is perfect agreement between the observed frequencies and the expected frequencies, the value of χ^2 is _____.

2. Any arrangement of data into a two-way classification is called a _____.

3. We use a chi-square test
   a. to compare differences between sample means   *b.* to find the standard error of the mean   *c.* to compare ratios of two or more binomial variables   *d.* only when one sample size is less than 30   *e.* none of these

4. When there is very little agreement between observed and expected frequencies, that is when the expected frequency is considerably different from the observed frequency, the chi-square test statistic will
   a. equal one   *b.* be very large   *c.* be close to zero   *d.* be negative   *e.* equal five

5. If a contingency table has *r* rows and *c* columns, the number of degrees of freedom is _____.

6. The number of acceptable and unacceptable parts resulting from two different manufacturing methods are shown below:

	Number of Acceptable Parts	Number of Unacceptable Parts
Process A	13	11
Process B	7	29

   If the two processes are equally effective, we would expect the number of acceptable parts made by process *A* to be   *a.* 8   *b.* 7   *c.* 11   *d.* 10   *e.* 12

7. Refer to question 6. We would expect the number of acceptable parts made by process B to be   *a.* 8   *b.* 7   *c.* 11   *d.* 10   *e.* 12

8. If the observed frequencies are

10	10
10	10

   what are the expected frequencies?

   a.

8	12
12	8

   b.

1.0	1.0
1.0	1.0

   c.

10	10
10	10

   d.

5	5
5	5

   e. none of these

9. A group of 300 children are asked if they know how to swim. The results follow.

	Swim	Do Not Swim
Boys	110	40
Girls	80	70

   What is the expected number of boys who swim?
   a. 95   *b.* 75   *c.* 90   *d.* 110   *e.* none of these

10. Refer to question 9. What is the expected number of girls who swim?
    a. 95   *b.* 75   *c.* 85   *d.* 80   *e.* none of these

Form B

1. A random survey of 400 of the cars parked in the student parking lot during day session classes at Staten Island Community College shows that 283 of them have air conditioning. A similar survey of 275 cars during evening session classes shows that 193 of them have air conditioning. Using a 5% level of significance, test the null hypothesis that there is no difference between the proportion of day session and evening session students whose cars have air conditioning.

2. A survey of 400 families in Tucson, Denver, and New Orleans taken to determine the proportion of families that own a color television shows the following results:

		Tucson	Denver	New Orleans
OWN COLOR TELEVISION?	Yes	90	70	80
	No	70	30	60

Using a 5% level of significance, test the null hypothesis that the true proportion of families that own color television sets is the same for all three cities.

3. A survey of 800 college students was taken to determine how many of them wear contact lenses. The following table indicates the results of the survey according to the sex of the student:

		SEX	
		Male	Female
WEAR CONTACT	Yes	69	123
LENSES?	No	331	277

Using a 1% level of significance, test the null hypothesis that there is no difference between the proportion of male and female students who wear contact lenses.

4. A state Consumer's Protection Agency investigates complaints about the guarantee work provided by four large companies. The following table indicates the results of the agency's survey involving customers of the four companies:

		COMPANY			
		A	B	C	D
SATISFIED?	Yes	79	68	54	93
	No	31	52	46	107

Using a 5% level of significance, test the null hypothesis that there is no difference between the corresponding true proportions.

5. Six hundred new car owners were asked to indicate what influenced their decision to buy a particular car. The following table indicates the sex of the car owner and the reason given for buying the car:

REASON FOR BUYING CAR

	Style	Economy of Operation
Male	108	192
Female	174	126

Using a 5% level of significance, test the null hypothesis that a decision is independent of the sex of the owner asked.

6. The following table indicates the sex of prospective jurors in three different counties in a state. Using a 5% level of significance, test the null hypothesis that the true proportion of prospective female jurors is the same for all three counties.

COUNTY

	A	B	C
Male	39	87	62
Female	51	78	78

7. A group of 111 men and 59 women were asked to indicate their preference between two brands of deodorant. The results are as follows:

	Brand A	Brand B
Men	62	49
Women	41	18

Using a 5% level of significance, test the null hypothesis that preference for a particular brand of deodorant is not related to sex.

8. The voters of a town were asked to indicate if they favor adding chlorine to the town's water supply. The following results were obtained:

	Opposed	No Opinion	In Favor
Male	88	239	494
Female	62	191	106

Using a 5% level of significance, test the null hypothesis that the decision was not influenced by the sex of the voter.

9. A newspaper reporter asked 450 randomly selected people if they were in favor of America selling nuclear reactors to foreign countries. The following table indicates their answers according to their sex:

	Male	Female
Yes	146	122
No	104	78

Using a 1% level of significance, test the null hypothesis that there is no difference between the corresponding true proportions.

10. Three hundred new workers were recently hired by the ABC Construction Company. The company claims that it does not discriminate in its hiring practices. The following table indicates the number of people who applied, their race, and the number of people hired:

	White	Black	Mexican and Indian
Hired	125	90	85
Not Hired	375	210	315
Total Number of Applicants	500	300	400

Using a 5% level of significance, can this company be accused of discrimination in its hiring practices?

11. The following table shows the total number of arrests during 1973 for various offenses:

Offense	Male	Female
Murder and Nonnegligent Manslaughter	11,760	2,077
Robbery	92,190	6,679
Aggravated Assault	126,717	19,306
Burglary	284,679	16,244
Larceny	420,049	193,885
Forgery	29,090	10,688
Narcotic Drug Laws	394,327	66,604

Source: Federal Bureau of Investigation. Uniform Crime Reports, 1973.

Using a 5% level of significance, test the null hypothesis that the sex of the offender is not related to the type of offense charged against the person. (Hint: Use a calculator.)

A more detailed representation of increased crime offenses by women within the past decade is presented in the clipping which follows.

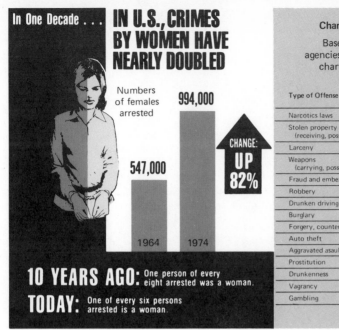

Changing Pattern of Crimes—a Sampling

Based on reports from over 1,800 local agencies, comprising a smaller sample than the chart but typical of nationwide trends—

Type of Offense	Number of Female Arrests 1960	1974	Change
Narcotics laws	3,607	34,646	Up 861%
Stolen property (receiving, possessing)	535	4,639	Up 767%
Larceny	24,769	124,838	Up 404%
Weapons (carrying, possessing)	1,156	5,672	Up 391%
Fraud and embezzlement	3,671	15,856	Up 332%
Robbery	1,247	5,059	Up 306%
Drunken driving	6,093	21,276	Up 249%
Burglary	2,952	10,212	Up 246%
Forgery, counterfeiting	2,634	7,025	Up 167%
Auto theft	1,571	4,109	Up 162%
Aggravated asault	5,739	13,414	Up 134%
Prostitution	13,895	26,393	Up 90%
Drunkenness	74,114	36,413	Down 51%
Vagrancy	6,645	2,655	Down 60%
Gambling	6,912	2,411	Down 65%

Source: Federal Bureau of Investigation

U.S. News & World Report, December 22, 1975. Reprinted from *U.S. News & World Report,* Copyright © 1975 U.S. News & World Report, Inc.

SUGGESTED READING

Anderson, R. L. and Bancroft, T. A. *Statistical Theory in Research.* New York: McGraw-Hill, 1952. chap. 12.

Chapman, D. G. and Schaufele, R. A. *Elementary Probability Models and Statistical Inference.* Lexington, Mass.: Xerox College Publishing Co., 1970. chap. 11.

Cochran, W. G. "The χ^2 Test of Goodness of Fit" in *Annals of Mathematical Statistics.* vol. 23, 1952, pp. 315–345.

Dixon, W. J. and Massey, F. J., Jr. *Introduction to Statistical Analysis.* 3rd ed. New York: McGraw-Hill, 1969.

Nie, N.; Bent, D. H.; and Hull, C. H. *Statistical Package for the Social Sciences.* New York: McGraw-Hill, 1970.

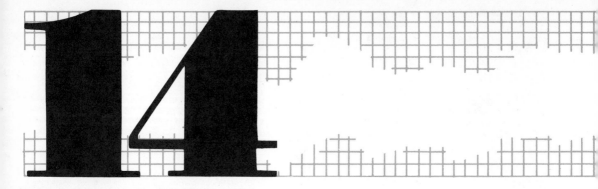

14

ANALYSIS
OF VARIANCE

HOW SPENDING PATTERNS CHANGED
IN THE INFLATION-PLAGUED RECESSION

PEOPLE SPENT MORE FOR...

	Share of All Consumer Outlays 1973	1975
Food and beverages	20.8%	21.7%
Housing	15.2%	15.5%
Utilities	3.9%	4.5%
Medical care	8.3%	8.6%
Personal business (brokers' and lawyers' fees, etc.)	5.0%	5.1%

PEOPLE SPENT LESS FOR...

	Share of All Consumer Outlays 1973	1975
Transportation	13.7%	12.8%
Clothes, accessories	8.9%	8.5%
Home furnishings, supplies	8.9%	8.5%
Foreign travel, religious activities, miscellaneous	1.9%	1.7%
Education	1.6%	1.5%
Personal care	1.6%	1.5%
Tobacco products	1.6%	1.5%

NO CHANGE IN...

	Share of All Consumer Outlays 1973	1975
Recreation	6.8%	6.8%
Telephone service	1.8%	1.8%

Note: Each tenth of 1 per cent represents nearly a billion dollars in purchases at 1975 prices.

Source: U.S. Dept of Commerce; 1975 estimate by USN&WR Economic Unit

U.S. News & World Report, April 5, 1976. Reprinted from *U.S. News & World Report,* Copyright © 1976 U.S. News & World Report, Inc.

By looking at the data given in the chart, can we say that the overall average amount of money spent for various items has changed significantly?

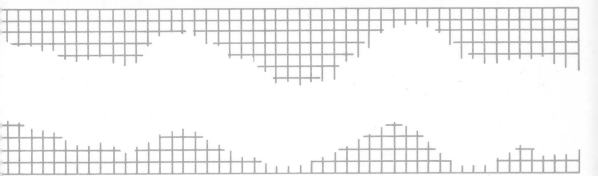

Preview

In this chapter we will discuss the following topics:

1. *Single Factor Analysis of Variance*
This is a technique that is used when comparing several sample means.

2. *Analysis of Variance Techniques*
We discuss how to set up ANOVA charts. We also develop formulas that we use when working with ANOVA charts.

3. *F-Distribution*
This is a distribution that we use when comparing variances. We apply this distribution to determine whether the differences in sample means are significant.

Introduction

Suppose a company is interested in determining whether changing the lighting conditions of its factory will have any effect on the number of items produced by a worker. It can arrange the lighting conditions in four different ways A, B, C, and D. To determine how lighting affects production, the factory supervisor decides to randomly arrange the lights under each of the four possible conditions for an equal number of days. He will then measure worker productivity under each of these conditions.

After all the data is collected, the manager calculates the number of items produced under each of these lighting conditions. He now wishes to test whether there is any difference between these sample means. How does he proceed?

The null hypothesis is H_0: $\mu_A = \mu_B = \mu_C = \mu_D$. This tells us that the mean number of items produced is the same for each lighting condition. The alternate hypothesis is H_1: $\mu_A \neq \mu_B \neq \mu_C \neq \mu_D$. This tells us that not all the means are the same. Thus the lighting condition does affect production. The

supervisor will reject the null hypothesis if one (or more) of the means is different from the others.

To determine if lighting affects production, he could use the techniques of Section 12.4 for testing differences between means. However, he would have to apply those techniques many times since each time he would be able to test only two means. Thus he would have to test the following null hypotheses:

$H_0: \mu_A = \mu_B \qquad H_0: \mu_A = \mu_D \qquad H_0: \mu_B = \mu_D$
$H_0: \mu_A = \mu_C \qquad H_0: \mu_B = \mu_C \qquad H_0: \mu_C = \mu_D$

In order to conclude that there is not a difference between the sample means, he would have to accept each of the six separate null hypotheses previously listed. Performing these tests involves a tremendous amount of computations. Furthermore, in doing it this way, the probability of making a Type-I error is quite large.

Since many of the problems that occur in applied statistics involve testing whether there is any significant difference between several means, statisticians use a special analysis of variance technique. This is abbreviated as **ANOVA.** The reason why these types of problems occur so often is because many companies often hire engineers to design new techniques or processes for producing products. The company must then compare the sample means of these new processes with the sample mean of the old process. Using the ANOVA technique, we have to test only one hypothesis in order to determine when to reject or to accept a null hypothesis.

ANOVA techniques can be applied to many different types of problems. In this chapter we will discuss one simple application of ANOVA techniques. This is the case in which the data is classified into groups on the basis of one single property.

14.1   SINGLE FACTOR ANOVA

Let us return to the factory supervisor problem discussed in the introduction. The manager repeated each of the lighting conditions on 5 different days. Table 14.1 indicates the number of items produced under each of the conditions.

TABLE 14.1

Single Factor
ANOVA

NUMBER OF ITEMS PRODUCED UNDER DIFFERENT LIGHTING
CONDITIONS

		DAY 1	DAY 2	DAY 3	DAY 4	DAY 5	AVERAGE
Lighting Conditions	A	12	10	15	12	13	12.4
	B	16	14	9	10	15	12.8
	C	11	15	8	12	10	11.2
	D	15	14	12	11	13	13

The means for each of these lighting conditions are 12.4, 12.8, 11.2, and 13. Since the average number of items produced is not the same for all the lighting conditions, the question becomes "Is the variation among individual sample means due purely to chance or are these differences due to the different lighting conditions?" The null hypothesis is H_0: $\mu_A = \mu_B = \mu_C = \mu_D$. If the null hypothesis is true, then the lighting condition does not affect production. We can then consider our results as a listing of the number of items produced on 20 randomly selected days under one of the lighting conditions.

In working with such problems we will assume that the number of items produced is normally distributed and that the variance is the same for each of the lighting conditions. We will also assume that the experiments with the different lighting conditions are independent of each other. Usually the experiments are conducted in random order so that we have independence.

Let us now apply the ANOVA techniques to this problem. Notice that we have included the row totals and the column totals in Table 14.2.

TABLE 14.2

NUMBER OF ITEMS PRODUCED UNDER DIFFERENT LIGHTING
CONDITIONS

		DAY 1	DAY 2	DAY 3	DAY 4	DAY 5	Row Total
Lighting Conditions	A	12	10	15	12	13	62
	B	16	14	9	10	15	64
	C	11	15	8	12	10	56
	D	15	14	12	11	13	65
Column Total		54	53	44	45	51	247 Grand Total

The first number that we calculate is called the **Total Sum of Squares** and is abbreviated as SS (total). To obtain the SS (total) we square each of the numbers in the table and add these squares together. Then we divide the square of the grand total (total of all the rows or the total of all the columns) by the number in the sample. (In our case there are 20 in the sample.) Subtracting this result from the sum of the squares, we get

$$(12^2 + 10^2 + 15^2 + 12^2 + 13^2 + 16^2 + 14^2 + 9^2 + 10^2 + 15^2 + 11^2 + 15^2 + 8^2$$
$$+ 12^2 + 10^2 + 15^2 + 14^2 + 12^2 + 11^2 + 13^2) - \frac{(247)^2}{20}$$

$$= 3149 - 3050.45$$
$$= 98.55$$

*Thus SS (total) = 98.55. (We will put an * in front of important results.)

To find the next important result we square each of the row totals and divide the sum of these squares by the number in each row, which is 5. Then we divide the square of the grand total by the number in the sample. We subtract this result from the sum of squares. We get

$$\frac{62^2 + 64^2 + 56^2 + 65^2}{5} - \frac{(247)^2}{20} = 3060.2 - 3050.45 = 9.75$$

This result is called the **Sum of Squares Due to the Factor.** The factor in our example is the lighting conditions. Thus

$$*SS(lighting) = 9.75$$

Now we calculate a number which is called the **Sum of Squares of the Error** abbreviated as SS(error). We first square each entry of the table and add these squares together. Then we square each row total and divide the sum of these squares by the number in each row. Subtracting this result from the sum of squares, we get

$$12^2 + 10^2 + 15^2 + 12^2 + \cdots + 11^2 + 13^2 \quad - \quad \frac{62^2 + 64^2 + 56^2 + 65^2}{5}$$

$$= 3149 - 3060.2$$
$$= 88.8$$

Thus

$$*SS(error) = 88.8$$

We enter these results along with the appropriate number of degrees of freedom, which we will determine shortly, in a table known as an **ANOVA table.** The general form of an ANOVA table is shown in Table 14.3.

TABLE 14.3

ANOVA TABLE

SOURCE OF VARIATION	SUM OF SQUARES (SS)	DEGREES OF FREEDOM (df)	MEAN SQUARE (MS)	F-RATIO
Factors of Experiment				
Error				
Total				

In our case we have the ANOVA table shown in Table 14.4. We have entered the important results that we obtained in the preceding paragraphs in the appropriate space of the table.

TABLE 14.4

ANOVA TABLE

SOURCE OF VARIATION	SUM OF SQUARES	DEGREES OF FREEDOM	MEAN SQUARE	F-RATIO
Factors (lighting)	9.75	3	3.25	
Error	88.8	16	5.55	$\dfrac{3.25}{5.55} = 0.59$
Total	98.55	19		

The degrees of freedom are obtained according to the following rules.

Rules
1. The number of degrees of freedom for the factor tested is one less than the number of possible levels at which the factor is tested. If there are r levels of the factor, the number of degrees of freedom is $r - 1$.
2. The number of degrees of freedom for the error is one less than the number of repetitions of each condition multiplied by the number of possible levels of the factor. If each condition is repeated c times, then the number of degrees of freedom is $r(c - 1)$.
3. Finally, the number of degrees of freedom for the total is one less than the total number in the sample. If the sample consists of n things, the number of degrees of freedom is $n - 1$.

In our case there are 4 experimental conditions and each condition is repeated 5 times so that $r = 4$, $c = 5$, and $n = 20$.

Thus

$$df(\text{factor}) = r - 1 = 4 - 1 = 3$$
$$df(\text{error}) = r(c - 1) = 4(5 - 1) = 4 \cdot 4 = 16$$
$$df(\text{total}) = n - 1 = 20 - 1 = 19$$

We enter these values in an ANOVA table as shown in Table 14.4.

Now we calculate the values that belong in the **Mean Square** column. We divide the sum of squares for the row by the number of degrees of freedom for that row. Thus,

$$MS(\text{lighting}) = \frac{SS(\text{lighting})}{df(\text{lighting})} = \frac{9.75}{3} = 3.25$$

$$MS(\text{error}) = \frac{SS(\text{error})}{df(\text{error})} = \frac{88.8}{16} = 5.55$$

Although we are comparing the sample means for the different lighting conditions, the ANOVA technique compares variances under the assumption that the variances among all the levels is 0. If the variance is 0, then all the means are the same.

When comparing variances we use the **F-distribution**. The test statistic is

Figure 14.1

$$F = \frac{MS(\text{lighting})}{MS(\text{error})}$$

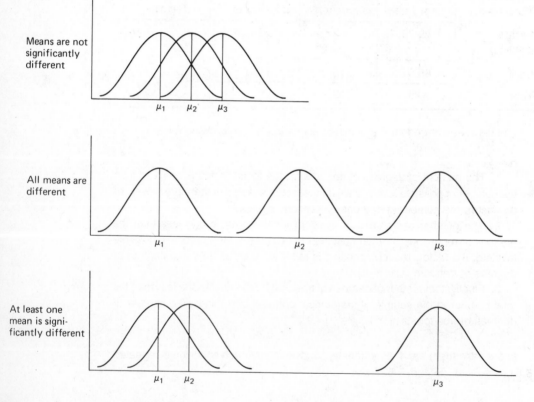

Means are not significantly different

All means are different

At least one mean is significantly different

When using the F-distribution we are testing whether the variances are zero (in which case the means are equal) or not. These situations are pictured in Figure 14.1

The values of the F-distribution depend on the number of degrees of freedom of the numerator, that of the denominator, and upon the level of significance. Table X in the Appendix gives us the different values corresponding to different degrees of freedom. From Table X in the Appendix we find that the F-value with 3 degrees of freedom for the numerator and 16 degrees of freedom for the denominator, at the 5% level of significance, is 3.24. We reject the null hypothesis if the F-value is greater than 3.24.

In our case the value of the F statistic is

$$\frac{MS(\text{lighting})}{MS(\text{error})} = \frac{3.25}{5.55} = 0.59$$

Since 0.59 is less than 3.24, we do not reject the null hypothesis. Thus the data does not indicate that the lighting condition affects production.

Let us summarize the procedure to be used when testing several sample means by the ANOVA technique. First draw an ANOVA table such as shown in Figure 14.5.

TABLE 14.5

ANOVA TABLE

SOURCE OF VARIATION	SUM OF SQUARES	DEGREES OF FREEDOM	MEAN SQUARE	F-RATIO
Factor of the Experiment	(1)	(4)	(7)	(9)
Error	(2)	(5)	(8)	
Total	(3)	(6)		

In this table we have placed numbers in parentheses in the various cells. The values that belong in each of these cells are obtained by using the following formulas:

cell (1) $\quad \dfrac{\Sigma(\text{each row total})^2}{\text{number in each row}} - \dfrac{(\text{grand total})^2}{\text{total sample size}}$

cell (2) $\quad \Sigma(\text{each original number})^2 - \dfrac{\Sigma(\text{each row total})^2}{\text{number in each row}}$

cell (3) $\Sigma(\text{each original number})^2 - \dfrac{(\text{grand total})^2}{\text{total sample size}}$

cell (4)   If there are r levels of the factor, $r - 1$

cell (5)   If there are c repetitions of each of the r levels, $r(c - 1)$

cell (6)   If the total sample consists of n things, $n - 1$

cell (7) $\dfrac{\text{cell (1) value}}{\text{cell (4) value}}$

cell (8) $\dfrac{\text{cell (2) value}}{\text{cell (5) value}}$

cell (9) $\dfrac{\text{cell (7) value}}{\text{cell (8) value}}$

Although these formulas seem complicated, they are easy to use as the following examples illustrate.

EXAMPLE 1

Three brands of cigarettes, six from each brand, were tested for tar content. Does the data shown below indicate that there is a significant difference in the average tar content for these three brands of cigarettes? Use a 5% level of significance.

Row Total

								Row Total
	X	14	16	12	18	11	13	84
Brand	Y	10	11	22	19	9	18	89
	Z	24	22	19	18	20	19	122
Column Total		48	49	53	55	40	50	295 Grand Total

SOLUTION

To solve this problem we set up the following ANOVA table:

SOURCE OF VARIATION	SUM OF SQUARES	DEGREES OF FREEDOM	MEAN SQUARE	F-RATIO
Cigarettes	(1)	(4)	(7)	(9)
Error	(2)	(5)	(8)	
Total	(3)	(6)		

Now we calculate the values that belong in each of the cells by using the formulas given on pages 415 and 416. We have

cell (1) $\dfrac{\Sigma(\text{each row total})^2}{\text{number in each row}} - \dfrac{(\text{grand total})^2}{\text{total sample size}}$

$= \dfrac{84^2 + 89^2 + 122^2}{6} - \dfrac{(295)^2}{18}$

$= 4976.83 - 4834.72$

$= 142.11$

cell (2) $\Sigma(\text{each original number})^2 - \dfrac{\Sigma(\text{each row total})^2}{\text{number in each row}}$

$= (14^2 + 16^2 + 12^2 + \cdots + 19^2) - \dfrac{(84^2 + 89^2 + 122^2)}{6}$

$= 5187 - 4976.83$

$= 210.17$

cell (3) $\Sigma(\text{each original number})^2 - \dfrac{(\text{grand total})^2}{\text{total sample size}}$

$= (14^2 + 16^2 + 12^2 + \cdots + 19^2) - \dfrac{(295)^2}{18}$

$= 5187 - 4834.72$

$= 352.28$

cell (4) $df = r - 1$
$\qquad = 3 - 1 = 2$

cell (5) $df = r(c - 1)$
$\qquad = 3(6 - 1) = 3(5) = 15$

cell (6) $df = n - 1$
$\qquad = 18 - 1 = 17$

cell (7) $\dfrac{\text{cell (1) value}}{\text{cell (4) value}} = \dfrac{142.11}{2} = 71.06$

cell (8) $\dfrac{\text{cell (2) value}}{\text{cell (5) value}} = \dfrac{210.17}{15} = 14.01$

cell (9) $\dfrac{\text{cell (7) value}}{\text{cell (8) value}} = \dfrac{71.06}{14.01} = 5.07$

Now we enter these numbers on the ANOVA table:

SOURCE OF VARIATION	SUM OF SQUARES	DEGREES OF FREEDOM	MEAN SQUARE	F-RATIO
Cigarettes	142.11 (1)	2 (4)	71.06 (7)	(9)
Error	210.17 (2)	15 (5)	14.01 (8)	5.07
Total	352.28 (3)	17 (6)		

From Table X in the Appendix we find that the F-value with 2 degrees of freedom for the numerator and 15 degrees of freedom for the denominator at the 5% level of significance is 3.68. Since we obtained an F-value of 5.07 which is larger than 3.68, we reject the null hypothesis that the sample mean tar content is the same for the three brands of cigarettes.

EXAMPLE 2

Four groups of five students each were taught a skill by four different teaching techniques. At the end of a specified time the students were tested and their scores recorded. Does the following data indicate that there is a significant difference in the mean achievement for the four teaching techniques? Use a 1% level of significance.

							Row Total
TEACHING TECHNIQUE	A	64	73	69	75	78	359
	B	73	82	71	69	74	369
	C	61	79	71	73	66	350
	D	63	69	68	74	75	349
Column Total		261	303	279	291	293	1427 Grand Total

SOLUTION

To solve this problem we set up the following ANOVA table:

SOURCE OF VARIATION	SUM OF SQUARES	DEGREES OF FREEDOM	MEAN SQUARE	F-RATIO
Teaching Method	52.15 (1)	3 (4)	17.38 (7)	(9)
Error	500.4 (2)	16 (5)	31.28 (8)	0.56
Total	552.55 (3)	19 (6)		

Now we calculate the values that belong in each of the cells by using the formulas given on pages 415 and 416. We have

cell (1) $\dfrac{\Sigma(\text{each row total})^2}{\text{number in each row}} - \dfrac{(\text{grand total})^2}{\text{total sample size}}$

$= \dfrac{359^2 + 369^2 + 350^2 + 349^2}{5} - \dfrac{(1427)^2}{20}$

$= 101,868.6 - 101,816.45$

$= 52.15$

cell (2) $\Sigma(\text{each original number})^2 - \dfrac{\Sigma(\text{each row total})^2}{\text{number in each row}}$

$= (64^2 + 73^2 + 69^2 + 75^2 + \cdots + 74^2 + 75^2) - \left(\dfrac{359^2 + 369^2 + 350^2 + 349^2}{5} \right)$

$= 102,369 - 101,868.6$

$= 500.4$

cell (3) $\Sigma(\text{each original number})^2 - \dfrac{(\text{grand total})^2}{\text{total sample size}}$

$= (64^2 + 73^2 + 69^2 + 75^2 + \cdots + 74^2 + 75^2) - \dfrac{(1427)^2}{20}$

$= 102,369 - 101,816.45$

$= 552.55$

cell (4) $df = r - 1 = 4 - 1 = 3$

cell (5) $df = r(c - 1) = 4(5 - 1) = 4(4) = 16$

cell (6) $df = n - 1 = 20 - 1 = 19$

cell (7) $\dfrac{\text{cell (1) value}}{\text{cell (4) value}} = \dfrac{52.15}{3} = 17.38$

cell (8) $\dfrac{\text{cell (2) value}}{\text{cell (5) value}} = \dfrac{500.4}{16} = 31.28$

cell (9) $\dfrac{\text{cell (7) value}}{\text{cell (8) value}} = \dfrac{17.38}{31.28} = 0.56$

We enter all these values on the ANOVA table as shown.
From Table X in the Appendix we find that the F-value with 3 degrees of freedom for the numerator and 16 degrees of freedom for the denominator, at the 1% level of significance, is 5.29. Since we obtained and F-value of 0.56 we do not reject the null hypothesis that there is a significant difference in the mean achievement for the four teaching techniques.

EXERCISES

1. A record was kept on the number of items completed by 3 workers on each day of the week. Does the following data indicate that there is a significant difference in the average production of these employees? Use a 5% level of significance.

	A	17	14	28	16	11
WORKER	B	12	19	18	16	24
	C	14	13	18	22	19

2. Twenty-four animals were divided into groups of 4 each. Each animal in a group was injected with a chemical to increase its weight. Each group was given a different chemical. Does the following data indicate that there is a significant difference in the average weight gain of an animal over a fixed length of time as a result of the different chemicals? Use a 1% level of significance.

	A	10	13	7	9
	B	6	14	8	11
CHEMICAL	C	13	6	8	11
	D	10	9	15	13
	E	18	7	8	14
	F	12	13	8	9

3. The number of cases handled by each of five judges each day of the week is as follows:

	A	22	17	32	16	19
	B	19	36	24	18	31
JUDGE	C	25	29	14	30	26
	D	20	33	26	11	17
	E	25	17	22	34	27

Does the data indicate that there is a significant difference in the average number of cases handled by these judges? Use a 5% level of significance.

4. A farmer divides his 20 acre farm into 4 groups of 5 acres each. He then treats each of the acres with different types of fertilizers. Does the data shown below indicate that the type of fertilizer used significantly affects the yield of an acre? Use a 1% level of significance.

FERTILIZER	A	39	27	41	36	30
	B	38	32	29	40	26
	C	45	26	22	31	32
	D	28	29	21	34	30

5. Hearing-aid batteries produced by four different companies were tested to determine which brand lasts longer. Five batteries of each brand were tested. Does the following data indicate that there is no significant difference between the average life of these batteries? Use a 1% level of significance.

BATTERIES	A	22	29	19	16	18
	B	33	12	17	26	17
	C	18	27	19	17	32
	D	34	17	19	32	29

6. Sixteen music students were divided into 4 groups of 4 students each. Each group was taught how to play a new song on a particular instrument, each by a different method. The following chart indicates the number of minutes needed by each student to learn to play the song. Does the data indicate that there is a significant difference in the average time needed by each student to learn to play the song by the different methods? Use a 5% level of significance.

METHOD	A	55	48	49	58
	B	42	39	59	50
	C	46	32	51	48
	D	38	47	54	39

SELF-STUDY GUIDE

In this chapter we briefly introduced the important statistical technique known as analysis of variance or ANOVA. This technique is used when we wish to test a hypothesis about the equality of several means. We limited our discussion to normal populations with equal variances.

When applying the ANOVA technique, we do not test the differences between the means directly. Instead we test the variances. If the variances are zero, there is no difference between the means. We therefore analyze the source of the variation. Is it due purely to chance or is the difference in the variation significant? When analyzing variation by means of analysis of variance techniques, we use an ANOVA table.

Although we discussed the simplest type of ANOVA — the case in which there was only one factor of classification — it is worth noting that ANOVA techniques can be applied to more complicated situations. However these are beyond the scope of this book.

The most important thing to remember when testing several means is how to set up an ANOVA table and how to find the appropriate values for each cell of the ANOVA table. Both of these ideas are summarized on pages 415 and 416.

You should now be able to demonstrate your knowledge of the following ideas presented in this chapter by giving definitions, descriptions, or specific examples. Page references are given for each term so that you can check your answer.

ANOVA (page 410)	ANOVA table (page 412)
Total sum of squares (page 412)	MS (factor) (page 414)
SS (total) (page 412)	MS (error) (page 414)
SS (due to factors) (page 412)	F-distribution (page 414)
SS (due to error) (page 412)	

The tests of the following section will be more useful if you take them after you have studied the examples and solved the exercises in this chapter.

MASTERY TESTS

Form A

1. True or false: The ANOVA techniques can be used only when we are comparing two or more sample means.
2. The word ANOVA is an abbreviation for the words _____.
3. True or false: The ANOVA technique compares variances under the assumption that all the variances are 0.
4. The F-distribution is used when we wish to compare _____.
5. To obtain the F-ratio we divide the Mean Squares (factor) by _____.
6. The number of degrees of freedom for the error is _____.
7. True or false: The Total Sum of Squares can be obtained by adding the sum of squares of the factor with its degrees of freedom.
8. True or false: The values of the F-distribution depend upon the number of degrees of freedom of the numerator and upon the level of significance.
9. True or false: The _only_ way to compare several means is to use the F-distribution.
10. In an ANOVA table the number of degrees of freedom for the total is _____.

Form *B*

1. There are three savings banks in Adams City. Ten depositors from each bank are randomly selected and asked how long they had to wait before obtaining service from a teller. Their answers are as follows:

Bank A	Bank B	Bank C
11	17	27
14	22	25
17	21	17
16	32	23
13	16	31
12	9	14
18	12	17
22	18	19
10	19	12
19	24	8

Does the data indicate that there is a significant difference between the average waiting time at these banks? Use a 5% level of significance.

2. Bill is interested in buying a calculator and is considering three equally rated brands. He priced these brands at different stores and obtained the following prices for the cost of the calculator:

Calculator A	Calculator B	Calculator C
$29	$36	$33
24	29	27
36	19	29
22	28	22
33	37	28

Does the data indicate that there is a significant difference in the average cost of these brands of calculators? Use a 1% level of significance.

3. The State Attorney General's office receives numerous complaints about the service rendered by the four electric companies within the state. In one week it received the following number of daily complaints:

	A	11	19	22	21	26
	B	21	23	22	26	19
COMPANY	C	10	9	12	8	17
	D	11	14	13	16	22

Does the data indicate that there is no significant difference in the average number of complaints about each of the four companies? Use a 5% level of significance.

4. Seven movie critics in three different cities were asked to rate a new movie on a scale of 1 to 75. Their ratings are shown below:

City A	52	54	49	62	45	47	35
City B	61	62	68	43	39	53	58
City C	73	66	48	56	49	59	67

Does the data indicate that there is no significant difference in the average ratings of the movie by the critics of the different cities? Use a 5% level of significance.

5. The price of a container of milk varies both according to the neighborhood in which it is purchased and, within the neighborhood, the store in which it is purchased as follows:

		Price (in cents)					
AREA OF TOWN	Upper Class	48	39	45	49	42	46
IN WHICH MILK	Middle Class	38	39	42	45	48	47
WAS PURCHASED	Lower Class	47	53	46	49	52	50

Does the data indicate that the average price of a container of milk is significantly higher in the lower class neighborhood? Use a 1% level of significance.

6. The number of deliveries made by four trucking companies in a town during a week are as follows:

	A	36	41	45	42	31
TRUCKING	B	49	42	35	48	40
COMPANY	C	29	37	42	41	47
	D	24	35	40	29	45

Does the data indicate that the average number of deliveries is not significantly different for these trucking companies? Use a 5% level of significance.

7. The number of children reported lost on a daily basis at the three city beaches during the four day July 4th weekend are as follows:

Beach 1	59	63	48	60
Beach 2	61	66	43	62
Beach 3	63	64	50	61

Does the data indicate that the average number of lost children is not significantly different for these beaches? Use a 1% level of significance.

8. A psychologist administered a drug to 5 cats, 5 dogs, 5 rats, and 5 guinea pigs to determine how the drug affects the reaction time to a particular stimulus. The results of her experiments are as follows:

Reaction Time (in seconds)

ANIMALS	Cat	5.7	6.3	6.1	6.0	5.8
	Dog	6.2	5.3	5.7	5.6	5.5
	Rats	3.7	3.2	3.9	4.0	3.6
	Guinea Pigs	5.4	5.0	6.0	5.2	4.9

Does the data indicate that the average reaction time, as a result of the drug, is not significantly different for any of these animals? Use a 5% level of significance.

9. Environmentalists took four water samples at each of three different locations in an effort to measure whether the quantity of dissolved oxygen varied from one location to another. The quantity of dissolved oxygen in the water is used to determine the extent of the water pollution. The results of the survey are as follows:

Location 1	Location 2	Location 3
8.4	5.8	5.9
7.6	6.5	6.8
4.3	7.3	8.9
7.9	6.6	7.3

Does the data indicate that the average quantity of dissolved oxygen is the same at all three locations? Use a 1% level of significance.

SUGGESTED READING

Brownlee, K. A. *Statistical Theory and Methodology in Science and Engineering.* 2nd ed. New York: Wiley, 1965.

Edwards, Allen L. *Experimental Design in Psychological Research.* New York: Holt, Rinehart and Winston, 1950.

Guenther, W. C. *Analysis of Variance.* Englewood Cliffs, N.J.: Prentice-Hall, 1964.

Li, J. C. R. *Introduction to Statistical Inference.* Ann Arbor, Mich.: J. W. Edwards Press, 1961.

Snedecor, George W. and Cochran, William G. *Statistical Methods.* 6th ed., Ames, Iowa: The Iowa State University Press, 1967.

Steel, Robert G. and Torrie, James H. *Principles and Procedures of Statistics.* New York: McGraw-Hill, 1960.

Winer, B. J. *Statistical Principles in Experimental Design.* 2nd ed., New York: McGraw-Hill, 1971.

Yamane, Taro. *Statistics: An Introductory Analysis.* 2nd ed., New York: Harper & Row, 1967.

NONPARAMETRIC
STATISTICS

Import Dependence of U.S. for raw materials needed for an industrial economy				
item	1950	1970	2000	U.S. use as % of world use, 1970
	%	%	%	
Chromium	99	100	100	19
Tin	99	100	100	24
Manganese	77	95	100	14
Nickel	99	93	92	38
Lead	64	57	83	25
Aluminum	40	80	97	42
Zinc	37	61	85	
Copper	57	39	75	33
Iron	43	49	77	28
Tungsten	40	40	93	
Petroleum	12	23	?	33

Numbers in bold indicate those for which we are over 50% import-dependent.

Reprinted from the *ZPG National Reporter* (Nov. 1973) with permission from Zero Population Growth, Inc.

Source of data: TOWARDS A NATIONAL MATERIALS POLICY, by the National Commission on Materials Policy, 1972.   MINERAL FACTS AND PROBLEMS, by U.S. Bureau of Mines; 1970.

Over the years the United States has been importing raw materials from various foreign countries. Based upon the data contained in the clipping, would you say that the average percentage of imports for all items mentioned increased significantly between 1950 and 1970?

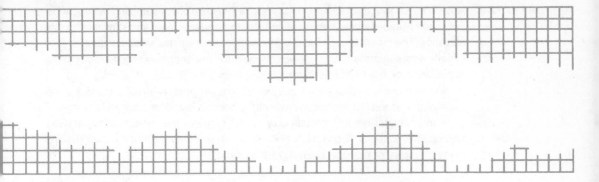

Preview

In this chapter we will discuss the following topics:

1. *Nonparametric Statistics or Distribution-Free Methods*
We analyze several tests which can be used when the assumptions about normally distributed populations or sample size cannot be satisfied.

2. *The Sign Test*
This is a test that is used in the before and after type study. We test whether or not $\mu_1 = \mu_2$ when we know that the samples are not independent.

3. *The Rank-Sum Test*
This alternative to the *standard* significance tests for the difference between two sample means is used when we have non-normally distributed populations. We can also use this test when the population variances are not equal.

4. *The Runs Test*
We use this test when we wish to test for randomness.

Introduction

In previous chapters we discussed procedures for testing various hypotheses involving means, proportions, variances, etc. Except for one case, all the tests discussed assumed that the populations from which the samples were taken were approximately normally distributed. Only when we applied the chi-square distribution in comparing observed frequencies with expected frequencies did we not specify the normal distribution.

Since there are many situations where this requirement cannot be satisfied, statisticians have developed techniques to be used in such cases. These techniques are known as **nonparametric statistics or distribution-free methods.** As the names imply, these methods are not dependent upon the distribution or parameters involved.

There are advantages and disadvantages associated with using nonparametric statistics. The advantages in using these methods as opposed to the *standard methods* are that (1) they are easier to understand, (2) they involve much less computation, and (3) they are less demanding in their assumptions about the nature of the sampling distributions. For these reasons many people often refer to nonparametric statistics as shortcut statistics. The disadvantages associated with nonparametric statistics are that they usually waste information, as we will see shortly, and that they tend to result in the acceptance of null hypotheses more often than they should.

Nonparametric statistical methods are often used when samples are small since most of the **standard tests** require that the sample sizes be reasonably large.

In this chapter we will discuss only briefly some of the more commonly used nonparametric tests. A complete discussion of all these methods would require many chapters or perhaps even several volumes.

15.1 THE SIGN TEST

By far the simplest of all the nonparametric tests is the **sign test.** It is used when we wish to test the null hypothesis that $\mu_1 = \mu_2$ and it is known that the samples are not independent. Usually the assumption of independence is not satisfied when measuring the same sample twice, as is done in the before and after type study.

To illustrate, suppose a college administrator is interested in knowing how a particular three week math mini-course affects a student's grade. Twenty students are selected and are given a math test. Then these students attend the mini-course and are retested. The administration would like to use the results of these two tests to determine whether the mini-course actually improves a student's score.

Table 15.1 contains the scores of the twenty students on the precourse test and the postcourse test. In this table we have taken each student's precourse test score and subtracted it from the student's postcourse test score to obtain the change score. An increase in score is assigned a plus (+) sign and a decrease in score is assigned a minus (−) sign. No sign is indicated when the two scores are identical.

The null hypothesis in this case is $\mu_1 = \mu_2$. This means that the mini-course does not significantly affect a student's score. Under this assumption we would expect an equal number of plus signs and minus signs. Thus if p is the proportion of plus signs, we would expect p to be around 0.5 (subject only to chance error).

Since we think that the mini-course does increase a student's score, the alternate hypothesis would be $\mu_2 > \mu_1$. Thus

$$H_0: \mu_1 = \mu_2$$
$$H_1: \mu_2 > \mu_1$$

We can now apply the methods discussed in Section 12.5 (pages 369–373) for testing a proportion. Recall that for testing a proportion we use the test statistic

$$z = \frac{p - \pi}{\sqrt{\dfrac{\pi(1 - \pi)}{N}}}$$

TABLE 15.1

SCORES OF TWENTY STUDENTS ON PRECOURSE AND POSTCOURSE TESTS

STUDENT	PRECOURSE SCORE	POSTCOURSE SCORE	SIGN OF DIFFERENCE
1	68	71	+
2	63	65	+
3	82	88	+
4	70	79	+
5	65	57	−
6	66	77	+
7	64	62	−
8	69	73	+
9	72	70	−
10	74	76	+
11	71	68	−
12	80	80	
13	59	71	+
14	85	80	−
15	57	65	+
16	83	87	+
17	43	48	+
18	94	94	
19	82	93	+
20	91	94	+

In applying this test statistic we let π be the true proportion of plus signs as specified in the null hypothesis. Thus if the mini-course does not affect a student's scores, we would expect as many plus signs as minus signs. There should be no more students obtaining higher scores than students obtaining lower scores as a result of this mini-course. Therefore, $\pi = 0.50$. We now count the number of plus signs. There are 13 plus signs out of a possible 18 sign changes. We ignore the cases which involve no change. So, $p = 13/18$, or 0.72, $N = 18$, and $\pi = 0.50$. Applying the test statistic, we have

$$z = \frac{p - \pi}{\sqrt{\dfrac{\pi(1 - \pi)}{N}}}$$

$$= \frac{0.72 - 0.50}{\sqrt{\dfrac{(0.5)(1 - 0.5)}{18}}}$$

$$= \frac{0.22}{\sqrt{0.01889}} = \frac{0.22}{0.118}$$

$$= 1.86$$

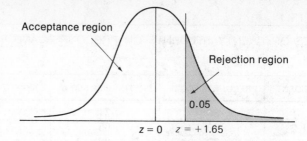

Figure 15.1

$z = 0$ $z = +1.65$

We use the one-tail rejection region of Figure 15.1. Since the value of $z = 1.86$ falls in the critical region, we reject the null hypothesis at the 5% level of significance. Thus the mini-course seems to have improved a student's score.

The following example will further illustrate the sign test technique.

EXAMPLE 1

A new weight reducing pill is given to 15 people once a week for 3 months to determine its effectiveness in reducing weight. The following data indicates the before and after weights in pounds of these 15 people:

Weight Before Taking Pill	131	127	116	153	178	202	192	183	171	182	169	155	163	171	208
Weight After Taking Pill	125	128	118	155	179	200	195	180	180	180	174	150	169	172	200

Using a 5% level of significance, test the null hypothesis that the pill is not effective in reducing weight.

SOLUTION

We arrange the data as follows:

Weight Before	Weight After	Sign of Difference
131	125	−
127	128	+
116	118	+
153	155	+
178	179	+
202	200	−
192	195	+
183	180	−
171	180	+
182	180	−
169	174	+
155	150	−
163	169	+
171	172	+
208	200	−

For each person we determine the change in weight. A plus sign indicates a gain and a minus sign indicates a loss. If the weight reducing pill is not effective, the average weight should be the same before and after taking this pill. Since we are testing whether a person's weight remains the same or is reduced, we have

$$H_0: \mu_1 = \mu_2$$
$$H_1: \mu_2 < \mu_1$$

Out of the 15 sign changes 6 are minus so that

$$p = \frac{6}{15} = 0.4 \quad \text{and} \quad N = 15$$

Also $\pi = 0.50$. Applying the test statistic, we get

$$z = \frac{p - \pi}{\sqrt{\dfrac{\pi(1 - \pi)}{N}}}$$

$$= \frac{0.4 - 0.5}{\sqrt{\dfrac{(0.5)(1 - 0.5)}{15}}}$$

$$= \frac{-0.1}{\sqrt{0.01667}} = \frac{-0.1}{0.129}$$

$$= -0.78$$

We use the one-tail rejection region of Figure 15.2. Since the value of $z = -0.78$ falls in the acceptance region, we cannot reject the null hypothesis. The weight reducing pill does not seem to be effective in reducing weight. It may even cause an increase in weight.

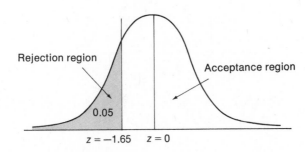

Rejection region

Acceptance region

0.05

$z = -1.65$ $z = 0$

Figure 15.2

Comment   We mentioned earlier that nonparametric methods are wasteful. A plus or minus sign merely tells us that a person gained weight or lost weight. It does not specify whether the gain was 1 pound, 10 pounds, or even 100 pounds. The same is true for the minus signs.

Comment   Since the sign test is so easy to use, many people use it even when the *standard* tests can be used. If a null hypothesis is rejected, we do not have to apply the *standard t*-test for comparing the means.

EXERCISES

1. A supermarket chain has noticed that shoplifting is on the rise. The manager decides to install closed circuit television cameras to combat the rise in shoplifting. The following table gives the before and after number of arrests per week in this supermarket chain over a 10 week period:

Number of Weekly Arrests Before Installation	10	9	8	6	5	9	5	9	5	7
Number of Weekly Arrests After Installation	9	8	8	7	3	9	4	6	4	2

   Using a 5% level of significance, test the null hypothesis that the closed circuit television cameras have no effect in reducing shoplifting.

2. Many people believe that the number of summonses issued by the police department has increased since the city threatened to layoff more police personnel. The following table gives the number of summonses issued per week by 12 policemen before and after the threat.

NUMBER OF SUMMONSES ISSUED

Policeman	Before Threat	After Threat
A	32	35
B	28	28
C	17	19
D	18	24
E	9	12
F	8	4
G	7	7
H	12	15
I	16	19
J	22	25
K	20	20
L	19	16

   Using a 5% level of significance, test the null hypothesis that the threat has no effect on the number of summonses issued.

3. A large electric company in the eastern United States decided to read electric meters monthly instead of every two months in an effort to decrease the daily number of complaints that it receives. The number of complaints that it received daily over a two week period before and after reading the meters monthly is as follows:

Day	Number of Complaints Before	Number of Complaints After
Monday	201	160
Tuesday	179	164
Wednesday	159	140
Thursday	195	184
Friday	177	174
Monday	170	142
Tuesday	184	191
Wednesday	182	188
Thursday	190	198
Friday	213	213

Using a 5% level of significance, test the null hypothesis that the monthly meter reading policy has not affected the number of complaints received.

4. A photocopying company has only one repairman. The company believes that this repairman is not competent since so many machines break down. He is fired and a new repairman is hired. The number of weekly repairs completed by these two workers over a twelve week period is as follows:

Repairman A	Repairman B
47	38
39	34
36	36
28	28
47	49
37	27
29	17
17	18
28	16
14	10
11	14
16	12

Using a 5% level of significance, test the null hypothesis that the number of breakdowns is not affected by the person who services the machines.

5. The State Environmental Protection Agency announced a new "get tough" policy against people who pollute the atmosphere. The number of daily cita-

tions, over a two week period, for polluting the air before and after instituting this new policy is as follows:

Before "get tough" Policy	After "get tough" Policy
32	35
37	31
35	39
28	35
41	24
44	44
35	38
31	39
34	38
47	57

Using a 5% level of significance, test the null hypothesis that the new "get tough" policy has not affected the number of citations issued.

15.2   THE RANK-SUM TEST

An important nonparametric test that is used as an alternative to the standard significance tests for the difference between two sample means is the **rank-sum test.** We can use this test when (a) the assumption about normality is not satisfied or when (b) the assumption that the samples are from populations with equal variances is not satisfied.

To illustrate how this test is used, we consider the following data on the number of minutes needed by two groups of music students to learn to play a particular song. Group A received special instruction whereas group B did not.

		Average
Group A	35, 39, 51, 63, 48, 31, 29, 41, 55	43.56
Group B	85, 28, 42, 37, 61, 54, 36, 57	50

The means of these two samples are 43.56 and 50. In this case we wish to decide whether the difference between the means is significant.

The two samples are arranged jointly, as if they were one sample, in order of increasing time. We get

Time	Group	Rank
28	B	1
29	A	2
31	A	3
35	A	4
36	B	5
37	B	6
39	A	7
41	A	8
42	B	9
48	A	10
51	A	11
54	B	12
55	A	13
57	B	14
61	B	15
63	A	16
85	B	17

We indicate each value whether it belongs to Group A or to Group B. Then we assign the ranks 1, 2, 3, 4, . . . , 17 to the scores, in this order, as shown.

Notice that the group A scores occupy the ranks of 2, 3, 4, 7, 8, 10, 11, 13, and 16. The group B scores occupy the ranks of 1, 5, 6, 9, 12, 14, 15, and 17. Now we sum the ranks of the group with the smaller sample size, in this case group B, getting

$$1 + 5 + 6 + 9 + 12 + 14 + 15 + 17 = 79$$

The sum of the ranks is denoted by R. In this case $R = 79$.

We always let n_1 and n_2 denote the sizes of the two samples where n_1 represents the smaller of the two sample sizes. Thus R represents the sum of the ranks of this smaller group. If both groups are of equal sizes, then either one is called n, and R represents the sum of the ranks of this group. Statistical theory tells us that if both n_1 and n_2 are large enough, each equal to eight or more, then the distribution of R can be approximated by a normal distribution. The test statistic is given by Formula 15.1.

FORMULA 15.1

$$z = \frac{R - \mu_R}{\sigma_R}$$

where

$$\mu_R = \frac{n_1(n_1 + n_2 + 1)}{2}$$

and

$$\sigma_R = \sqrt{\frac{n_1 n_2(n_1 + n_2 + 1)}{12}}$$

435

Using a 5% level of significance, we reject the null hypothesis of equal means if $z > 1.96$ or $z < -1.96$.

In our case $R = 79$, $n_1 = 8$, and $n_2 = 9$ so that

$$\mu_R = \frac{n_1(n_1 + n_2 + 1)}{2}$$

$$= \frac{8(8 + 9 + 1)}{2}$$

$$= 72$$

and

$$\sigma_R = \sqrt{\frac{n_1 n_2(n_1 + n_2 + 1)}{12}}$$

$$= \sqrt{\frac{8(9)(8 + 9 + 1)}{12}} = \sqrt{108}$$

$$= 10.39$$

The test statistic then becomes $\quad z = \dfrac{R - \mu_R}{\sigma_R} = \dfrac{79 - 72}{10.39} = 0.67$

Since the value of $z = 0.67$ falls in the acceptance region of Figure 15.3, we *do not* reject the null hypothesis. There is no significant difference between the means of these two groups.

Comment The method which we have just described is called the **Mann–Whitney test.** Statisticians have constructed tables which give the appropriate critical values when both sample sizes, n_1 and n_2, are smaller than 8. The interested reader can find such tables in many books on nonparametric statistics. The corresponding exact statistic is called the **Mann–Whitney U test.**

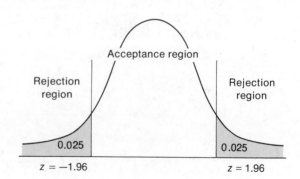

Figure 15.3

EXAMPLE 1   The Rank-Sum Test

An animal trainer in a circus is teaching 20 lions to perform a special trick. The lions have been divided into two groups, A and B. Group A gets positive reinforcement of food and favorable comments during the learning session whereas group B does not. The following table indicates the number of days needed by each lion to learn the trick:

Group A	78	95	82	69	111	65	73	84	92	110
Group B	121	132	101	79	94	88	102	93	98	127

Using a 5% level of significance, test the null hypothesis that the mean time for both groups is the same.

SOLUTION

The two samples are first arranged jointly, as if they were one large sample, in order of increasing size. We get

Days	Group	Rank
65	A	1
69	A	2
73	A	3
78	A	4
79	B	5
82	A	6
84	A	7
88	B	8
92	A	9
93	B	10
94	B	11
95	A	12
98	B	13
101	B	14
102	B	15
110	A	16
111	A	17
121	B	18
127	B	19
132	B	20

Since both groups are of equal size, we will work with group A. The sum of the ranks of group A is

$$1 + 2 + 3 + 4 + 6 + 7 + 9 + 12 + 16 + 17 = 77$$

Thus $R = 77$. Now we apply Formula 15.1. We have $R = 77$, $n_1 = 10$, and $n_2 = 10$ so that

$$\mu_R = \frac{n_1(n_1 + n_2 + 1)}{2}$$

$$= \frac{10(10 + 10 + 1)}{2}$$

$$= 105$$

and

$$\sigma_R = \sqrt{\frac{n_1 n_2(n_1 + n_2 + 1)}{12}}$$

$$= \sqrt{\frac{10 \cdot 10(10 + 10 + 1)}{12}}$$

$$= \sqrt{175}$$
$$= 13.23$$

The test statistic then becomes

$$z = \frac{R - \mu_R}{\sigma_R}$$

$$= \frac{77 - 105}{13.23}$$

$$= -2.12$$

We use the two-tail rejection of Figure 15.3. Since the value of $z = -2.12$ falls in the rejection region, we reject the null hypothesis and conclude that the number of minutes needed by each group is not the same. Positive reinforcement affects learning time.

Comment   In all examples discussed so far there were no ties for any rank. If a tie does come up, each of the tied observations is assigned the mean of the ranks which they occupy.

EXERCISES

In each of the following use the appropriate 5% level of significance.
1. Samples of water were taken from a river at ten different locations to determine the quantity of dissolved oxygen in the water. Such samples determine the

extent of water pollution. Samples of water were then taken from a second river. The results of these surveys follow:

River 1	4.7	6.3	7.2	6.8	7.5	8.4	9.3	8.7	7.9	5.4
River 2	8.6	5.7	5.9	6.5	7.1	7.8	6.6	4.9	5.8	

Test the null hypothesis that both rivers are equally polluted.

2. The number of pints of liquor sold by two liquor stores per day during a two week period is shown below. One store is closed on Sundays.

Store A	56	68	73	41	39	55	81	75	60	45	38	54	46	52
Store B	47	44	53	61	72	50	59	65	76	58	71	74		

Test the null hypothesis that the average number of bottles sold is the same.

3. The number of guests that attended ten wedding receptions at Bob's and the number of guests that attended eight wedding receptions at Bill's is shown below:

Bob's	143	157	180	168	110	112	156	148	131	120
Bill's	132	149	144	103	172	169	147	153		

Test the null hypothesis that the average number of guests that attended the receptions at both restaurants is the same.

4. The number of pounds of newspapers collected for recycling during one month from houses on the north and south side of Main Street is as follows:

North Side	312	278	464	222	378	279	300		
South Side	270	391	324	262	404	396	412	340	310

Test the null hypothesis that the average number of pounds collected from houses on both sides of Main Street is the same.

5. The ages of six cashiers on the sixth floor and seven cashiers on the seventh floor are as follows:

Sixth Floor	22	17	35	49	31	28	
Seventh Floor	18	25	29	36	42	26	38

Test the null hypothesis that the average age of a cashier on the sixth and seventh floors is the same.

6. The number of students registered in eight sections of Calculus I, which is offered during the fall semester, and in nine sections of the same course offered during the spring semester is as follows:

Fall Semester	45	48	42	38	29	53	36	32	
Spring Semester	44	40	39	35	46	50	41	34	30

Test the null hypothesis that the average number of students that register for Calculus I courses during both semesters is the same.

15.3   THE RUNS TEST

All the samples discussed so far in this book were assumed to be random samples. How does one test randomness?

In recent years mathematicians have developed a **runs test** for determining the randomness of samples. This test is based on the order in which the observations are made. For example, suppose 25 people are waiting in line for admission to a theater and they are arranged as follows where m denotes male and f denotes female:

f, f, f, f, m, f, m, m, m, m, m, f, f, f, f, f, f, m, m, m, m, f, m, m, f, m

Is this a random arrangement of the m's and f's?

In order to answer this question, statisticians use the **theory of runs**. We first have the following definition.

DEFINITION 15.1

A **run** is a succession of identical letters or symbols which is followed and preceded by a different letter or by no letter at all.

There are 10 runs in the above sequence of m's and f's. These are

Run	Letters
1	ffff
2	m
3	f
4	mmmm
5	ffffff
6	mmmm
7	f
8	mm
9	f
10	m

Many runs would indicate that the data occur in definite cycles according to some pattern. The same is true for data with few runs. In either case we do not have a random sample. We still need some way of determining when the number of runs is reasonable.

When using the runs test, note that the length of each individual run is not important. What is important is the number of times that each letter appears in the entire sequence of letters. Thus, in our example there are 25 people waiting in line; 13 are female and 12 are male, so that f appears 13 times and m appears 12 times. We have n_1 samples of one kind and n_2 samples of another kind. We now wish to test whether this sample is random.

Table XI in the Appendix gives us the critical values for the total number of runs. To use this table we first determine the larger of n_1 and n_2. In our case there are 13 f's and 12 m's so that the larger is 13 and the smaller is 12. We now move across the top of the chart until we reach 13. Then we move down until we get to the 12 row. Notice that there are two numbers in the box corresponding to larger 13, smaller 12. These are the numbers 8 and 19. These are also the critical values. If the number of runs is between 8 and 19, we do not reject the null hypothesis. This would mean that we have a random sample. If the number of runs is less than 8 or more than 19, we no longer have a random sample. In our case since we had 10 runs, we do not reject the null hypotheses of randomness.

Let us further illustrate the runs test with several examples.

EXAMPLE 1

Twenty people are waiting on line in a bank. These people will either deposit or withdraw money. Let d represent a customer that makes a deposit and let w represent a customer who is making a withdrawal. If the people are arranged in the following order, test for randomness. Use a 5% level of significance.

d, d, d, w, w, w, w, w, d, d, d, d, d, d, d, w, w, w, d, d

SOLUTION

There are 5 runs as shown in the following chart:

Run	Letters
1	ddd
2	wwwww
3	ddddddd
4	www
5	dd

There are also 12 d's and 8 w's in this succession of letters, so that the larger is 12 and the smaller is 8. From Table XI we find that the critical values, where the larger number is 12 and the smaller number is 8, are 6 and 16 runs. Since we obtained only 5 runs, we reject the null hypotheses and conclude that these people are not arranged in random order.

EXAMPLE 2

Thirty dresses are arranged on a rack as follows. Note that r represents a red dress and b represents a blue dress. Using a 5% level of significance, determine if the dresses are arranged in a random order.

r, b, r, b, r, r, b, r, b, b, r, r, r, r, r, b, r, b, b, r, b, r, b, r, b, b, b, b, r, r

SOLUTION

There are 16 r's and 14 b's so that the larger number is 16 and the smaller number is 14. There are also 19 runs as shown.

Run	Letters
1	r
2	b
3	r
4	b
5	rr
6	b
7	r
8	bb
9	rrrrr
10	b
11	r
12	bb
13	r
14	b
15	r
16	b
17	r
18	bbbb
19	rr

From Table XI we note that the critical values, where the larger number is 16 and the smaller number is 14, are 10 and 22. Since we obtained 19 runs, we do not reject the null hypothesis of randomness.

The theory of runs can also be applied to any set of numbers to determine whether these numbers appear in a random order or not. In such cases we first calculate the median of these numbers (see page 55), since approximately one-half of the numbers are below the median and one-half of the numbers are above the median. We then go through the sequence of numbers and replace each number with the letter a if it is above the median and with the letter b if it is below the median. We omit any values which equal the median. Once we have a sequence of a's and b's, we proceed in exactly the same way as we did in Examples 1 and 2 of this section.

EXAMPLE 3

The number of defective items produced by a machine per day over a period of a month is:

13, 17, 14, 20, 18, 16, 14, 19, 21, 20, 14, 17, 12, 14, 19, 20, 17, 18, 14, 20, 17, 19, 17, 14, 19, 21, 16, 12, 15, 22

Using a 5% level of significance, test for randomness.

SOLUTION

We first arrange the numbers in order, from smallest to largest, to determine the median. We get 12, 12, 13, 14, 14, 14, 14, 14, 14, 15, 16, 16, 17, 17, 17, 17, 17, 18, 18, 19, 19, 19, 19, 20, 20, 20, 20, 21, 21, 22. The median of these numbers is 17. We now replace all the numbers of the original sequence with the letter b if the number is below 17 and with the letter a if the number is above 17. We do not replace the 17 with any letter. The new sequence then becomes

b, b, a, a, b, b, a, a, a, b, b, b, a, a, a, b, a, a, b, a, a, b, b, b, a

There are now 13 a's and 12 b's, and there are 12 runs as follows:

Run	Letters
1	bb
2	aa
3	bb
4	aaa
5	bbb
6	aaa
7	b
8	aa
9	b
10	aa
11	bbb
12	a

From Table XI we note that the critical values, where the larger number is 13 and the smaller number is 12, are 8 and 19. Since we obtained 12 runs, we do not reject the null hypothesis.

Comment   Table XI can be used only when n_1 and n_2 are not greater than 20. If either is larger than 20, we use the normal curve approximation. In this case the test statistic is

$$z = \frac{X - \mu_R}{\sigma_R}$$

where

$$\mu_R = \frac{2n_1 \cdot n_2}{n_1 + n_2} + 1$$

and

$$\sigma_R = \sqrt{\frac{2n_1 n_2 (2n_1 n_2 - n_1 - n_2)}{(n_1 + n_2)^2 (n_1 + n_2 - 1)}}$$

1. The sex of the first 25 babies born on Monday at Maternity Hospital is

   b, b, g, g, g, b, b, b, b, g, g, b, g, g, g, b, g, g, b, b, g, g, g, g, b

   where b = boy and g = girl. Test for randomness.

2. From 9:00 P.M. to midnight 23 vehicles passed through the toll gate as shown by the following. Letting T = truck and C = car, test for randomness.

   T, C, T, T, C, C, C, T, C, C, T, T, C, C, T, T, T, T, C, T, C, T, T

3. An interviewer asks people, as they are getting off the escalator, whether or not they smoke. They answer Yes (Y) or No (N) as follows:

   N, Y, N, Y, Y, N, N, N, Y, Y, N, Y, N, Y, N,
   N, N, N, N, Y, N, Y, Y, N, N, N, Y, Y, N

   Test for randomness.

4. As they register for a physical education course, students are asked to indicate their preference for either hockey (H) or football (F). Their preferences are as follows:

F, F, H, H, H, F, H, F, H, H, H, F, H, F, F, F, F,
F, F, H, F, F, H, F, H, F, F, F, H, F, F, H

Test for randomness.

5. The family income of the applicants for a scholarship, shown in the order in which the applications were received, is as follows:

$7800, $9100, $14,100, $17,080, $12,063, $8400, $13,721, $6291, $4400, $6800, $17,031, $11,096, $9500, $8276, $12,039, $16,321, $6246, $5963, $17,468, $15,820, $17,306, $7100, $6583

Test for randomness.

6. A newspaper reporter covering a rally asks students if they are undergraduates (U) or graduates (G) at the university. Their answers are as follows:

G, G, U, G, U, G, G, U, U, U, U, G, U, U,
U, G, U, G, U, U, G, U, U, G, G, U, G, U

Test for randomness.

7. Control tower personnel at a large airport submitted the following report on the type of plane which landed at the airport on July 5. It was either a cargo plane (C) or a passenger plane (P).

C, C, P, P, C, C, P, P, P, C, P, P, P, C, C, C, P,
C, P, P, C, C, P, P, P, C, P, P, C, P, P, P, C

Test for randomness.

8. During the last month the photocopying machine on the 13th floor made the following number of copies:

31, 68, 52, 39, 59, 64, 28, 64, 53, 76, 23,
73, 56, 42, 57, 40, 39, 76, 26, 19, 17, 63, 79

Test for randomness.

9. The number of patients coming to a dentist each day over the last few weeks is

36, 42, 26, 33, 12, 17, 39, 15, 23, 31, 43, 9, 16, 18, 23, 19

Test for randomness.

10. The midterm grades on the statistics exam were

86, 91, 100, 68, 42, 73, 78, 84, 53, 98, 71, 90, 48,
63, 78, 82, 55, 100, 92, 62, 60, 73, 94, 97, 63, 98

Test for randomness.

SELF-STUDY GUIDE

In this chapter we discussed several of the nonparametric statistical methods that are often used when we cannot use the standard tests. By far the easiest and most popular of these methods is the sign test. This test is used when we wish to compare two sample means and we know that the samples are not independent. Because of its simplicity the sign test is used by many people even when a standard test can be used. However, this method is very wasteful of information.

Another important nonparametric test is the rank–sum test (Mann–Whitney test) which is used when the normality assumption is not satisfied or when the variances are not equal. The sum of the ranks is normally distributed when the sample size is large enough, in which case we can use an appropriate z statistic.

Finally we discussed the runs test which is used to test for randomness or a lack of it. In determining whether we have a random sample or not, we use Table XI in the Appendix to find the appropriate critical values.

You should be able to identify each symbol in the following formula, understand the relationships among the symbols expressed in the formula, understand the significance of the formula, and use the formula in solving problems.

Test statistic of rank–sum test is

$$z = \frac{R - \mu_R}{\sigma_R} \quad \text{where} \quad \mu_R = \frac{n_1(n_1 + n_2 + 1)}{2}, \quad \sigma_R = \sqrt{\frac{n_1 n_2 (n_1 + n_2 + 1)}{12}}$$

and n_1 is the smaller sample size.

You should now be able to demonstrate your knowledge of the following ideas presented in this chapter by giving definitions, descriptions, or specific examples. Page references are given for each term so that you can check your answer.

Nonparametric statistics (page 427)   Mann–Whitney test (page 436)
Distribution-free methods (page 427)   Mann–Whitney U test (page 436)
Standard tests (page 428)   Runs test (page 440)
Sign-test (page 428)   Run (page 441)
Rank–sum test (page 434)   Testing randomness (page 441)

The tests of the following section will be more useful if you take them after you have studied the examples and solved the exercises in this chapter.

MASTERY TESTS

Form *A*

1. Nonparametric tests are often known as _____ methods.
2. The sign test is used when we wish to test the null hypothesis $\mu_1 = \mu_2$ and when _____.
3. True or false: Nonparametric tests do not require that the populations from which samples are taken be normally distributed.
4. When taking repeated measurements of the same sample as in the before and after type study, the nonparametric test that we use is the _____.
5. The rank–sum test is used when we wish to test _____.
6. Another name for the Mann–Whitney test is the _____ test.
7. To test for randomness we can use the _____ test.
8. When using the rank–sum test, what happens when a tie occurs?
9. True or false: A succession of identical letters followed by and preceded by a different letter or no letter at all is called a sequence.
10. True or false: The length of each individual run is crucial when we use the runs test.

In each of the following use the appropriate 5% level of significance.

1. The weights of ten people before they stopped drinking and three months after they stopped drinking are as follows:

Before	153	146	182	176	133	153	161	120	110	120
After	158	132	179	184	145	162	178	131	110	131

Using the sign test, test the null hypothesis that drinking does not have any effect on a person's weight.

2. Two groups of students are given a special reading comprehension exam. Group A students are from middle-class neighborhood schools. Group B students are from lower-class neighborhood schools. The following chart indicates the results of the exam:

Group A	38	42	56	39	27	45	36	62	73		
Group B	52	46	37	63	55	48	59	61	70	35	28

Using the rank–sum test, test the null hypothesis that the mean score for both groups is the same.

3. At a certain ice-cream parlor ice-cream cones are available with only vanilla (V) or strawberry (S) ice cream. Thirty children come in and order the following ice-cream cones:

> V, V, S, S, S, V, V, V, V, S, S, S, V, S, V,
> S, S, V, V, V, S, V, V, S, S, S, V, V, S, V

Test for randomness.

4. At a recent party the ladies wore either pantsuits (P) or dresses (D). As each lady entered, the hostess noted how they were dressed. She noted the following:

> P, P, D, D, D, P, D, P, P, D, P, P, D, D, D,
> D, P, P, P, D, D, P, D, D, P, P, P, P, D

Test for randomness.

5. The following data indicates the number of minutes needed by two groups of students to complete the proficiency exam. Using the rank–sum test, test the null hypothesis that the mean time for both groups is the same.

Group A	75	80	72	69	83	84	77	68	78	73	68	63
Group B	65	85	81	73	77	76	82	89	74	79	70	75	78	82

6. The following chart indicates the number of trials required by two groups of elephants to learn to perform a certain act. Group A received a special method of training whereas group B did not.

Group A	20	28	32	17	24	19	25	21	23	27	
Group B	34	36	33	29	19	28	24	23	27	29	31	32

Using the rank–sum test, test the null hypothesis that the mean number of trials for both groups is the same.

7. The number of cars parked daily in the student parking lot during the first few weeks of the semester is

$$75, 62, 94, 108, 83, 95, 62, 78, 84, 80, 95, 83, 60,$$
$$72, 123, 111, 85, 78, 63, 69, 76, 85, 93, 100$$

Test for randomness.

8. A dispatcher keeps records on whether an arriving train is an express (E) or a local (L). The following is his list for March 13:

E, E, L, L, L, L, L, E, E, L, E, L, E, E, E, L, L, L, L, E, E, L, L, E, L, L, L, E, E, L

Test for randomness.

9. A new method of teaching biology produced the following results:

Preteaching Score	65	81	84	92	95	93	78	69	78	76	78	92	94	93	80
Postteaching Score	73	85	80	89	98	90	84	76	78	83	82	96	97	90	80

Using the sign test, test the null hypothesis that there is no difference in performance as a result of this teaching method.

10. Ten men were given a special blood-pressure reducing pill. These are the results:

Pressure Before	116	121	113	145	163	172	162	182	192	111
Pressure After	114	118	110	140	157	177	162	175	173	109

Using the sign test, test the null hypothesis that the blood-pressure reducing pill is not effective.

11. A photographer arranges 25 first grade boys and girls for a group picture in the following order (b = boy, g = girl):

b, g, g, g, g, b, b, b, g, b, b, g, g, b, g, b, b, b, g, g, g, g, b, b, b

Test for randomness.

12. The number of speeding tickets issued daily by the 28 members of the Highway Patrol to motorists during the month of February was as follows:

$$6, 18, 5, 8, 17, 12, 19, 22, 15, 17, 16, 14, 11, 5,$$
$$9, 8, 12, 12, 3, 4, 7, 0, 8, 17, 29, 34, 28, 17$$

Test for randomness.

13. The following chart indicates the reaction time, in seconds, to two different stimuli:

A	12	14	15	10	9	8	13	12	16	17	19	16	18	17	16	19
B	11	10	17	10	7	6	11	14	13	15	23	17	14	17	23	15

Using the sign test, test the null hypothesis that there is no difference in the reaction time to both stimuli.

14. A die was rolled many times. If a 1, 3, or 5 came up, we labeled this as O for an odd number. If a 2, 4, or 6 came up, we labeled this as E for an even number. When the die was tossed 20 times, the following sequence of odd and even numbers occurred.

$$O, E, O, O, E, O, O, E, E, E, O, O, E, E, E, O, O, E, E, O$$

Test for randomness.

15. Twelve female and ten male students are asked to rate an actor according to a particular scale. Their ratings are:

Female	57	62	68	53	61	55	49	68	72	60	58	64	65
Male	48	53	46	70	62	58	55	69	73	61			

Using the rank sum test, test the null hypothesis that the mean ratings of both groups is the same.

SUGGESTED READING

Conover, W. J. Practical Nonparametric Statistics. New York: Wiley, 1971.

Noether, G. E. Elements of Nonparametric Statistics. New York: Wiley, 1967.

Siegel, S. Nonparametric Statistics for the Behavioral Sciences. New York: McGraw-Hill, 1956.

Wald, A. and Wolfowitz, J. "On a Test of Whether Two Samples are from the Same Population." Annals of Mathematical Statistics., vol. 2, 1940, pp. 147–162.

Appendix

STATISTICAL TABLES

SUGGESTIONS FOR USING THE SQUARE ROOT TABLE

To find the square root of any number between 1.00 and 10.00, we use the third column. For example, $\sqrt{3.21}$ is 1.79165 and $\sqrt{6.58}$ is 2.56515. The fourth column enables us to find the square root of all numbers between 10.0 and 100.0, since $\sqrt{10N}$ gives us 10(1.00), and 10(10.00), or 10.0 to 100.0. For example, from the fourth column we find that $\sqrt{32.1}$ is 5.66569 and $\sqrt{65.8}$ is 8.11172. Thus, using the third and fourth columns, we can find the square root of all numbers between 1.00 and 100.0.

Suppose we want $\sqrt{321}$. We can find $\sqrt{3.21}$ and $\sqrt{32.1}$. Which should we use? We locate the decimal point. (If no decimal point is indicated, it is understood to be at the end of the number.) Then move the decimal point an even number of places to the right or to the left until a number greater than or equal to 1 but less than 100 is reached. If the resulting number is less than 10, go to the $\sqrt{N}$ column; if it is between 10 and 100, go to the $\sqrt{10N}$ column. The next question is, where do we put the decimal? Since we moved the decimal point an even number of places to the left or to the right to get a number between 1 and 100, the decimal point of the answer obtained from Table I is moved half as many places in the *opposite* direction.

Thus, to find $\sqrt{321}$, we move the decimal two places to the left, to $\sqrt{3.21}$. From Table I, we get 1.79165. Since we moved the decimal two places to the left to arrive at this answer, we now move the decimal half as many places in the opposite direction. Therefore $\sqrt{321} = 17.9165$. In the same way, $\sqrt{3210}$ becomes $\sqrt{32.10}$, which is 5.66569, so $\sqrt{3210} = 56.6569$. $\sqrt{0.000321}$ becomes $\sqrt{0003.21}$, which is 1.79165, so $\sqrt{0.000321} = 0.0179165$.

Table I. Squares, square roots, and reciprocals

N	N^2	$\sqrt{N}$	$\sqrt{10N}$	N	N^2	$\sqrt{N}$	$\sqrt{10N}$
1.00	1.0000	1.00000	3.16228	**1.50**	2.2500	1.22474	3.87298
1.01	1.0201	1.00499	3.17805	1.51	2.2801	1.22882	3.88587
1.02	1.0404	1.00995	3.19374	1.52	2.3104	1.23288	3.89872
1.03	1.0609	1.01489	3.20936	1.53	2.3409	1.23693	3.91152
1.04	1.0816	1.01980	3.22490	1.54	2.3716	1.24097	3.92428
1.05	1.1025	1.02470	3.24037	1.55	2.4025	1.24499	3.93700
1.06	1.1236	1.02956	3.25576	1.56	2.4336	1.24900	3.94968
1.07	1.1449	1.03441	3.27109	1.57	2.4649	1.25300	3.96232
1.08	1.1664	1.03923	3.28634	1.58	2.4964	1.25698	3.97492
1.09	1.1881	1.04403	3.30151	1.59	2.5281	1.26095	3.98748
1.10	1.2100	1.04881	3.31662	**1.60**	2.5600	1.26491	4.00000
1.11	1.2321	1.05357	3.33167	1.61	2.5921	1.26886	4.01248
1.12	1.2544	1.05830	3.34664	1.62	2.6244	1.27279	4.02492
1.13	1.2769	1.06301	3.36155	1.63	2.6569	1.27671	4.03733
1.14	1.2996	1.06771	3.37639	1.64	2.6896	1.28062	4.04969
1.15	1.3225	1.07238	3.39116	1.65	2.7225	1.28452	4.06202
1.16	1.3456	1.07703	3.40588	1.66	2.7556	1.28841	4.07431
1.17	1.3689	1.08167	3.42053	1.67	2.7889	1.29228	4.08656
1.18	1.3924	1.08628	3.43511	1.68	2.8224	1.29615	4.09878
1.19	1.4161	1.09087	3.44964	1.69	2.8561	1.30000	4.11096
1.20	1.4400	1.09545	3.46410	**1.70**	2.8900	1.30384	4.12311
1.21	1.4641	1.10000	3.47851	1.71	2.9241	1.30767	4.13521
1.22	1.4884	1.10454	3.49285	1.72	2.9584	1.31149	4.14729
1.23	1.5129	1.10905	3.50714	1.73	2.9929	1.31529	4.15933
1.24	1.5376	1.11355	3.52136	1.74	3.0276	1.31909	4.17133
1.25	1.5625	1.11803	3.53553	1.75	3.0625	1.32288	4.18330
1.26	1.5876	1.12250	3.54965	1.76	3.0976	1.32665	4.19524
1.27	1.6129	1.12694	3.56371	1.77	3.1329	1.33041	4.20714
1.28	1.6384	1.13137	3.57771	1.78	3.1684	1.33417	4.21900
1.29	1.6641	1.13578	3.59166	1.79	3.2041	1.33791	4.23084
1.30	1.6900	1.14018	3.60555	**1.80**	3.2400	1.34164	4.24264
1.31	1.7161	1.14455	3.61939	1.81	3.2761	1.34536	4.25441
1.32	1.7424	1.14891	3.63318	1.82	3.3124	1.34907	4.26615
1.33	1.7689	1.15326	3.64692	1.83	3.3489	1.35277	4.27785
1.34	1.7956	1.15758	3.66060	1.84	3.3856	1.35647	4.28952
1.35	1.8225	1.16190	3.67423	1.85	3.4225	1.36015	4.30116
1.36	1.8496	1.16619	3.68782	1.86	3.4596	1.36382	4.31277
1.37	1.8769	1.17047	3.70135	1.87	3.4969	1.36748	4.32435
1.38	1.9044	1.17473	3.71484	1.88	3.5344	1.37113	4.33590
1.39	1.9321	1.17898	3.72827	1.89	3.5721	1.37477	4.34741
1.40	1.9600	1.18322	3.74166	**1.90**	3.6100	1.37840	4.35890
1.41	1 9881	1.18743	3.75500	1.91	3.6481	1.38203	4.37035
1.42	2.0164	1.19164	3.76829	1.92	3.6864	1.38564	4.38178
1.43	2.0449	1.19583	3.78153	1.93	3.7249	1.38924	4.39318
1.44	2.0736	1.20000	3.79473	1.94	3.7636	1.39284	4.40454
1.45	2.1025	1.20416	3.80789	1.95	3.8025	1.39642	4.41588
1.46	2.1316	1.20830	3.82099	1.96	3.8416	1.40000	4.42719
1.47	2.1609	1.21244	3.83406	1.97	3.8809	1.40357	4.43847
1.48	2.1904	1.21655	3.84708	1.98	3.9204	1.40712	4.44972
1.49	2.2201	1.22066	3.86005	1.99	3.9601	1.41067	4.46094
1.50	2.2500	1.22474	3.87298	**2.00**	4.0000	1.41421	4.47214
N	N^2	$\sqrt{N}$	$\sqrt{10N}$	N	N^2	$\sqrt{N}$	$\sqrt{10N}$

N	N²	√N	√10N		N	N²	√N	√10N
2.00	4.0000	1.41421	4.47214		**2.50**	6.2500	1.58114	5.00000
2.01	4.0401	1.41774	4.48330		2.51	6.3001	1.58430	5.00999
2.02	4.0804	1.42127	4.49444		2.52	6.3504	1.58745	5.01996
2.03	4.1209	1.42478	4.50555		2.53	6.4009	1.59060	5.02991
2.04	4.1616	1.42829	4.51664		2.54	6.4516	1.59374	5.03984
2.05	4.2025	1.43178	4.52769		2.55	6.5025	1.59687	5.04975
2.06	4.2436	1.43527	4.53872		2.56	6.5536	1.60000	5.05964
2.07	4.2849	1.43875	4.54973		2.57	6.6049	1.60312	5.06952
2.08	4.3264	1.44222	4.56070		2.58	6.6564	1.60624	5.07937
2.09	4.3681	1.44568	4.57165		2.59	6.7081	1.60935	5.08920
2.10	4.4100	1.44914	4.58258		**2.60**	6.7600	1.61245	5.09902
2.11	4.4521	1.45258	4.59347		2.61	6.8121	1.61555	5.10882
2.12	4.4944	1.45602	4.60435		2.62	6.8644	1.61864	5.11859
2.13	4.5369	1.45945	4.61519		2.63	6.9169	1.62173	5.12835
2.14	4.5796	1.46287	4.62601		2.64	6.9696	1.62481	5.13809
2.15	4.6225	1.46629	4.63681		2.65	7.0225	1.62788	5.14782
2.16	4.6656	1.46969	4.64758		2.66	7.0756	1.63095	5.15752
2.17	4.7089	1.47309	4.65833		2.67	7.1289	1.63401	5.16720
2.18	4.7524	1.47648	4.66905		2.68	7.1824	1.63707	5.17687
2.19	4.7961	1.47986	4.67974		2.69	7.2361	1.64012	5.18652
2.20	4.8400	1.48324	4.69042		**2.70**	7.2900	1.64317	5.19615
2.21	4.8841	1.48661	4.70106		2.71	7.3441	1.64621	5.20577
2.22	5.9284	1.48997	4.71169		2.72	7.3984	1.64924	5.21536
2.23	4.9729	1.49332	4.72229		2.73	7.4529	1.65227	5.22494
2.24	5.0176	1.49666	4.73286		2.74	7.5076	1.65529	5.23450
2.25	5.0625	1.50000	4.74342		2.75	7.5625	1.65831	5.24404
2.26	5.1076	1.50333	4.75395		2.76	7.6176	1.66132	5.25357
2.27	5.1529	1.50665	4.76445		2.77	7.6729	1.66433	5.26308
2.28	5.1984	1.50997	4.77493		2.78	7.7284	1.66733	5.27257
2.29	5.2441	1.51327	4.78539		2.79	7.7841	1.67033	5.28205
2.30	5.2900	1.51658	4.79583		**2.80**	7.8400	1.67332	5.29150
2.31	5.3361	1.51987	4.80625		2.81	7.8961	1.67631	5.30094
2.32	5.3824	1.52315	4.81664		2.82	7.9524	1.67929	5.31037
2.33	5.4289	1.52643	4.82701		2.83	8.0089	1.68226	5.31977
2.34	5.4756	1.52971	4.83735		2.84	8.0656	1.68523	5.32917
2.35	5.5225	1.53297	4.84768		2.85	8.1225	1.68819	5.33854
2.36	5.5696	1.53623	4.85798		2.86	8.1796	1.69115	5.34790
2.37	5.6169	1.53948	4.86826		2.87	8.2369	1.69411	5.35724
2.38	5.6644	1.54272	4.87852		2.88	8.2944	1.69706	5.36656
2.39	5.7121	1.54596	4.88876		2.89	8.3521	1.70000	5.37587
2.40	5.7600	1.54919	4.89898		**2.90**	8.4100	1.70294	5.38516
2.41	5.8081	1.55252	4.90918		2.91	8.4681	1.70587	5.39444
2.42	5.8564	1.55563	4.91935		2.92	8.5264	1.70880	5.40370
2.43	5.9049	1.55885	4.92950		2.93	8.5849	1.71172	5.41295
2.44	5.9536	1.56205	4.93964		2.94	8.6436	1.71464	5.42218
2.45	6.0025	1.56525	4.94975		2.95	8.7025	1.71756	5.43139
2.46	6.0516	1.56844	4.95984		2.96	8.7616	1.72047	5.44059
2.47	6.1009	1.57162	4.96991		2.97	8.8209	1.72337	5.44977
2.48	6.1054	1.57480	4.97996		2.98	8.8804	1.72627	5.45894
2.49	6.2001	1.57797	4.98999		2.99	8.9401	1.72916	5.46809
2.50	6.2500	1.58114	5.00000		**3.00**	9.0000	1.73205	5.47723
N	N²	√N	√10N		N	N²	√N	√10N

N	N²	√N	√10N		N	N²	√N	√10N
3.00	9.0000	1.73205	5.47723		**3.50**	12.2500	1.87083	5.91608
3.01	9.0601	1.73494	5.48635		3.51	12.3201	1.87350	5.92453
3.02	9.1204	1.73781	5.49545		3.52	12.3904	1.87617	5.93296
3.03	9.1809	1.74069	5.50454		3.53	12.4609	1.87883	5.94138
3.04	9.2416	1.74356	5.51362		3.54	12.5316	1.88149	5.94979
3.05	9.3025	1.74642	5.52268		3.55	12.6025	1.88414	5.95819
3.06	9.3636	1.74929	5.53173		3.56	12.6736	1.88680	5.96657
3.07	9.4249	1.75214	5.54076		3.57	12.7449	1.88944	5.97495
3.08	9.4864	1.75499	5.54977		3.58	12.8164	1.89209	5.98331
3.09	9.5481	1.75784	5.55878		3.59	12.8881	1.89473	5.99166
3.10	9.6100	1.76068	5.56776		**3.60**	12.9600	1.89737	6.00000
3.11	9.6721	1.76352	5.57674		3.61	13.0321	1.90000	6.00833
3.12	9.7344	1.76635	5.58570		3.62	13.1044	1.90263	6.01664
3.13	9.7969	1.76918	5.59464		3.63	13.1769	1.90526	6.02495
3.14	9.8596	1.77200	5.60357		3.64	13.2496	1.90788	6.03324
3.15	9.9225	1.77482	5.61249		3.65	13.3225	1.91050	6.04152
3.16	9.9856	1.77764	5.62139		3.66	13.3956	1.91311	6.04949
3.17	10.0489	1.78045	5.63028		3.67	13.4689	1.91572	6.05805
3.18	10.1124	1.78326	5.63915		3.68	13.5424	1.91833	6.06630
3.19	10.1761	1.78606	5.64801		3.69	13.6161	1.92094	6.07454
3.20	10.2400	1.78885	5.65685		**3.70**	13.6900	1.92354	6.08276
3.21	10.3041	1.79165	5.66569		3.71	13.7641	1.92614	6.09098
3.22	10.3684	1.79444	5.67450		3.72	13.8384	1.92873	6.09918
3.23	10.4329	1.79722	5.68331		3.73	13.9129	1.93132	6.10737
3.24	10.4976	1.80000	5.69210		3.74	13.9876	1.93391	6.11555
3.25	10.5625	1.80278	5.70088		3.75	14.0625	1.93649	6.12372
3.26	10.6276	1.80555	5.70964		3.76	14.1376	1.93907	6.13188
3.27	10.6929	1.80831	5.71839		3.77	14.2129	1.94165	6.14003
3.28	10.7584	1.81108	5.72713		3.78	14.2884	1.94422	6.14817
3.29	10.8241	1.81384	5.73585		3.79	14.3641	1.94679	6.15630
3.30	10.8900	1.81659	5.74456		**3.80**	14.4400	1.94936	6.16441
3.31	10.9561	1.81934	5.75326		3.81	14.5161	1.95192	6.17252
3.32	10.0224	1.82209	5.76194		3.82	14.5924	1.95448	6.18061
3.33	11.0889	1.82483	5.77062		3.83	14.6689	1.95704	6.18870
3.34	11.1556	1.82757	5.77927		3.84	14.7456	1.95959	6.19677
3.35	11.2225	1.83030	5.78792		3.85	14.8225	1.96214	6.20484
3.36	11.2896	1.83303	5.79655		3.86	14.8996	1.96469	6.21289
3.37	11.3569	1.83576	5.80517		3.87	14.9769	1.96723	6.22093
3.38	11.4244	1.83848	5.81378		3.88	15.0544	1.96977	6.22896
3.39	11.4921	1.84120	5.82237		3.89	15.1321	1.97231	6.23699
3.40	11.5600	1.84391	5.83095		**3.90**	51.2100	1.97484	6.24500
3.41	11.6281	1.84662	5.83952		3.91	15.2881	1.97737	6.25300
3.42	11.6964	1.84932	5.84808		3.92	15.3664	1.97990	6.26099
3.43	11.7649	1.85203	5.85662		3.93	15.4449	1.98242	6.26897
3.44	11.8336	1.85472	5.86515		3.94	15.5236	1.98494	6.27694
3.45	11.9025	1.85742	5.87367		3.95	15.6025	1.98746	6.28490
3.46	11.9716	1.86011	5.88218		3.96	15.6816	1.98997	6.29285
3.47	12.0409	1.86279	5.89067		3.97	15.7609	1.99249	6.30079
3.48	12.1104	1.86548	5.89915		3.98	15.8404	1.99499	6.30872
3.49	12.1801	1.86815	5.90762		3.99	15.9201	1.99750	6.31644
3.50	12.2500	1.87083	5.91608		**4.00**	16.0000	2.00000	6.32456
N	N²	√N	√10N		N	N²	√N	√10N

N	N²	√N	√10N
4.00	16.0000	2.00000	6.32456
4.01	16.0801	2.00250	6.33246
4.02	16.1604	2.00499	6.34035
4.03	16.2409	2.00749	6.34823
4.04	16.3216	2.00998	6.35610
4.05	16.4025	2.01246	6.36396
4.06	16.4836	2.01494	6.37181
4.07	16.5649	2.01742	6.37966
4.08	16.6464	2.01990	6.38749
4.09	16.7281	2.02237	6.39531
4.10	16.8100	2.02485	6.40312
4.11	16.8921	2.02731	6.41093
4.12	16.9744	2.02978	6.41872
4.13	17.0569	2.03224	6.42651
4.14	17.1396	2.03470	6.43428
4.15	17.2225	2.03715	6.44205
4.16	17.3056	2.03961	6.44981
4.17	17.3889	2.04206	6.45755
4.18	17.4724	2.04450	6.46529
4.19	17.5561	2.04695	6.47302
4.20	17.6400	2.04939	6.48074
4.21	17.7241	2.05183	6.48845
4.22	17.8084	2.05426	6.49615
4.23	17.8929	2.05670	6.50384
4.24	17.9776	2.05913	6.51153
4.25	18.0625	2.06155	6.51920
4.26	18.1476	2.06398	6.52687
4.27	18.2329	2.06640	6.53452
4.28	18.3184	2.06882	6.54217
4.29	18.4041	2.07123	6.54981
4.30	18.4900	2.07364	6.55744
4.31	18.5761	2.07605	6.66506
4.32	18.6624	2.07846	6.57267
4.33	18.7489	2.08087	6.58027
4.34	18.8356	2.08327	6.58787
4.35	18.9225	2.08567	6.59545
4.36	19.0096	2.08806	6.60303
4.37	19.0969	2.09045	6.61060
4.38	19.1844	2.09284	6.61816
4.39	19.2721	2.09523	6.62571
4.40	19.3600	2.09762	6.63325
4.41	19.4481	2.10000	6.64078
4.42	19.5364	2.10238	6.64831
4.43	19.6249	2.10476	6.65582
4.44	19.7136	2.10713	6.66333
4.45	19.8025	2.10950	6.67083
4.46	19.8916	2.11187	6.67832
4.47	19.9809	2.11424	6.68581
4.48	20.0704	2.11660	6.69328
4.49	20.1601	2.11896	6.70075
4.50	20.2500	2.12132	6.70820
N	N²	√N	√10N

N	N²	√N	√10N
4.50	20.2500	2.12132	6.70820
4.51	20.3401	2.12368	6.71565
4.52	20.4304	2.12603	6.72309
4.53	20.5209	2.12838	6.73053
4.54	20.6116	2.13073	6.73795
4.55	20.7025	2.13307	6.74537
4.56	20.7936	2.13542	6.75278
4.57	20.8849	2.13776	6.76018
4.58	20.9764	2.14009	6.76757
4.59	21.0681	2.14243	6.77495
4.60	21.1600	2.14476	6.78233
4.61	21.2521	2.14709	6.78970
4.62	21.3444	2.14942	6.79706
4.63	21.4369	2.15174	6.80441
4.64	21.5296	2.15407	6.81175
4.65	21.6225	2.15639	6.81909
4.66	21.7156	2.15870	6.82642
4.67	21.8089	2.16102	6.83374
4.68	21.9024	2.16333	6.84105
4.69	21.9961	2.16564	6.84836
4.70	22.0900	2.16795	6.85565
4.71	22.1841	2.17025	6.86294
4.72	22.2784	2.17256	6.87023
4.73	22.3729	2.17486	6.87750
4.74	22.4676	2.17715	6.88477
4.75	22.5625	2.17945	6.89202
4.76	22.6576	2.18174	6.89928
4.77	22.7529	2.18403	6.90652
4.78	22.8484	2.18632	6.91375
4.79	22.9441	2.18861	6.92098
4.80	23.0400	2.19089	6.92820
4.81	23.1361	2.19317	6.93542
4.82	23.2324	2.19545	6.94262
4.83	23.3289	2.19773	6.94982
4.84	23.4256	2.20000	6.95701
4.85	23.5225	2.20227	6.96419
4.86	23.6196	2.20454	6.97137
4.87	23.7169	2.20681	6.97854
4.88	23.8144	2.20907	6.98570
4.89	23.9121	2.21133	6.99285
4.90	24.0100	2.21359	7.00000
4.91	24.1081	2.21585	7.00714
4.92	24.2064	2.21811	7.01427
4.93	24.3049	2.22036	7.02140
4.94	24.4036	2.22261	7.02851
4.95	24.5025	2.22486	7.03562
4.96	24.6016	2.22711	7.04273
4.97	24.7009	2.22935	7.04982
4.98	24.8004	2.23159	7.05691
4.99	24.9001	2.23383	7.06399
5.00	25.0000	2.23607	7.07107
N	N²	√N	√10N

N	N²	√N	√10N		N	N²	√N	√10N
5.00	25.0000	2.23607	7.07107		**5.50**	30.2500	2.34521	7.41620
5.01	25.1001	2.23830	7.07814		5.51	30.3601	2.34734	7.42294
5.02	25.2004	2.24054	7.08520		5.52	30.4704	2.34947	7.42967
5.03	25.3009	2.24277	7.09225		5.53	30.5809	2.35160	7.43640
5.04	25.4016	2.24499	7.09930		5.54	30.6916	2.35372	7.44312
5.05	25.5025	2.24722	7.10634		5.55	30.8025	2.35584	7.44983
5.06	25.6036	2.24944	7.11337		5.56	30.9136	2.35797	7.45654
5.07	25.7049	2.25167	7.12039		5.57	31.0249	2.36008	7.46324
5.08	25.8064	2.25389	7.12741		5.58	31.1364	2.36220	7.46994
5.09	25.9081	2.25610	7.13442		5.59	31.2481	2.36432	7.47663
5.10	26.0100	2.25832	7.14143		**5.60**	31.3600	2.36643	7.48331
5.11	26.1121	2.26053	7.14843		5.61	31.4721	2.36854	7.48999
5.12	26.2144	2.26274	7.15542		5.62	31.5844	2.37065	7.49667
5.13	26.3169	2.26495	7.16240		5.63	31.6969	2.37276	7.50333
5.14	26.4196	2.26716	7.16938		5.64	31.8096	2.37487	7.50999
5.15	26.5225	2.26936	7.17635		5.65	31.9225	2.37697	7.51665
5.16	26.6256	2.27156	7.18331		5.66	32.0356	2.37908	7.52330
5.17	26.7289	2.27376	7.19027		5.67	32.1489	2.38118	7.52994
5.18	26.8324	2.27596	7.19722		5.68	32.2624	2.38328	7.53658
5.19	26.9361	2.27816	7.20417		5.69	32.3761	2.38537	7.54321
5.20	27.0400	2.28035	7.21110		**5.70**	32.4900	2.38747	7.54983
5.21	27.1441	2.28254	7.21803		5.71	32.6041	2.38956	7.55645
5.22	27.2484	2.28473	7.22496		5.72	32.7184	2.39165	7.56307
5.23	27.3529	2.28692	7.23187		5.73	32.8329	2.39374	7.56968
5.24	27.4576	2.28910	7.23838		5.74	32.9476	2.39583	7.57628
5.25	27.5625	2.29129	7.24569		5.75	33.0625	2.39792	7.58288
5.26	27.6676	2.29347	7.25259		5.76	33.1776	2.40000	7.58947
5.27	27.7729	2.29565	7.25948		5.77	33.2929	2.40208	7.59605
5.28	27.8784	2.29783	7.26636		5.78	33.4084	2.40416	7.60263
5.29	27.9841	2.30000	7.27324		5.79	33.5241	2.40624	7.60920
5.30	28.0900	2.30217	7.28011		**5.80**	33.6400	2.40832	7.61577
5.31	28.1961	2.30434	7.28697		5.81	33.7561	2.41039	7.62234
5.32	28.3024	2.30651	7.29383		5.82	33.8724	2.41247	7.62889
5.33	28.4089	2.30868	7.30068		5.83	33.9889	2.41454	7.63544
5.34	28.5156	2.31084	7.30753		5.84	34.1056	2.41661	7.64199
5.35	28.6225	2.31301	7.31437		5.85	34.2225	2.41868	7.64853
5.36	28.7296	2.31517	7.32120		5.86	34.3396	2.42074	7.65506
5.37	28.8369	2.31733	7.32803		5.87	34.4569	2.42281	7.66159
5.38	28.9444	2.31948	7.33485		5.88	34.5744	2.42487	7.66812
5.39	29.0521	2.32164	7.34166		5.89	34.6921	2.42693	7.67463
5.40	29.1600	2.32379	7.34847		**5.90**	34.8100	2.42899	7.68115
5.41	29.2681	2.32594	7.35527		5.91	34.9281	2.43105	7.68765
5.42	29.3764	2.32809	7.36206		5.92	35.0464	2.43311	7.69415
5.43	29.4849	2.33024	7.36885		5.93	35.1649	2.43516	7.70065
5.44	29.5936	2.33238	7.37564		5.94	35.2836	2.43721	7.70714
5.45	29.7025	2.33452	7.38241		5.95	35.4025	2.43926	7.71362
5.46	29.8116	2.33666	7.38918		5.96	35.5216	2.44131	7.72010
5.47	29.9209	2.33880	7.39594		5.97	35.6409	2.44336	7.72658
5.48	30.0304	2.34094	7.40270		5.98	35.7604	2.44540	7.73305
5.49	30.1401	2.34307	7.40945		5.99	35.8801	2.44745	7.73951
5.50	30.2500	2.34521	7.41620		**6.00**	36.0000	2.44949	7.74597
N	N²	√N	√10N		N	N²	√N	√10N

N	N²	√N	√10N	N	N²	√N	√10N
6.00	36.0000	2.44949	7.74597	**6.50**	42.2500	2.54951	8.06226
6.01	36.1201	2.45153	7.75242	6.51	42.3801	2.55147	8.06846
6.02	36.2404	2.45357	7.75887	6.52	42.5104	2.55343	8.07465
6.03	36.3609	2.45561	7.76531	6.53	42.6409	2.55539	8.08084
6.04	36.4816	2.45764	7.77174	6.54	42.7716	2.55734	8.08703
6.05	36.6025	2.45967	7.77817	6.55	42.9025	2.55930	8.09321
6.06	36.7236	2.46171	7.78460	6.56	43.0336	2.56125	8.09938
6.07	36.8449	2.46374	7.79102	6.57	43.1649	2.56320	8.10555
6.08	36.9664	2.46577	7.79744	6.58	43.2964	2.56515	8.11172
6.09	37.0881	2.46779	7.80385	6.59	43.4281	2.56710	8.11788
6.10	37.2100	2.46982	7.81025	**6.60**	43.5600	2.56905	8.12404
6.11	37.3321	2.47184	7.81665	6.61	43.6921	2.57099	8.13019
6.12	37.4544	2.47386	7.82304	6.62	43.8244	2.57294	8.13634
6.13	37.5769	2.47588	7.82943	6.63	43.9569	2.57488	8.14248
6.14	37.6996	2.47790	7.83582	6.64	44.0896	2.57682	8.14862
6.15	37.8225	2.47992	7.84219	6.65	44.2225	2.57876	8.15475
6.16	37.9456	2.48193	7.84857	6.66	44.3556	2.58070	8.16088
6.17	38.0689	2.48395	7.85493	6.67	44.4889	2.58263	8.16701
6.18	38.1924	2.48596	7.86130	6.68	44.6224	2.58457	8.17313
6.19	38.3161	2.48797	7.86766	6.69	44.7561	2.58650	8.17924
6.20	38.4400	2.48998	7.87401	**6.70**	44.8900	2.58844	8.18535
6.21	38.5641	2.49199	7.88036	6.71	45.0241	2.59037	8.19146
6.22	38.6884	2.49399	7.88670	6.72	45.1584	2.59230	8.19756
6.23	38.8129	2.49600	7.89303	6.73	45.2929	2.59422	8.20366
6.24	38.9376	2.49800	7.89937	6.74	45.4276	2.59615	8.20975
6.25	39.0625	2.50000	7.90569	6.75	45.5625	2.59808	8.21584
6.26	39.1876	2.50200	7.91202	6.76	45.6976	2.60000	8.22192
6.27	39.3129	2.50400	7.91833	6.77	45.8329	2.60192	8.22800
6.28	39.4384	2.50599	7.92465	6.78	45.9684	2.60384	8.23408
6.29	39.5641	2.50799	7.93095	6.79	46.1041	2.60576	8.24015
6.30	39.6900	2.50998	7.93725	**6.80**	46.2400	2.60768	8.24621
6.31	39.8161	2.51197	7.94355	6.81	46.3761	2.60960	8.25227
6.32	39.9424	2.51396	7.94984	6.82	46.5124	2.61151	8.25833
6.33	40.0689	2.51595	7.95613	6.83	46.6489	2.61343	8.26438
6.34	40.1956	2.51794	7.96241	6.84	46.7856	2.61534	8.27043
6.35	40.3225	2.51992	7.96869	6.85	46.9225	2.61725	8.27647
6.36	40.4496	2.52190	7.97496	6.86	47.0596	2.61916	8.28251
6.37	40.5769	2.52389	7.98123	6.87	47.1969	2.62107	8.28855
6.38	40.7044	2.52587	7.98749	6.88	47.3344	2.62298	8.29458
6.39	40.8321	2.52784	7.99375	6.89	47.4721	2.62488	8.30060
6.40	40.9600	2.52982	8.00000	**6.90**	47.6100	2.62679	8.30662
6.41	41.0881	2.53180	8.00625	6.91	47.7481	2.62869	8.31264
6.42	41.2164	2.53377	8.01249	6.92	47.8864	2.63059	8.31865
6.43	41.3449	2.53574	8.01873	6.93	48.0249	2.63249	8.32466
6.44	41.4736	2.53772	8.02496	6.94	48.1636	2.63439	8.33067
6.45	41.6025	2.53969	8.03119	6.95	48.3025	2.63629	8.33667
6.46	41.7316	2.54165	8.03741	6.96	48.4416	2.63818	8.34266
6.47	41.8609	2.54362	8.04363	6.97	48.5809	2.64008	8.34865
6.48	41.9904	2.54558	8.04984	6.98	48.7204	2.64197	8.35464
6.49	42.1201	2.54755	8.05605	6.99	48.8601	2.64386	8.36062
6.50	42.2500	2.54951	8.06226	**7.00**	49.0000	2.64575	8.36660
N	N²	√N	√10N	N	N²	√N	√10N

N	N²	√N̄	√10N̄		N	N²	√N̄	√10N̄
7.00	49.0000	2.64575	8.36660		**7.50**	56.2500	2.73861	8.66025
7.01	49.1401	2.64764	8.37257		7.51	56.4001	2.74044	8.66603
7.02	49.2804	2.64953	8.37854		7.52	56.5504	2.74226	8.67179
7.03	49.4209	2.65141	8.38451		7.53	56.7009	2.74408	8.67756
7.04	49.5616	2.65330	8.39047		7.54	56.8516	2.74591	8.68332
7.05	49.7025	2.65518	8.39643		7.55	57.0025	2.74773	8.68907
7.06	49.8436	2.65707	8.40238		7.56	57.1536	2.74955	8.69483
7.07	49.9849	2.65895	8.40833		7.57	57.3049	2.75136	8.70057
7.08	50.1264	2.66083	8.41427		7.58	57.4564	2.75318	8.70632
7.09	50.2681	2.66271	8.42021		7.59	57.6081	2.75500	8.71206
7.10	50.4100	2.66458	8.42615		**7.60**	57.7600	2.75681	8.71780
7.11	50.5521	2.66646	8.43208		7.61	57.9121	2.75862	8.72353
7.12	50.6944	2.66833	8.43801		7.62	58.0644	2.76043	8.72926
7.13	50.8369	2.67021	8.44393		7.63	58.2169	2.76225	8.73499
7.14	50.9796	2.67208	8.44985		7.64	58.3696	2.76405	8.74071
7.15	51.1225	2.67395	8.45577		7.65	58.5225	2.76586	8.74643
7.16	51.2656	2.67582	8.46168		7.66	58.6756	2.76767	8.75214
7.17	51.4089	2.67769	8.46759		7.67	58.8289	2.76948	8.75785
7.18	51.5524	2.67955	8.47349		7.68	58.9824	2.77128	8.76356
7.19	51.6961	2.68142	8.47939		7.69	59.1361	2.77308	8.76926
7.20	51.8400	2.68328	8.48528		**7.70**	59.2900	2.77489	8.77496
7.21	51.9841	2.68514	8.49117		7.71	59.4441	2.77669	8.78066
7.22	52.1284	2.68701	8.49706		7.72	59.5984	2.77849	8.78635
7.23	52.2729	2.68887	8.50294		7.73	59.7529	2.78029	8.79204
7.24	52.4176	2.69072	8.50882		7.74	59.9076	2.78209	8.79773
7.25	52.5625	2.69258	8.51469		7.75	60.0625	2.78388	8.80341
7.26	52.7076	2.69444	8.52056		7.76	60.2176	2.78568	8.80909
7.27	52.8529	2.69629	8.52643		7.77	60.3729	2.78747	8.81476
7.28	52.9984	2.69815	8.53229		7.78	60.5284	2.78927	8.82043
7.29	53.1441	2.70000	8.53815		7.79	60.6841	2.79106	8.82610
7.30	53.2900	2.70185	8.54400		**7.80**	60.8400	2.79285	8.83176
7.31	53.4361	2.70370	8.54985		7.81	60.9961	2.79464	8.83742
7.32	53.5824	2.70555	8.55570		7.82	61.1524	2.79643	8.84308
7.33	53.7289	2.70740	8.56154		7.83	61.3089	2.79821	8.84873
7.34	53.8756	2.70924	8.56738		7.84	61.4656	2.80000	8.85438
7.35	54.0225	2.71109	8.57321		7.85	61.6225	2.80179	8.86002
7.36	54.1696	2.71293	8.57904		7.86	61.7796	2.80357	8.86566
7.37	54.3169	2.71477	8.58487		7.87	61.9369	2.80535	8.87130
7.38	54.4644	2.71662	8.59069		7.88	62.0944	2.80713	8.87694
7.39	54.6121	2.71846	8.59651		7.89	62.2521	2.80891	8.88257
7.40	54.7600	2.72029	8.60233		**7.90**	62.4100	2.81069	8.88819
7.41	54.9081	2.72213	8.60814		7.91	62.5681	2.81247	8.89382
7.42	55.0564	2.72397	8.61394		7.92	62.7264	2.81425	8.89944
7.43	55.2049	2.72580	8.61974		7.93	62.8849	2.81603	8.90505
7.44	55.3536	2.72764	8.62554		7.94	63.0436	2.81780	8.91067
7.45	55.5025	2.72947	8.63134		7.95	63.2025	2.81957	8.91628
7.46	55.6516	2.73130	8.63713		7.96	63.3616	2.82135	8.92188
7.47	55.8009	2.73313	8.64292		7.97	63.5209	2.82312	8.92749
7.48	55.9504	2.73496	8.64870		7.98	63.6804	2.82489	8.93308
7.49	56.1001	2.73679	8.65448		7.99	63.8401	2.82666	8.93868
7.50	56.2500	2.73861	8.66025		**8.00**	64.0000	2.82843	8.94427
N	N²	√N̄	√10N̄		N	N²	√N̄	√10N̄

N	N²	√N	√10N	N	N²	√N	√10N
8.00	64.0000	2.82843	8.94427	**8.50**	72.2500	2.91548	9.21954
8.01	64.1601	2.83019	8.94986	8.51	72.4201	2.91719	9.22497
8.02	64.3204	2.83196	8.95545	8.52	72.5904	2.91890	9.23038
8.03	64.4809	2.83373	8.96103	8.53	72.7609	2.92062	9.23580
8.04	64.6416	2.83549	8.96660	8.54	72.9316	2.92233	9.24121
8.05	64.8025	2.83725	8.97218	8.55	73.1025	2.92404	9.24662
8.06	64.9636	2.83901	8.97775	8.56	73.2736	2.92575	9.25203
8.07	65.1249	2.84077	8.98332	8.57	73.4449	2.92746	9.25743
8.08	65.2864	2.84253	8.98888	8.58	73.6164	2.92916	9.26283
8.09	65.4481	2.84429	8.99444	8.59	73.7881	2.93087	9.26823
8.10	65.6100	2.84605	9.00000	**8.60**	73.9600	2.93258	9.27362
8.11	65.7721	2.84781	9.00555	8.61	74.1321	2.93428	9.27901
8.12	65.9344	2.84956	9.01110	8.62	74.3044	2.93598	9.28440
8.13	66.0969	2.85132	9.01665	8.63	74.4769	2.93769	9.28978
8.14	66.2596	2.85307	9.02219	8.64	74.6496	2.93939	9.29516
8.15	66.4225	2.85482	9.02774	8.65	74.8225	2.94109	9.30054
8.16	66.5856	2.85657	9.03327	8.66	74.9956	2.94279	9.30591
8.17	66.7489	2.85832	9.03881	8.67	75.1689	2.94449	9.31128
8.18	66.9124	2.86007	9.04434	8.68	75.3424	2.94618	9.31665
8.19	67.0761	2.86182	9.04986	8.69	75.5161	2.94788	9.32202
8.20	67.2400	2.86356	9.05539	**8.70**	75.6900	2.94958	9.32738
8.21	67.4041	2.86531	9.06091	8.71	75.8641	2.95127	9.33274
8.22	67.5684	2.86705	9.06642	8.72	76.0384	2.95296	9.33809
8.23	67.7329	2.86880	9.07193	8.73	76.2129	2.95466	9.34345
8.24	67.8976	2.87054	9.07744	8.74	76.3876	2.95635	9.34880
8.25	68.0625	2.87228	9.08295	8.75	76.5625	2.95804	9.35414
8.26	68.2276	2.87402	9.08845	8.76	76.7376	2.95973	9.35949
8.27	68.3929	2.87576	9.09395	8.77	76.9129	2.96142	9.36483
8.28	68.5584	2.87750	9.09945	8.78	77.0884	2.96311	9.37017
8.29	68.7241	2.87924	9.10494	8.79	77.2641	2.96479	9.37550
8.30	68.8900	2.88097	9.11045	**8.80**	77.4400	2.96648	9.38083
8.31	69.0561	2.88271	9.11592	8.81	77.6161	2.96816	9.38616
8.32	69.2224	2.88444	9.12140	8.82	77.7924	2.96985	9.39149
8.33	69.3889	2.88617	9.12688	8.83	77.9689	2.97153	9.39681
8.34	69.5556	2.88791	9.13236	8.84	78.1456	2.97321	9.40213
8.35	69.7225	2.88964	9.13783	8.85	78.3225	2.97489	9.40744
8.36	69.8896	2.89137	9.14330	8.86	78.4996	2.97658	9.41276
8.37	70.0569	2.89310	9.14877	8.87	78.6769	2.97825	9.41807
8.38	70.2244	2.89482	9.15423	8.88	78.8544	2.97993	9.42338
8.39	70.3921	2.89655	9.15969	8.89	79.0321	2.98161	9.42868
8.40	70.5600	2.89828	9.16515	**8.90**	79.2100	2.98329	9.43398
8.41	70.7281	2.90000	9.17061	8.91	79.3881	2.98496	9.43928
8.42	70.8964	2.90172	9.17606	8.92	79.5664	2.98664	9.44458
8.43	71.0649	2.90345	9.18150	8.93	79.7449	2.98831	9.44987
8.44	71.2336	2.90517	9.18695	8.94	79.9236	2.98998	9.45516
8.45	71.4025	2.90689	9.19239	8.95	80.1025	2.99166	9.46044
8.46	71.5716	2.90861	9.19783	8.96	80.2816	2.99333	9.46573
8.47	71.7409	2.91033	9.20326	8.97	80.4609	2.99500	9.47101
8.48	71.9104	2.91204	9.20869	8.98	80.6404	2.99666	9.47629
8.49	72.0801	2.91376	9.21412	8.99	80.8201	2.99833	9.48156
8.50	72.2500	2.91548	9.21954	**9.00**	81.0000	3.00000	9.48683
N	N²	√N	√10N	N	N²	√N	√10N

N	N²	√N	√10N		N	N²	√N	√10N
9.00	81.0000	3.00000	9.48683		**9.50**	90.2500	3.08221	9.74679
9.01	81.1801	3.00167	9.49210		9.51	90.4401	3.08383	9.75192
9.02	81.3604	3.00333	9.49737		9.52	90.6304	3.08545	9.75705
9.03	81.5409	3.00500	9.50263		9.53	90.8209	3.08707	9.76217
9.04	81.7216	3.00666	9.50789		9.54	91.0116	3.08869	9.76729
9.05	81.9025	3.00832	9.51315		9.55	91.2025	3.09031	9.77241
9.06	82.0836	3.00998	9.51840		9.56	91.3936	3.09192	9.77753
9.07	82.2649	3.01164	9.52365		9.57	91.5849	3.09354	9.78264
9.08	82.4464	3.01330	9.52890		9.58	91.7764	3.09516	9.78775
9.09	82.6281	3.01496	9.53415		9.59	91.9681	3.09677	9.79285
9.10	82.8100	3.01662	9.53939		**9.60**	92.1600	3.09839	9.79796
9.11	82.9921	3.01828	9.54463		9.61	92.3521	3.10000	9.80306
9.12	83.1744	3.01993	9.54987		9.62	92.5444	3.10161	9.80816
9.13	83.3569	3.02159	9.55510		9.63	92.7369	3.10322	9.81326
9.14	83.5396	3.02324	9.56033		9.64	92.9296	3.10483	9.81835
9.15	83.7225	3.02490	9.56556		9.65	93.1225	3.10644	9.82344
9.16	83.9056	3.02655	9.57079		9.66	93.3156	3.10805	9.82853
9.17	84.0889	3.02820	9.57601		9.67	93.5089	3.10966	9.83362
9.18	84.2724	3.02985	9.58123		9.68	93.7024	3.11127	9.83870
9.19	84.4561	3.03150	9.58645		9.69	93.8961	3.11288	9.84378
9.20	84.6400	3.03315	9.59166		**9.70**	94.0900	3.11448	9.84886
9.21	84.8241	3.03480	9.59687		9.71	94.2841	3.11609	9.85393
9.22	85.0084	3.03645	9.60208		9.72	94.4784	3.11769	9.85901
9.23	85.1929	3.03809	9.60729		9.73	94.6729	3.11929	9.86408
9.24	85.3776	3.03974	9.61249		9.74	94.8676	3.12090	9.86914
9.25	85.5625	3.04138	9.61769		9.75	95.0625	3.12250	9.87421
9.26	85.7476	3.04302	9.62289		9.76	95.2576	3.12410	9.87927
9.27	85.9329	3.04467	9.62808		9.77	95.4529	3.12570	9.88433
9.28	86.1184	3.04631	9.63328		9.78	95.6484	3.12730	9.88939
9.29	86.3041	3.04795	9.63846		9.79	95.8441	3.12890	9.89444
9.30	86.4900	3.04959	9.64365		**9.80**	96.0400	3.13050	9.89949
9.31	86.6761	3.05123	9.64883		9.81	96.2361	3.13209	9.90454
9.32	86.8624	3.05287	9.65401		9.82	96.4324	3.13369	9.90959
9.33	87.0489	3.05450	9.65919		9.83	96.6289	3.13528	9.91464
9.34	87.2356	3.05614	9.66437		9.84	96.8256	3.13688	9.91968
9.35	87.4225	3.05778	9.66954		9.85	97.0025	3.13847	9.92472
9.36	87.6096	3.05941	9.67471		9.86	97.2196	3.14006	9.92974
9.37	87.7969	3.06105	9.67988		9.87	97.4169	3.14166	9.93479
9.38	87.9844	3.06268	9.68504		9.88	97.6144	3.14325	9.93982
9.39	88.1721	3.06431	9.69020		9.89	97.8121	3.14484	9.94485
9.40	88.3600	3.06594	9.69536		**9.90**	98.0100	3.14643	9.94987
9.41	88.5481	3.06757	9.70052		9.91	98.2081	3.14802	9.95490
9.42	88.7364	3.06920	9.70567		9.92	98.4064	3.14960	9.95992
9.43	88.9249	3.07083	9.71082		9.93	98.6049	3.15119	9.96494
9.44	89.1136	3.07246	9.71597		9.94	98.8036	3.15278	9.96995
9.45	89.3025	3.07409	9.72111		9.95	99.0025	3.15436	9.97497
9.46	89.4916	3.07571	9.72625		9.96	99.2016	3.15595	9.97998
9.47	89.6809	3.07734	9.73139		9.97	99.4009	3.15753	9.98499
9.48	89.8704	3.07896	9.73653		9.98	99.6004	3.15911	9.98999
9.49	90.0601	3.08058	9.74166		9.99	99.8001	3.16070	9.99500
9.50	90.2500	3.08221	9.74679		**10.0**	100.000	3.16228	10.0000
N	N²	√N	√10N		N	N²	√N	√10N

Table II. Factorials

n	$n!$
0	1
1	1
2	2
3	6
4	24
5	120
6	720
7	5,040
8	40,320
9	362,880
10	3,628,800
11	39,916,800
12	479,001,600
13	6,227,020,800
14	87,178,291,200
15	1,307,674,368,000
16	20,922,789,888,000
17	355,687,428,096,000
18	6,402,373,705,728,000
19	121,645,100,408,832,000
20	2,432,902,008,176,640,000

Table III. Binomial coefficients $\dfrac{n!}{x!(n-x)!}$

n \ x	2	3	4	5	6	7	8	9	10
2	1								
3	3	1							
4	6	4	1						
5	10	10	5	1					
6	15	20	15	6	1				
7	21	35	35	21	7	1			
8	28	56	70	56	28	8	1		
9	36	84	126	126	84	36	9	1	
10	45	120	210	252	210	120	45	10	1
11	55	165	330	462	462	330	165	55	11
12	66	220	495	792	924	792	495	220	66
13	78	286	715	1,287	1,716	1,716	1,287	715	286
14	91	364	1,001	2,002	3,003	3,432	3,003	2,002	1,001
15	105	455	1,365	3,003	5,005	6,435	6,435	5,005	3,003
16	120	560	1,820	4,368	8,008	11,440	12,870	11,440	8,008
17	136	680	2,380	6,188	12,376	19,448	24,310	24,310	19,448
18	153	816	3,060	8,568	18,564	31,824	43,758	48,620	43,758
19	171	969	3,876	11,628	27,132	50,388	75,582	92,378	92,378
20	190	1,140	4,845	15,504	38,760	77,520	125,970	167,960	184,756

Table IV. Binomial probabilities

							p					
n	x	0.05	0.1	0.2	0.3	0.4	0.5	0.6	0.7	0.8	0.9	0.95
2	0	0.902	0.810	0.640	0.490	0.360	0.250	0.160	0.090	0.040	0.010	0.002
	1	0.095	0.180	0.320	0.420	0.480	0.500	0.480	0.420	0.320	0.180	0.095
	2	0.002	0.010	0.040	0.090	0.160	0.250	0.360	0.490	0.640	0.810	0.902
3	0	0.857	0.729	0.512	0.343	0.216	0.125	0.064	0.027	0.008	0.001	
	1	0.135	0.243	0.384	0.441	0.432	0.375	0.288	0.189	0.096	0.027	0.007
	2	0.007	0.027	0.096	0.189	0.288	0.375	0.432	0.441	0.384	0.243	0.135
	3		0.001	0.008	0.027	0.064	0.125	0.216	0.343	0.512	0.729	0.857
4	0	0.815	0.656	0.410	0.240	0.130	0.062	0.026	0.008	0.002		
	1	0.171	0.292	0.410	0.412	0.346	0.250	0.154	0.076	0.026	0.004	
	2	0.014	0.049	0.154	0.265	0.346	0.375	0.346	0.265	0.154	0.049	0.014
	3		0.004	0.026	0.076	0.154	0.250	0.346	0.412	0.410	0.292	0.171
	4			0.002	0.008	0.026	0.062	0.130	0.240	0.410	0.656	0.815
5	0	0.774	0.590	0.328	0.168	0.078	0.031	0.010	0.002			
	1	0.204	0.328	0.410	0.360	0.259	0.156	0.077	0.028	0.006		
	2	0.021	0.073	0.205	0.309	0.346	0.312	0.230	0.132	0.051	0.008	0.001
	3	0.001	0.008	0.051	0.132	0.230	0.312	0.346	0.309	0.205	0.073	0.021
	4			0.006	0.028	0.077	0.156	0.259	0.360	0.410	0.328	0.204
	5				0.002	0.010	0.031	0.078	0.168	0.328	0.590	0.774
6	0	0.735	0.531	0.262	0.118	0.047	0.016	0.004	0.001			
	1	0.232	0.354	0.393	0.303	0.187	0.094	0.037	0.010	0.002		
	2	0.031	0.098	0.246	0.324	0.311	0.234	0.138	0.060	0.015	0.001	
	3	0.002	0.015	0.082	0.185	0.276	0.312	0.276	0.185	0.082	0.015	0.002
	4		0.001	0.015	0.060	0.138	0.234	0.311	0.324	0.246	0.098	0.031
	5			0.002	0.010	0.037	0.094	0.187	0.303	0.393	0.354	0.232
	6				0.001	0.004	0.016	0.047	0.118	0.262	0.531	0.735
7	0	0.698	0.478	0.210	0.082	0.028	0.008	0.002				
	1	0.257	0.372	0.367	0.247	0.131	0.055	0.017	0.004			
	2	0.041	0.124	0.275	0.318	0.261	0.164	0.077	0.025	0.004		
	3	0.004	0.023	0.115	0.227	0.290	0.273	0.194	0.097	0.029	0.003	
	4		0.003	0.029	0.097	0.194	0.273	0.290	0.227	0.115	0.023	0.004
	5			0.004	0.025	0.077	0.164	0.261	0.318	0.275	0.124	0.041
	6				0.004	0.017	0.055	0.131	0.247	0.367	0.372	0.257
	7					0.002	0.008	0.028	0.082	0.210	0.478	0.698
8	0	0.663	0.430	0.168	0.058	0.017	0.004	0.001				
	1	0.279	0.383	0.336	0.198	0.090	0.031	0.008	0.001			
	2	0.051	0.149	0.294	0.296	0.209	0.109	0.041	0.010	0.001		
	3	0.005	0.033	0.147	0.254	0.279	0.219	0.124	0.047	0.009		
	4		0.005	0.046	0.136	0.232	0.273	0.232	0.136	0.046	0.005	
	5			0.009	0.047	0.124	0.219	0.279	0.254	0.147	0.033	0.005
	6			0.001	0.010	0.041	0.109	0.209	0.296	0.294	0.149	0.051
	7				0.001	0.008	0.031	0.090	0.198	0.336	0.383	0.279
	8					0.001	0.004	0.017	0.058	0.168	0.430	0.663

(continued)

n	x	0.05	0.1	0.2	0.3	0.4	0.5	0.6	0.7	0.8	0.9	0.95
9	0	0.630	0.387	0.134	0.040	0.010	0.002					
	1	0.299	0.387	0.302	0.156	0.060	0.018	0.004				
	2	0.063	0.172	0.302	0.267	0.161	0.070	0.021	0.004			
	3	0.008	0.045	0.176	0.267	0.251	0.164	0.074	0.021	0.003		
	4	0.001	0.007	0.066	0.172	0.251	0.246	0.167	0.074	0.017	0.001	
	5		0.001	0.017	0.074	0.167	0.246	0.251	0.172	0.066	0.007	0.001
	6			0.003	0.021	0.074	0.164	0.251	0.267	0.176	0.045	0.008
	7				0.004	0.021	0.070	0.161	0.267	0.302	0.172	0.063
	8					0.004	0.018	0.060	0.156	0.302	0.387	0.299
	9						0.002	0.010	0.040	0.134	0.387	0.630
10	0	0.599	0.349	0.107	0.028	0.006	0.001					
	1	0.315	0.387	0.268	0.121	0.040	0.010	0.002				
	2	0.075	0.194	0.302	0.233	0.121	0.044	0.011	0.001			
	3	0.010	0.057	0.201	0.267	0.215	0.117	0.042	0.009	0.001		
	4	0.001	0.011	0.088	0.200	0.251	0.205	0.111	0.037	0.006		
	5		0.001	0.026	0.103	0.201	0.246	0.201	0.103	0.026	0.001	
	6			0.006	0.037	0.111	0.205	0.251	0.200	0.088	0.011	0.001
	7			0.001	0.009	0.042	0.117	0.215	0.267	0.201	0.057	0.010
	8				0.001	0.011	0.044	0.121	0.233	0.302	0.194	0.075
	9					0.002	0.010	0.040	0.121	0.268	0.387	0.315
	10						0.001	0.006	0.028	0.107	0.349	0.599
11	0	0.569	0.314	0.086	0.020	0.004						
	1	0.329	0.384	0.236	0.093	0.027	0.005	0.001				
	2	0.087	0.213	0.295	0.200	0.089	0.027	0.005	0.001			
	3	0.014	0.071	0.221	0.257	0.177	0.081	0.023	0.004			
	4	0.001	0.016	0.111	0.220	0.236	0.161	0.070	0.017	0.002		
	5		0.002	0.039	0.132	0.221	0.226	0.147	0.057	0.010		
	6			0.010	0.057	0.147	0.226	0.221	0.132	0.039	0.002	
	7			0.002	0.017	0.070	0.161	0.236	0.220	0.111	0.016	0.001
	8				0.004	0.023	0.081	0.177	0.257	0.221	0.071	0.014
	9				0.001	0.005	0.027	0.089	0.200	0.295	0.213	0.087
	10					0.001	0.005	0.027	0.093	0.236	0.384	0.329
	11						0.004	0.020	0.086	0.314	0.569
12	0	0.540	0.282	0.069	0.014	0.002						
	1	0.341	0.377	0.206	0.071	0.017	0.003					
	2	0.099	0.230	0.283	0.168	0.064	0.016	0.002				
	3	0.017	0.085	0.236	0.240	0.142	0.054	0.012	0.001			
	4	0.002	0.021	0.133	0.231	0.213	0.121	0.042	0.008	0.001		
	5		0.004	0.053	0.158	0.227	0.193	0.101	0.029	0.003		
	6			0.016	0.079	0.177	0.226	0.177	0.079	0.016		
	7			0.003	0.029	0.101	0.193	0.227	0.158	0.053	0.004	
	8			0.001	0.008	0.042	0.121	0.213	0.231	0.133	0.021	0.002
	9				0.001	0.012	0.054	0.142	0.240	0.236	0.085	0.017
	10					0.002	0.016	0.064	0.168	0.283	0.230	0.099
	11						0.003	0.017	0.071	0.206	0.377	0.341
	12							0.002	0.014	0.069	0.282	0.540

						p						
n	x	0.05	0.1	0.2	0.3	0.4	0.5	0.6	0.7	0.8	0.9	0.95
13	0	0.513	0.254	0.055	0.010	0.001						
	1	0.351	0.367	0.179	0.054	0.011	0.002					
	2	0.111	0.245	0.268	0.139	0.045	0.010	0.001				
	3	0.021	0.100	0.246	0.218	0.111	0.035	0.006	0.001			
	4	0.003	0.028	0.154	0.234	0.184	0.087	0.024	0.003			
	5		0.006	0.069	0.180	0.221	0.157	0.066	0.014	0.001		
	6		0.001	0.023	0.103	0.197	0.209	0.131	0.044	0.006		
	7			0.006	0.044	0.131	0.209	0.197	0.103	0.023	0.001	
	8			0.001	0.014	0.066	0.157	0.221	0.180	0.069	0.006	
	9				0.003	0.024	0.087	0.184	0.234	0.154	0.028	0.003
	10				0.001	0.006	0.035	0.111	0.218	0.246	0.100	0.021
	11					0.001	0.010	0.045	0.139	0.268	0.245	0.111
	12						0.002	0.011	0.054	0.179	0.367	0.351
	13							0.001	0.010	0.055	0.254	0.513
14	0	0.488	0.229	0.044	0.007	0.001						
	1	0.359	0.356	0.154	0.041	0.007	0.001					
	2	0.123	0.257	0.250	0.113	0.032	0.006	0.001				
	3	0.026	0.114	0.250	0.194	0.085	0.022	0.003				
	4	0.004	0.035	0.172	0.229	0.155	0.061	0.014	0.001			
	5		0.008	0.086	0.196	0.207	0.122	0.041	0.007			
	6		0.001	0.032	0.126	0.207	0.183	0.092	0.023	0.002		
	7			0.009	0.062	0.157	0.209	0.157	0.062	0.009		
	8			0.002	0.023	0.092	0.183	0.207	0.126	0.032	0.001	
	9				0.007	0.041	0.122	0.207	0.196	0.086	0.008	
	10				0.001	0.014	0.061	0.155	0.229	0.172	0.035	0.004
	11					0.003	0.022	0.085	0.194	0.250	0.114	0.026
	12					0.001	0.006	0.032	0.113	0.250	0.257	0.123
	13						0.001	0.007	0.041	0.154	0.356	0.359
	14							0.001	0.007	0.044	0.229	0.488
15	0	0.463	0.206	0.035	0.005							
	1	0.366	0.343	0.132	0.031	0.005						
	2	0.135	0.267	0.231	0.092	0.022	0.003					
	3	0.031	0.129	0.250	0.170	0.063	0.014	0.002				
	4	0.005	0.043	0.188	0.219	0.127	0.042	0.007	0.001			
	5	0.001	0.010	0.103	0.206	0.186	0.092	0.024	0.003			
	6		0.002	0.043	0.147	0.207	0.153	0.061	0.012	0.001		
	7			0.014	0.081	0.177	0.196	0.118	0.035	0.003		
	8			0.003	0.035	0.118	0.196	0.177	0.081	0.014		
	9			0.001	0.012	0.061	0.153	0.207	0.147	0.043	0.002	
	10				0.003	0.024	0.092	0.186	0.206	0.103	0.010	0.001
	11				0.001	0.007	0.042	0.127	0.219	0.188	0.043	0.005
	12					0.002	0.014	0.063	0.170	0.250	0.129	0.031
	13						0.003	0.022	0.092	0.231	0.267	0.135
	14							0.005	0.031	0.132	0.343	0.366
	15								0.005	0.035	0.206	0.463

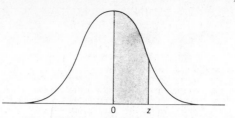

The entries in Table V are the probabilities that a random variable having the standard normal distribution takes on a value between 0 and z; they are given by the area under the curve shaded in the above diagram.

Table V. The standard normal distribution

z	.00	.01	.02	.03	.04	.05	.06	.07	.08	.09
0.0	.0000	.0040	.0080	.0120	.0160	.0199	.0239	.0279	.0319	.0359
0.1	.0398	.0438	.0478	.0517	.0557	.0596	.0636	.0675	.0714	.0753
0.2	.0793	.0832	.0871	.0910	.0948	.0987	.1026	.1064	.1103	.1141
0.3	.1179	.1217	.1255	.1293	.1331	.1368	.1406	.1443	.1480	.1517
0.4	.1554	.1591	.1628	.1664	.1700	.1736	.1772	.1808	.1844	.1879
0.5	.1915	.1950	.1985	.2019	.2054	.2088	.2123	.2157	.2190	.2224
0.6	.2257	.2291	.2324	.2357	.2389	.2422	.2454	.2486	.2517	.2549
0.7	.2580	.2611	.2642	.2673	.2704	.2734	.2764	.2794	.2823	.2852
0.8	.2881	.2910	.2939	.2967	.2995	.3023	.3051	.3078	.3106	.3133
0.9	.3159	.3186	.3212	.3238	.3264	.3289	.3315	.3340	.3365	.3389
1.0	.3413	.3438	.3461	.3485	.3508	.3531	.3554	.3577	.3599	.3621
1.1	.3643	.3665	.3686	.3708	.3729	.3749	.3770	.3790	.3810	.3830
1.2	.3849	.3869	.3888	.3907	.3925	.3944	.3962	.3980	.3997	.4015
1.3	.4032	.4049	.4066	.4082	.4099	.4115	.4131	.4147	.4162	.4177
1.4	.4192	.4207	.4222	.4236	.4251	.4265	.4279	.4292	.4306	.4319
1.5	.4332	.4345	.4357	.4370	.4382	.4394	.4406	.4418	.4429	.4441
1.6	.4452	.4463	.4474	.4484	.4495	.4505	.4515	.4525	.4535	.4545
1.7	.4554	.4564	.4573	.4582	.4591	.4599	.4608	.4616	.4625	.4633
1.8	.4641	.4649	.4656	.4664	.4671	.4678	.4686	.4693	.4699	.4706
1.9	.4713	.4719	.4726	.4732	.4738	.4744	.4750	.4756	.4761	.4767
2.0	.4772	.4778	.4783	.4788	.4793	.4798	.4803	.4808	.4812	.4817
2.1	.4821	.4826	.4830	.4834	.4838	.4842	.4846	.4850	.4854	.4857
2.2	.4861	.4864	.4868	.4871	.4875	.4878	.4881	.4884	.4887	.4890
2.3	.4893	.4896	.4898	.4901	.4904	.4906	.4909	.4911	.4913	.4916
2.4	.4918	.4920	.4922	.4925	.4927	.4929	.4931	.4932	.4934	.4936
2.5	.4938	.4940	.4941	.4943	.4945	.4946	.4948	.4949	.4951	.4952
2.6	.4953	.4955	.4956	.4957	.4959	.4960	.4961	.4962	.4963	.4964
2.7	.4965	.4966	.4967	.4968	.4969	.4970	.4971	.4972	.4973	.4974
2.8	.4974	.4975	.4976	.4977	.4977	.4978	.4979	.4979	.4980	.4981
2.9	.4981	.4982	.4982	.4983	.4984	.4984	.4985	.4985	.4986	.4986
3.0	.4987	.4987	.4987	.4988	.4988	.4989	.4989	.4989	.4990	.4990

Anne-Marie Brazoura   Feb. 6

Quiz 3

(8)

For the following frequency distribution: (1) calculate the class marks
(2) construct a histogram (bar graph)
for the distribution
(3) construct the frequency polygon
for the distribution

Observation	Class Mark	Frequency
150-154	152	10
145-149	147	6
140-144	142	14
135-139	137	22
130-134	132	17
125-129	127	11
120-124	122	8

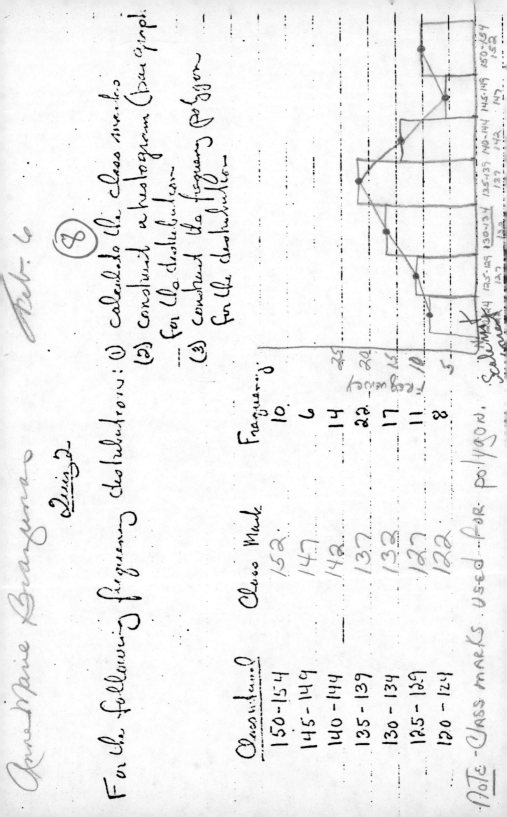

FREQUENCY

25
20
15
10
5

Scale Mark 4 125-129 130-134 135-139 140-144 145-149 150-154
122 127 132 137 142 147 152

NOTE - CLASS MARKS USED FOR POLYGON.

Table VI. Critical values of r

n	$r_{0.025}$	$r_{0.005}$	n	$r_{0.025}$	$r_{0.005}$
3	0.997		18	0.468	0.590
4	0.950	0.999	19	0.456	0.575
5	0.878	0.959	20	0.444	0.561
6	0.811	0.917	21	0.433	0.549
7	0.754	0.875	22	0.423	0.537
8	0.707	0.834	27	0.381	0.487
9	0.666	0.798	32	0.349	0.449
10	0.632	0.765	37	0.325	0.418
11	0.602	0.735	42	0.304	0.393
12	0.576	0.708	47	0.288	0.372
13	0.553	0.684	52	0.273	0.354
14	0.532	0.661	62	0.250	0.325
15	0.514	0.641	72	0.232	0.302
16	0.497	0.623	82	0.217	0.283
17	0.482	0.606	92	0.205	0.267

This table is abridged from Table VI of R. A. Fisher and F. Yates: *Statistical Tables for Biological, Agricultural, and Medical Research*, published by Longman Group, Ltd., London (previously published by Oliver & Boyd, Edinburgh), by permission of the authors and publishers.

Table VII. Table of random digits.

COL. LINE	(1)	(2)	(3)	(4)	(5)	(6)
1	10480	15011	01536	02011	81647	91646
2	22368	46573	25595	85393	30995	89198
3	24130	48360	22527	97265	76393	64809
4	42167	93093	06243	61680	07856	16376
5	37570	39975	81837	16656	06121	91782
6	77921	06907	11008	42751	27756	53498
7	99562	72905	56420	69994	98872	31016
8	96301	91977	05463	07972	18876	20922
9	89579	14342	63661	10281	17453	18103
10	85475	36857	53342	53988	53060	59533
11	28918	69578	88231	33276	70997	79936
12	63553	40961	48235	03427	49626	69445
13	09429	93969	52636	92737	88974	33488
14	10365	61129	87529	85689	48237	52267
15	07119	97336	71048	08178	77233	13916
16	51085	12765	51821	51259	77452	16308
17	02368	21382	52404	60268	89368	19885
18	01011	54092	33362	94904	31273	04146
19	52162	53916	46369	58586	23216	14513
20	07056	97628	33787	09998	42698	06691
21	48663	91245	85828	14346	09172	30168
22	54164	58492	22421	74103	47070	25306
23	32639	32363	05597	24200	13363	38005
24	29334	27001	87637	87308	58731	00256
25	02488	33062	28834	07351	19731	92420
26	81525	72295	04839	96423	24878	82651
27	29676	20591	68086	26432	46901	20849
28	00742	57392	39064	66432	84673	40027
29	05366	04213	25669	26422	44407	44048
30	91921	26418	64117	94305	26766	25940
31	00582	04711	87917	77341	42206	35126
32	00725	69884	62797	56170	86324	88072
33	69011	65795	95876	55293	18988	27354
34	25976	57948	29888	88604	67917	48708
35	09763	83473	73577	12908	30883	18317
36	91567	42595	27958	30134	04024	86385
37	17955	56349	90999	49127	20044	59931
38	46503	18584	18845	49618	02304	51038
39	92157	89634	94824	78171	84610	82834
40	14577	62765	35605	81263	39667	47358
41	98427	07523	33362	64270	01638	92477
42	34914	63976	88720	82765	34476	17032
43	70060	28277	39475	46473	23219	53416
44	53976	54914	06990	67245	68350	82948
45	76072	29515	40980	07391	58745	25774
46	90725	52210	83974	29992	65831	38857
47	64364	67412	33339	31926	14883	24413
48	08962	00358	31662	25388	61642	34072
49	95012	68379	93526	70765	10592	04542
50	15664	10493	20492	38391	91132	21999

(7)	(8)	(9)	(10)	(11)	(12)	(13)	(14)
69179	14194	62590	36207	20969	99570	91291	90700
27982	53402	93965	34095	52666	19174	39615	99505
15179	24830	49340	32081	30680	19655	63348	58629
39440	53537	71341	57004	00849	74917	97758	16379
60468	81305	49684	60672	14110	06927	01263	54613
18602	70659	90655	15053	21916	81825	44394	42880
71194	18738	44013	48840	63213	21069	10634	12952
94595	56869	69014	60045	18425	84903	42508	32307
57740	84378	25331	12566	58678	44947	05585	56941
38867	62300	08158	17983	16439	11458	18593	64952
56865	05859	90106	31595	01547	85590	91610	78188
18663	72695	52180	20847	12234	90511	33703	90322
36320	17617	30015	08272	84115	27156	30613	74952
67689	93394	01511	26358	85104	20285	29975	89868
47564	81056	97735	85977	29372	74461	28551	90707
60756	92144	49442	53900	70960	63990	75601	40719
55322	44819	01188	65255	64835	44919	05944	55157
18594	29852	71585	85030	51132	01915	92747	64951
83149	98736	23495	64350	94738	17752	35156	35749
76988	13602	51851	46104	88916	19509	25625	58104
90229	04734	59193	22178	30421	61666	99904	32812
76468	26384	58151	06646	21524	15227	96909	44592
94342	28728	35806	06912	17012	64161	18296	22851
45834	15398	46557	41135	10367	07684	36188	18510
60952	61280	50001	67658	32586	86679	50720	94953
66566	14778	76797	14780	13300	87074	79666	95725
89768	81536	86645	12659	92259	57102	80428	25280
32832	61362	98947	96067	64760	64584	96096	98253
37937	63904	45766	66134	75470	66520	34693	90449
39972	22209	71500	64568	91402	42416	07844	69618
74087	99547	81817	42607	43808	76655	62028	76630
76222	36086	84637	93161	76038	65855	77919	88006
26575	08625	40801	59920	29841	80150	12777	48501
18912	82271	65424	69774	33611	54262	85963	03547
28290	35797	05998	41688	34952	37888	38917	88050
29880	99730	55536	84855	29080	09250	79656	73211
06115	20542	18059	02008	73708	83517	36103	42791
20655	58727	28168	15475	56942	53389	20562	87338
09922	25417	44137	48413	25555	21246	35509	20468
56873	56307	61607	49518	89656	20103	77490	18062
66969	98420	04880	45585	46565	04102	46880	45709
87589	40836	32427	70002	70663	88863	77775	69348
94970	25832	69975	94884	19661	72828	00102	66794
11398	42878	80287	88267	47363	46634	06541	97809
22987	80059	39911	96189	41151	14222	60697	59583
50490	83765	55657	14361	31720	57375	56228	41546
59744	92351	97473	89286	35931	04110	23726	51900
81249	35648	56891	69352	48373	45578	78547	81788
76463	54328	02349	17247	28865	14777	62730	92277
59516	81652	27195	48223	46751	22923	32261	85653

Page 1 of *Table of 105,000 Random Decimal Digits*, Statement No. 4914, May, 1949, File No. 261-A-1, Interstate Commerce Commission, Washington, D.C.

Table VIII. The *t* distribution

df	$t_{0.050}$	$t_{0.025}$	$t_{0.010}$	$t_{0.005}$	df
1	6.314	12.706	31.821	63.657	1
2	2.920	4.303	6.965	9.925	2
3	2.353	3.182	4.541	5.841	3
4	2.132	2.776	3.747	4.604	4
5	2.015	2.571	3.365	4.032	5
6	1.943	2.447	3.143	3.707	6
7	1.895	2.365	2.998	3.499	7
8	1.860	2.306	2.896	3.355	8
9	1.833	2.262	2.821	3.250	9
10	1.812	2.228	2.764	3.169	10
11	1.796	2.201	2.718	3.106	11
12	1.782	2.179	2.681	3.055	12
13	1.771	2.160	2.650	3.012	13
14	1.761	2.145	2.624	2.977	14
15	1.753	2.131	2.602	2.947	15
16	1.746	2.120	2.583	2.921	16
17	1.740	2.110	2.567	2.898	17
18	1.734	2.101	2.552	2.878	18
19	1.729	2.093	2.539	2.861	19
20	1.725	2.086	2.528	2.845	20
21	1.721	2.080	2.518	2.831	21
22	1.717	2.074	2.508	2.819	22
23	1.714	2.069	2.500	2.807	23
24	1.711	2.064	2.492	2.797	24
25	1.708	2.060	2.485	2.787	25
26	1.706	2.056	2.479	2.779	26
27	1.703	2.052	2.473	2.771	27
28	1.701	2.048	2.467	2.763	28
29	1.699	2.045	2.462	2.756	29
inf.	1.645	1.960	2.326	2.576	inf.

This table is abridged from Table IV of R. A. Fisher and F. Yates: *Statistical Tables for Biological, Agricultural, and Medical Research*, published by Longman Group, Ltd., London (previously published by Oliver & Boyd, Edinburgh), by permission of the authors and publishers.

Table IX. The χ² distribution

df	$\chi^2_{0.05}$	$\chi^2_{0.01}$	df
1	3.841	6.635	1
2	5.991	9.210	2
3	7.815	11.345	3
4	9.488	13.277	4
5	11.070	15.086	5
6	12.592	16.812	6
7	14.067	18.475	7
8	15.507	20.090	8
9	16.919	21.666	9
10	18.307	23.209	10
11	19.675	24.725	11
12	21.026	26.217	12
13	22.362	27.688	13
14	23.685	29.141	14
15	24.996	30.578	15
16	26.296	32.000	16
17	27.587	33.409	17
18	28.869	34.805	18
19	30.144	36.191	19
20	31.410	37.566	20
21	32.671	38.932	21
22	33.924	40.289	22
23	35.172	41.638	23
24	36.415	42.980	24
25	37.652	44.314	25
26	38.885	45.642	26
27	40.113	46.963	27
28	41.337	48.278	28
29	42.557	49.588	29
30	43.773	50.892	30

Table X. Critical values of the F-distribution ($\alpha = 0.05$)

		Degrees of freedom for numerator									
		1	2	3	4	5	6	7	8	9	10
Degrees of freedom for denominator	1	161	200	216	225	230	234	237	239	241	242
	2	18.5	19.0	19.2	19.2	19.3	19.3	19.4	19.4	19.4	19.4
	3	10.1	9.55	9.28	9.12	9.01	8.94	8.89	8.85	8.81	8.79
	4	7.71	6.94	6.59	6.39	6.26	6.16	6.09	6.04	6.00	5.96
	5	6.61	5.79	5.41	5.19	5.05	4.95	4.88	4.82	4.77	4.74
	6	5.99	5.14	4.76	4.53	4.39	4.28	4.21	4.15	4.10	4.06
	7	5.59	4.74	4.35	4.12	3.97	3.87	3.79	3.73	3.68	3.64
	8	5.32	4.46	4.07	3.84	3.69	3.58	3.50	3.44	3.39	3.35
	9	5.12	4.26	3.86	3.63	3.48	3.37	3.29	3.23	3.18	3.14
	10	4.96	4.10	3.71	3.48	3.33	3.22	3.14	3.07	3.02	2.98
	11	4.84	3.98	3.59	3.36	3.20	3.09	3.01	2.95	2.90	2.85
	12	4.75	3.89	3.49	3.26	3.11	3.00	2.91	2.85	2.80	2.75
	13	4.67	3.81	3.41	3.18	3.03	2.92	2.83	2.77	2.71	2.67
	14	4.60	3.74	3.34	3.11	2.96	2.85	2.76	2.70	2.65	2.60
	15	4.54	3.68	3.29	3.06	2.90	2.79	2.71	2.64	2.59	2.54
	16	4.49	3.63	3.24	3.01	2.85	2.74	2.66	2.59	2.54	2.49
	17	4.45	3.59	3.20	2.96	2.81	2.70	2.61	2.55	2.49	2.45
	18	4.41	3.55	3.16	2.93	2.77	2.66	2.58	2.51	2.46	2.41
	19	4.38	3.52	3.13	2.90	2.74	2.63	2.54	2.48	2.42	2.38
	20	4.35	3.49	3.10	2.87	2.71	2.60	2.51	2.45	2.39	2.35
	21	4.32	3.47	3.07	2.84	2.68	2.57	2.49	2.42	2.37	2.32
	22	4.30	3.44	3.05	2.82	2.66	2.55	2.46	2.40	2.34	2.30
	23	4.28	3.42	3.03	2.80	2.64	2.53	2.44	2.37	2.32	2.27
	24	4.26	3.40	3.01	2.78	2.62	2.51	2.42	2.36	2.30	2.25
	25	4.24	3.39	2.99	2.76	2.60	2.49	2.40	2.34	2.28	2.24
	30	4.17	3.32	2.92	2.69	2.53	2.42	2.33	2.27	2.21	2.16
	40	4.08	3.23	2.84	2.61	2.45	2.34	2.25	2.18	2.12	2.08
	60	4.00	3.15	2.76	2.53	2.37	2.25	2.17	2.10	2.04	1.99
	120	3.92	3.07	2.68	2.45	2.29	2.18	2.09	2.02	1.96	1.91
	∞	3.84	3.00	2.60	2.37	2.21	2.10	2.01	1.94	1.88	1.83

		12	15	20	24	30	40	60	120	∞
					Degrees of freedom for numerator					
	1	6,106	6,157	6,209	6,235	6,261	6,287	6,313	6,339	6,366
	2	99.4	99.4	99.4	99.5	99.5	99.5	99.5	99.5	99.5
	3	27.1	26.9	26.7	26.6	26.5	26.4	26.3	26.2	26.1
	4	14.4	14.2	14.0	13.9	13.8	13.7	13.7	13.6	13.5
	5	9.89	9.72	9.55	9.47	9.38	9.29	9.20	9.11	9.02
	6	7.72	7.56	7.40	7.31	7.23	7.14	7.06	6.97	6.88
	7	6.47	6.31	6.16	6.07	5.99	5.91	5.82	5.74	5.65
	8	5.67	5.52	5.36	5.28	5.20	5.12	5.03	4.95	4.86
	9	5.11	4.96	4.81	4.73	4.65	4.57	4.48	4.40	4.31
	10	4.71	4.56	4.41	4.33	4.25	4.17	4.08	4.00	3.91
Degrees of freedom for denominator	11	4.40	4.25	4.10	4.02	3.94	3.86	3.78	3.69	3.60
	12	4.16	4.01	3.86	3.78	3.70	3.62	3.54	3.45	3.36
	13	3.96	3.82	3.66	3.59	3.51	3.43	3.34	3.25	3.17
	14	3.80	3.66	3.51	3.43	3.35	3.27	3.18	3.09	3.00
	15	3.67	3.52	3.37	3.29	3.21	3.13	3.05	2.96	2.87
	16	3.55	3.41	3.26	3.18	3.10	3.02	2.93	2.84	2.75
	17	3.46	3.31	3.16	3.08	3.00	2.92	2.83	2.75	2.65
	18	3.37	3.23	3.08	3.00	2.92	2.84	2.75	2.66	2.57
	19	3.30	3.15	3.00	2.92	2.84	2.76	2.67	2.58	2.49
	20	3.23	3.09	2.94	2.86	2.78	2.69	2.61	2.52	2.42
	21	3.17	3.03	2.88	2.80	2.72	2.64	2.55	2.46	2.36
	22	3.12	2.98	2.83	2.75	2.67	2.58	2.50	2.40	2.31
	23	3.07	2.93	2.78	2.70	2.62	2.54	2.45	2.35	2.26
	24	3.03	2.89	2.74	2.66	2.58	2.49	2.40	2.31	2.21
	25	2.99	2.85	2.70	2.62	2.53	2.45	2.36	2.27	2.17
	30	2.84	2.70	2.55	2.47	2.39	2.30	2.21	2.11	2.01
	40	2.66	2.52	2.37	2.29	2.20	2.11	2.02	1.92	1.80
	60	2.50	2.35	2.20	2.12	2.03	1.94	1.84	1.73	1.60
	120	2.34	2.19	2.03	1.95	1.86	1.76	1.66	1.53	1.38
	∞	2.18	2.04	1.88	1.79	1.70	1.59	1.47	1.32	1.00

Critical Values of the F Distribution ($\alpha = 0.01$)

		Degrees of freedom for numerator								
	1	2	3	4	5	6	7	8	9	10
1	4,052	5,000	5,403	5,625	5,764	5,859	5,928	5,982	6,023	6,056
2	98.5	99.0	99.2	99.2	99.3	99.3	99.4	99.4	99.4	99.4
3	34.1	30.8	29.5	28.7	28.2	27.9	27.7	27.5	27.3	27.2
4	21.2	18.0	16.7	16.0	15.5	15.2	15.0	14.8	14.7	14.5
5	16.3	13.3	12.1	11.4	11.0	10.7	10.5	10.3	10.2	10.1
6	13.7	10.9	9.78	9.15	8.75	8.47	8.26	8.10	7.98	7.87
7	12.2	9.55	8.45	7.85	7.46	7.19	6.99	6.84	6.72	6.62
8	11.3	8.65	7.59	7.01	6.63	6.37	6.18	6.03	5.91	5.81
9	10.6	8.02	6.99	6.42	6.06	5.80	5.61	5.47	5.35	5.26
10	10.0	7.56	6.55	5.99	5.64	5.39	5.20	5.06	4.94	4.85
11	9.65	7.21	6.22	5.67	5.32	5.07	4.89	4.74	4.63	4.54
12	9.33	6.93	5.95	5.41	5.06	4.82	4.64	4.50	4.39	4.30
13	9.07	6.70	5.74	5.21	4.86	4.62	4.44	4.30	4.19	4.10
14	8.86	6.51	5.56	5.04	4.70	4.46	4.28	4.14	4.03	3.94
15	8.68	6.36	5.42	4.89	4.56	4.32	4.14	4.00	3.89	3.80
16	8.53	6.23	5.29	4.77	4.44	4.20	4.03	3.89	3.78	3.69
17	8.40	6.11	5.19	4.67	4.34	4.10	3.93	3.79	3.68	3.59
18	8.29	6.01	5.09	4.58	4.25	4.01	3.84	3.71	3.60	3.51
19	8.19	5.93	5.01	4.50	4.17	3.94	3.77	3.63	3.52	3.43
20	8.10	5.85	4.94	4.43	4.10	3.87	3.70	3.56	3.46	3.37
21	8.02	5.78	4.87	4.37	4.04	3.81	3.64	3.51	3.40	3.31
22	7.95	5.72	4.82	4.31	3.99	3.76	3.59	3.45	3.35	3.26
23	7.88	5.66	4.76	4.26	3.94	3.71	3.54	3.41	3.30	3.21
24	7.82	5.61	4.72	4.22	3.90	3.67	3.50	3.36	3.26	3.17
25	7.77	5.57	4.68	4.18	3.86	3.63	3.46	3.32	3.22	3.13
30	7.56	5.39	4.51	4.02	3.70	3.47	3.30	3.17	3.07	2.98
40	7.31	5.18	4.31	3.83	3.51	3.29	3.12	2.99	2.89	2.80
60	7.08	4.98	4.13	3.65	3.34	3.12	2.95	2.82	2.72	2.63
120	6.85	4.79	3.95	3.48	3.17	2.96	2.79	2.66	2.56	2.47
∞	6.63	4.61	3.78	3.32	3.02	2.80	2.64	2.51	2.41	2.32

Degrees of freedom for denominator

		\multicolumn{9}{c}{Degrees of freedom for numerator}
		12	15	20	24	30	40	60	120	∞
	1	244	246	248	249	250	251	252	253	254
	2	19.4	19.4	19.4	19.5	19.5	19.5	19.5	19.5	19.5
	3	8.74	8.70	8.66	8.64	8.62	8.59	8.57	8.55	8.53
	4	5.91	5.86	5.80	5.77	5.75	5.72	5.69	5.66	5.63
	5	4.68	4.62	4.56	4.53	4.50	4.46	4.43	4.40	4.37
	6	4.00	3.94	3.87	3.84	3.81	3.77	3.74	3.70	3.67
	7	3.57	3.51	3.44	3.41	3.38	3.34	3.30	3.27	3.23
	8	3.28	3.22	3.15	3.12	3.08	3.04	3.01	2.97	2.93
	9	3.07	3.01	2.94	2.90	2.86	2.83	2.79	2.75	2.71
	10	2.91	2.85	2.77	2.74	2.70	2.66	2.62	2.58	2.54
	11	2.79	2.72	2.65	2.61	2.57	2.53	2.49	2.45	2.40
	12	2.69	2.62	2.54	2.51	2.47	2.43	2.38	2.34	2.30
	13	2.60	2.53	2.46	2.42	2.38	2.34	2.30	2.25	2.21
	14	2.53	2.46	2.39	2.35	2.31	2.27	2.22	2.18	2.13
	15	2.48	2.40	2.33	2.29	2.25	2.20	2.16	2.11	2.07
	16	2.42	2.35	2.28	2.24	2.19	2.15	2.11	2.06	2.01
	17	2.38	2.31	2.23	2.19	2.15	2.10	2.06	2.01	1.96
	18	2.34	2.27	2.19	2.15	2.11	2.06	2.02	1.97	1.92
	19	2.31	2.23	2.16	2.11	2.07	2.03	1.98	1.93	1.88
	20	2.28	2.20	2.12	2.08	2.04	1.99	1.95	1.90	1.84
	21	2.25	2.18	2.10	2.05	2.01	1.96	1.92	1.87	1.81
	22	2.23	2.15	2.07	2.03	1.98	1.94	1.89	1.84	1.78
	23	2.20	2.13	2.05	2.01	1.96	1.91	1.86	1.81	1.76
	24	2.18	2.11	2.03	1.98	1.94	1.89	1.84	1.79	1.73
	25	2.16	2.09	2.01	1.96	1.92	1.87	1.82	1.77	1.71
	30	2.09	2.01	1.93	1.89	1.84	1.79	1.74	1.68	1.62
	40	2.00	1.92	1.84	1.79	1.74	1.69	1.64	1.58	1.51
	60	1.92	1.84	1.75	1.70	1.65	1.59	1.53	1.47	1.39
	120	1.83	1.75	1.66	1.61	1.55	1.50	1.43	1.35	1.25
	∞	1.75	1.67	1.57	1.52	1.46	1.39	1.32	1.22	1.00

Degrees of freedom for denominator

Table XI. Critical values for total number of runs (table shows critical values for two-tailed test at $\alpha = 0.05$)

The larger of n_1 and n_2

The smaller of n_1 and n_2	5	6	7	8	9	10	11	12	13	14	15	16	17	18	19	20
2								2/6	2/6	2/6	2/6	2/6	2/6	2/6	2/6	2/6
3		2/8	2/8	2/8	2/8	2/8	2/8	2/8	2/8	2/8	3/8	3/8	3/8	3/8	3/8	3/8
4	2/9	2/9	2/10	3/10	3/10	3/10	3/10	3/10	3/10	3/10	3/10	4/10	4/10	4/10	4/10	4/10
5	2/10	3/10	3/11	3/11	3/12	3/12	4/12	4/12	4/12	4/12	4/12	4/12	4/12	5/12	5/12	5/12
6		3/11	3/12	3/12	4/13	4/13	4/13	4/13	5/14	5/14	5/14	5/14	5/14	5/14	6/14	6/14
7			3/13	4/13	4/14	5/14	5/14	5/14	5/15	5/15	5/15	6/16	6/16	6/16	6/16	6/16
8				4/14	5/14	5/15	5/15	6/16	6/16	6/16	6/16	6/17	7/17	7/17	7/17	7/17
9					5/15	5/16	6/16	6/16	6/17	7/17	7/18	7/18	7/18	8/18	8/18	8/18
10						6/16	6/17	7/17	7/18	7/18	7/18	8/19	8/19	8/19	8/20	9/20
11							7/17	7/18	7/19	8/19	8/19	8/20	9/20	9/20	9/21	9/21
12								7/19	8/19	8/20	8/20	9/21	9/21	9/21	10/22	10/22
13									8/20	9/20	9/21	9/21	10/22	10/22	10/23	10/23
14										9/21	9/22	10/22	10/23	10/23	11/23	11/24
15											10/22	10/23	11/23	11/24	11/24	12/25
16												11/23	11/24	11/25	12/25	12/25
17													11/25	12/25	12/26	13/26
18														12/26	13/26	13/27
19															13/27	13/27
20																14/28

From C. Eisenhart and F. Swed, "Tables for testing randomness of grouping in a sequence of alternatives," *The Annals of Statistics*, 14(1943), 66–87. Reprinted by permission.

Answers

Chapter 1
EXERCISES (Pages 11–12)

1. *a.* No. Fifty-percent fewer cavities than whom?
   b. No. There is no reason to believe there is any relationship between the two factors.
   c. No. The ad does not specify the conditions needed to obtain the increased mileage.
   d. No. The people may have been healthy anyway.
2. *a.* Only descriptive statistics is needed.
   b. Only descriptive statistics is needed.
   c. Statistical inference is needed.
3. No. There is no reason to believe that there is a cause and effect relationship.
4. Most newspapers use descriptive statistics.
5. Probably descriptive statistics.
8, 9. See Chapter 9. Use a table of random digits.
10. No.
11. No.

MASTERY TESTS

Form *A*	Form *B*	
1. True	1. Not necessarily	6. No
2. True	2. No	7. No
3. False	4. No	8. One penny
4. False	5. No	9. Yes
5. False		

EXERCISES, SECTION 2.1 (Pages 24–27)

1.

Height	Tally	Frequency
58–59			1
60–61			1
62–63		0
64–65	|||||	5
66–67	||	2
68–69	||||| ||||	9
70–71	||||| ||	7
72–73	|||||	5
74–75	||||	4
76–77	||	2
	Total =	36

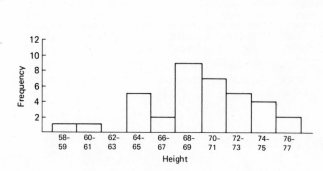

3.

Number of Cigarettes	Tally	Frequency
15–18	|||	3
19–22			1
23–26	||	2
27–30	||||| |||	8
31–34	|||	3
35–38	|||||	5
39–42	|||	3
43–46	|||	3
47–50	||	2
51–54			1
	Total =	31

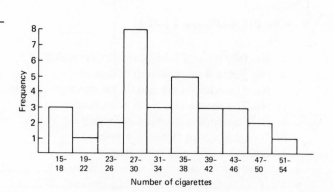

5.

Speed	Tally	Frequency
44–45			1
46–47	|||	3
48–49	|||||		6
50–51	|||||		6
52–53	|||	3
54–55	||||	4
56–57	||||| ||	7
58–59	|||||	5
60–61	|||||		6
62–63	||||	4
	Total =	45

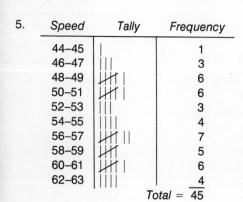

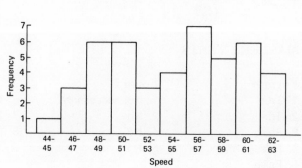

7. *a.* 10 intervals

IQ *Scores*	*Tally*	*Frequency*
74–81	⁴⁵⁶	7
82–88		11
89–95		12
96–102		14
103–109		19
110–116		13
117–123		12
124–130		6
131–137		3
138–144		3
	Total =	100

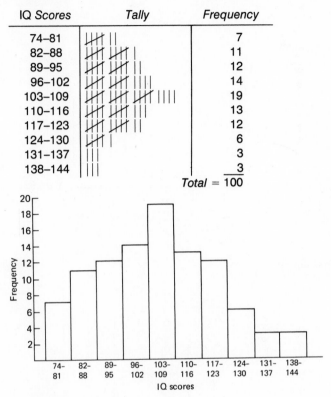

b. 5 intervals

IQ *Scores*	*Tally*	*Frequency*
74–88		18
89–102		26
103–116		32
117–130		18
131–144		6
	Total =	100

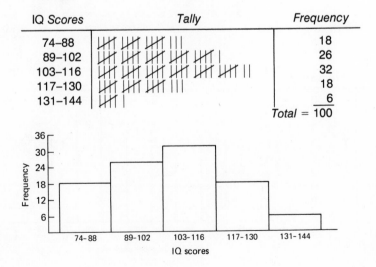

ANSWERS

c. The histogram of part *a* yields more information. By using fewer intervals we group too many IQ scores together.

9. *a.*

FACTORY A

Number of Products	Tally	Frequency
0–3		0
4–7	|||	3
8–11	|||	3
12–15			1
16–19	|||	3
20–23	||	2
24–27	|||	3
28–31	||||	4
32–35			1
36–39		0
	Total =	20

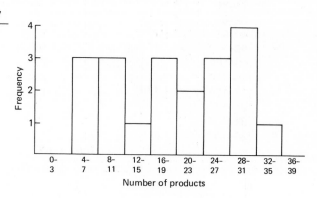

FACTORY B

Number of Products	Tally	Frequency
0–3		0
4–7	||	2
8–11	||	2
12–15	||	2
16–19	|||	3
20–23		0
24–27	||||	5
28–31	||||	4
32–35			1
36–39			1
	Total =	20

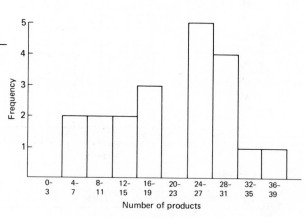

b.

Number of Products	Tally	Frequency
0–3		0
4–7	||||	5
8–11	||||	5
12–15	|||	3
16–19	||||		6
20–23	||	2
24–27	|||| |||	8
28–31	|||| |||	8
32–35	||	2
36–39			1
	Total =	40

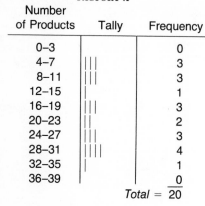

482

10. *a.*

Number of Products	Tally	Frequency
0–3		0
4–7		0
8–11		1
12–15		2
16–19		6
20–23		4
24–27		2
28–31		3
32–35		1
36–39		1
	Total =	20

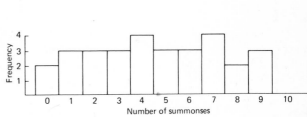

b. Production increased under the new conditions.

11.

PTL. HANNON

Summonses	Tally	Frequency
0		2
1		3
2		3
3		3
4		4
5		3
6		3
7		4
8		2
9		3
10		0
	Total =	30

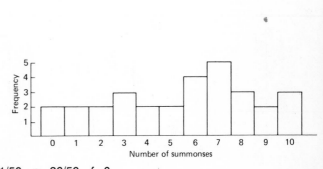

SGT. THURSDAY

Summonses	Tally	Frequency
0		2
1		2
2		2
3		3
4		2
5		2
6		4
7		5
8		3
9		2
10		3
	Total =	30

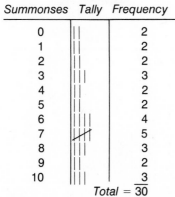

13. *a.* 12/50 *b.* 19/50 *c.* 19/50 *d.* 31/50 *e.* 36/50 *f.* 0

Chapter 2
EXERCISES, SECTION 2.2 (Pages 39–44)

1.

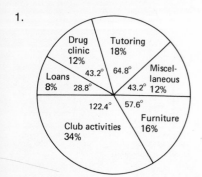

3.

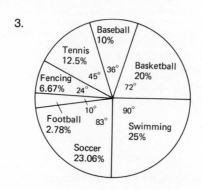

4. Married = 0.761 × 8000 = 6088
   Widowed = 0.077 × 8000 = 616
   Single = 0.083 × 8000 = 664
   Divorced = 0.079 × 8000 = 632

5. *a.* 440 *b.* 760 *c.* 200
6. *a.* The vertical scales are different.
   b. It depends upon the situation.

7.

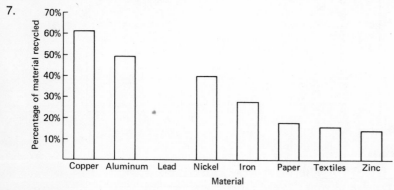

8.

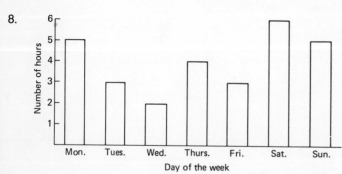

9.

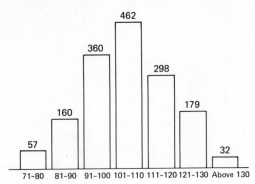

11. *a.* 1200 *b.* 900 *c.* 3000
12. *a.* 105 million
    b. From 125 million to 200 million, or by 75 million.

13.

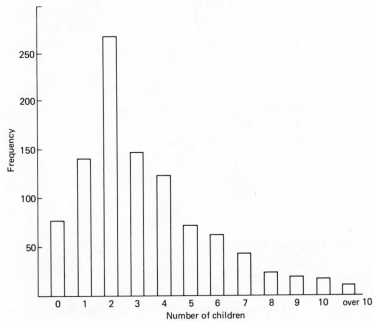

14. *a.* Normally distributed
    b. Normally distributed
    c. Normally distributed
    d. Probably normally distributed
    e. Probably not normally distributed
    f. Not normally distributed

MASTERY TESTS

Form *A*

1. *c*
2. *a*
3. 80
4. 105
5. 28
6. 2 P.M.–6 P.M.
7. 2 A.M.–4 A.M.
8. 1970
9. 1961

ANSWERS

Form *B*

2.

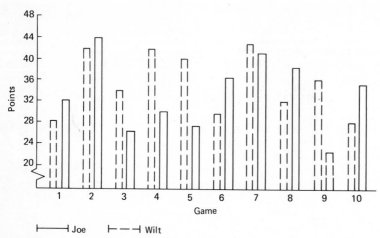

Points / Game

⊢————⊣ Joe   ⊢ — ⊣ Wilt

3. *a.* Between the last quarter of 1973 and the first quarter of 1974 there was a 2.9% increase.
   b. Between the third and fourth quarter of 1974.
5. There are more people earning less money than more money.

Chapter 3

EXERCISES, SECTION 3.1 (Pages 64–67)

1. Median = $14,130
   Mean = $14,550
   Mode = $14,130
2. Mean = 3919/28 = 139.96
   Median = 139
   Mode = 153
   Probably the mode.
3. Mean = 85.42
   Median = 88.5
   Mode = 76
4. $18,160
5. Mean = 1028/28 = 36.71
   Median = 36
   Mode = 34 and 36

6. 80.03
7. $8.00
8. *a.* Manufacturer A is using the mean, manufacturer B is using the mode, and manufacturer C is using the median.
   b. Probably from Manufacturer B.
9. 26.84
10. 150
11. Each is increased by $1500.
12. *a.* 15
    b. 14.37
    c. Probably the mode.
13. 1288
14. She gained 2 pounds.

EXERCISES, SECTION 3.4 (Pages 75–76)

1. Range is $26,000 – $7000 = $19,000.
2. Range is 98 – 50 = 48.
3. Range is 9 – 0 = 9. Standard deviation is $\sqrt{9.25}$, or 3.04.
4. Variance = 104. Standard deviation = $\sqrt{104}$, or 10.20.
5. Mean = 30. Variance = 173. Standard deviation = $\sqrt{173}$. Average deviation = 11.33
6. Mean = 13. Variance = 29.2. Standard deviation = $\sqrt{29.2}$, or 5.40. Average deviation = 4.8.
7. Mean = 9. Variance = 17.2. Standard deviation = $\sqrt{17.2}$, or 4.15. Average deviation = 3.6.
8. Mean = 27. Variance = 154.8. Standard deviation = $\sqrt{154.8}$, or 12.44. Average deviation = 10.8.
9. a. b. Mean = 84. Variance = 187.2. Standard deviation = $\sqrt{187.2}$, or 13.68.
10. Mean = $2900. Standard deviation = $450.
11. Variance = $\sqrt{584}$. Standard deviation = $\sqrt{584}$, or 24.17.

MASTERY TESTS

Form *A*

1. No. Both classes may not have the same number of students.
2. Probably the mean, although the mean could also be used.
3. Yes, if all the terms are equal.
4. No. The pool may be deeper than 5 feet at certain points.
5. Mean of Y is 10 times as great as the mean of X. Standard deviation of Y is 100 times as great as the mean of X.
6. Choice *a*
7. 300
8. The standard deviation of town B is larger.
9. Statement *c*
10. Probably the mode.

Form *B*

1. Probably economics since the standard deviation is smaller.
2. Brand B because it has a higher mean and a smaller standard deviation.
3. *a.* 856 *b.* 3600 *c.* No. The answers to *a* and *b* are not the same.
4. They are really the same. The weights assigned to the numbers are the same as the frequencies for each.
7. Yes, the standard deviation for dating service B is smaller than the standard deviation of dating service A.
9. Mode = 105. Median = 105. We cannot calculate the mean from the given information.

ANSWERS

10. *b.*

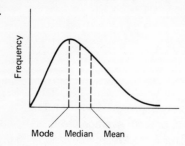

c.

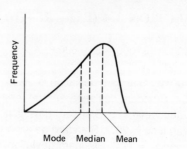

Chapter 4

EXERCISES, SECTION 4.2 (Pages 91–94)

1. Percentile rank of Doris: 45. Percentile rank of Jimmy: 65.
2. 26th percentile
3. *a.* 21st percentile
   b. 79th percentile
4. *a.* 25 percent   *d.* 50 percent
   b. 15 percent   *e.* 60 percent
   c. 75 percent
5. *a.* 2   *b.* −1.5   *c.* 3   *d.* 0
6. *a.* Jean, Sydney, Isaac, Sherry, Mabel
   b. Jean, Sydney, Isaac
   c. Sherry, Mabel
7. *a.* 32.82   *b.* 41.45   *c.* 27.12
8. *a.* 75   *b.* 34 percent   *c.* 47.4
9. Vera. Her *z*-score of 2.38 was higher than Judy's *z*-score of 1.94.
10. Lou, 160.5 pounds; Drew, 145.97 pounds; Sue, 156.22 pounds.
11. *a.* *z*-score for writing = −3.24   *z*-score for medicine = 2.50
   z-score for acting = 0.13   *z*-score for law = −1.70
   b. Medicine
   c. Writing

MASTERY TESTS

Form *A*

1. Mean   6. Choice *a*
2. Mean   7. 0
3. Choice *a*   8. Choice *a*
4. 0   9. Choice *b*
5. False   10. 20 percent

Form *B*

1. No   *b.* 16 percent
2. Yes   *c.* 97.73rd percentile
4. 19.41 accidents   7. *a.* 554
5. No, see exercise 4 on page 92.   *b.* 612
6. *a.* 15.87th percentile

Chapter 5

EXERCISES, SECTION 5.1 (Pages 109–113)

1. *a.* 1, 2, 3, 4, 5, 6   *b.* 1/2   *c.* 2/6 = 1/3   *d.* 1/6
2. *a.* 1/6   *b.* 1/36   *c.* 18/36 = 1/2   *d.* 10/36 = 5/18   *e.* The numbers may be both 2, 4 or 6. These numbers can be obtained in 9/36 ways. Thus, prob. = 9/36 = 1/4.
3. *a.* The only sums possible are 4, 5, 6, 7, 8, 9, 10, 11, 12. *b.* 0 *c.* 1 *d.* 20/36 = 5/9
4. *a.* 1/52 *b.* 4/52 = 1/13 *c.* 13/52 = 1/4 *d.* 24/52 = 6/13
5. 197/200
6. *a.* 4/20 = 1/5 *b.* 7/20 *c.* 3/20 *d.* 0
7. 6/36 = 1/6
8. There are 6 possible outcomes as shown below where E stands for envelope and L stands for letter.

Envelope 1	Envelope 2	Envelope 3
L_1	L_2	L_3
L_1	L_3	L_2
L_2	L_1	L_3
L_2	L_3	L_1
L_3	L_2	L_1
L_3	L_1	L_2

Only 1 of these outcomes is favorable, so that prob. = 1/6.
9. *a.* 16,900/95,700 = 0.18 *b.* 6400/95,700 = 0.07 *c.* 0.
10. *a.* 1/5 *b.* 1/5 *c.* 2/5 *d.* 3/5 *e.* 0
11. −1/2, 1.32, −1
12. *a.* 6/55 *b.* 8/55 *c.* 13/55 *d.* 26/55 *e.* 0.
13. 6/16, or 3/8
14. There are 24 possible ways in which they may land. The probability that they land in the specified order is 1/24.
15. *a.* *ccc, ccw, cwc, cww, wcc, wcw, wwc,* and *www*
    b. 1/8 *c.* 7/8 *d.* 1/8
16. *a.* 000, 001, 002, 010, 011, 012, 020, 021, 022, 100, 101, 102, 110, 111, 112, 120, 121, 122, 210, 211, 212, 220, 221, 222, 200, 201, 202
    b. 1/27 *c.* 1/27 *d.* 26/27
17. 1/10

EXERCISES, SECTION 5.2 (Pages 117–119)

1. 3 × 2 × 2 = 12 possible ways
2. 4 × 3 × 5 = 60 possible meals
3. 6 × 3 × 6 = 108 possible ways
4. *a.* 5 × 5 × 5 × 5 × 5 = 3125 *b.* 5 × 4 × 3 × 2 × 1 = 120

ANSWERS

5.

Child 1 Child 2   Child 3   Child 4   Child 5

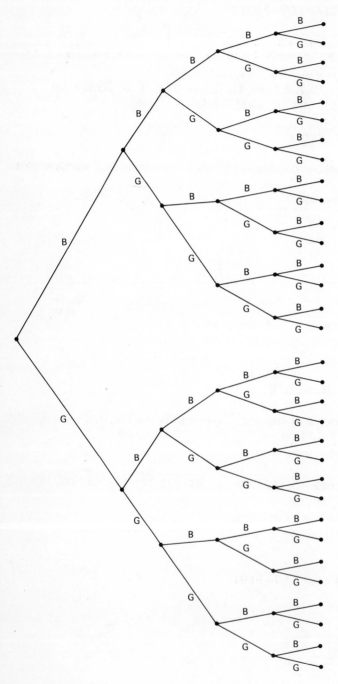

7. $5 \times 4 \times 3 \times 2 \times 1 = 120$ possible ways
8. $3 \times 3 \times 2 \times 1 = 18$ possible numbers
9. *a.* $1 \times 10 \times 10 \times 10 \times 10 \times 26 = 260{,}000$
   b. $10 \times 9 \times 8 \times 7 \times 6 \times 26 = 786{,}240$
   c. $10 \times 10 \times 10 \times 10 \times 10 \times 26 = 2{,}600{,}000$

10.

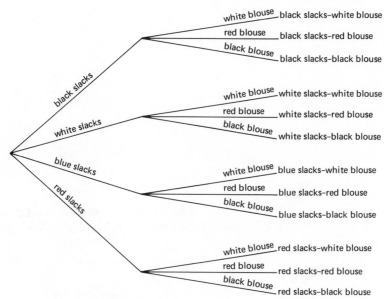

Outfit consists of

white blouse — black slacks–white blouse
red blouse — black slacks–red blouse
black blouse — black slacks–black blouse

white blouse — white slacks–white blouse
red blouse — white slacks–red blouse
black blouse — white slacks–black blouse

black slacks

white slacks

blue slacks

red slacks

white blouse — blue slacks–white blouse
red blouse — blue slacks–red blouse
black blouse — blue slacks–black blouse

white blouse — red slacks–white blouse
red blouse — red slacks–red blouse
black blouse — red slacks–black blouse

11. $9 \times 8 \times 12 \times 3 = 2592$
12. *a.* $10 \times 10 \times 10 \times 10 \times 10 \times 10 \times 10 \times 10 \times 10 = 1{,}000{,}000{,}000$
    b. $10 \times 10 \times 10 \times 10 \times 10 \times 10 \times 10 \times 10 \times 10 \times 10 = 10{,}000{,}000{,}000$

EXERCISES, SECTION 5.3 (Pages 125–126)

1. *a.* 5040   *d.* 1   *g.* 35   *j.* 120   *m.* 5040   *p.* 1
   b. 720   *e.* 1/3   *h.* 20   *k.* 30   *n.* 2
   c. 2   *f.* 6   *i.* 5040   *l.* 6   *o.* 1

2. $_7P_7 = 7! = 5040$

3. $_5P_3 = \dfrac{5!}{(5-3)!} = 60$

4. $_7P_3 = \dfrac{7!}{(7-3)!} = 210$

5. $_7P_7 = 7! = 5040$

6. $_8P_8 = 8! = 40{,}320$

7. $_7P_4 = \dfrac{7!}{(7-4)!} = 840$

ANSWERS

8. a. $_5P_5 = 5! = 120$

   b. $\dfrac{_4P_4}{_5P_5} = \dfrac{4!}{5!} = \dfrac{24}{120} = 0.2$ since Felix will be scheduled first.

9. a. $_8P_4 = \dfrac{8!}{(8-4)!} = 1680$ b. $\dfrac{_7P_3}{_8P_4} = \dfrac{210}{1680} = \dfrac{1}{8}$ c. $4 \times \dfrac{_7P_3}{_8P_4} = 4 \times \dfrac{210}{1680} = \dfrac{1}{2}$

10. a. $\dfrac{9!}{3!} = \dfrac{362{,}880}{6} = 60{,}480$ c. $\dfrac{8!}{2!2!} = 10{,}080$

    b. $\dfrac{9!}{2!2!} = \dfrac{362{,}880}{2.2} = 90{,}720$ d. $\dfrac{11!}{4!4!2!} = 34{,}650$

11. $_{17}P_{17} = 17!$

12. $2 \times 6 \times 5 \times 4 \times 3 \times 2 \times 1 = 1440$

13. a. $1 \times 10 \times 10 \times 10 \times 10 = 10{,}000$

    b. $10 \times 9 \times 8 \times 7 \times 6 = 30{,}240$

    c. $10 \times 10 \times 10 \times 10 \times 10 = 100{,}000$

14. $_9P_3 = \dfrac{9!}{(9-3)!} = 504$

15. a. $_{12}P_{12} = 12! = 479{,}001{,}600$

    b. $_4P_4 \cdot {_3P_3} \cdot {_2P_2} \cdot {_3P_3} = 24 \cdot 6 \cdot 2 \cdot 6 = 1728.$ We now multiply 1728 by 4! since there are 4 different subject matter books which can also be arranged in different orders. There are 1728×24, or 41,472, possible ways.

16. $_4P_4 = 4! = 24$

EXERCISES, SECTION 5.4 (Pages 135–136)

1. a. $\dfrac{7!}{6!(7-6)!} = 7$ b. $\dfrac{6!}{4!(6-4)!} = 15$ c. $\dfrac{8!}{3!(8-3)!} = 56$ d. $\dfrac{10!}{10!(10-0)!} = 1$

   e. $\dfrac{5!}{1!(5-1)!} = 5$ f. $\dfrac{9!}{6!(9-6)!} = 84$ g. $\dfrac{7!}{2!(7-2)!} = 21$ h. $\dfrac{8!}{4!(8-4)!} = 70$

   i. $_7C_8$ cannot be done   j. $\dfrac{7!}{7!(7-7)!} = 1$

2. $_{10}C_4 = \dfrac{10!}{4!(10-4)!} = 210$ 3. $_6C_2 = \dfrac{6!}{2!(6-2)!} = 15$ 4. $_{15}C_5 = \dfrac{15!}{5!(15-5)!} = 3003$

5. $_7C_3 \cdot {_8C_4} = \dfrac{7!}{3!4!} \cdot \dfrac{8!}{4!4!} = 2450$ 6. $_{10}C_3 = \dfrac{10!}{3!(10-3)!} = 120$

7. a. $_6C_2 = \dfrac{6!}{2!(6-2)!} = 15$ b. $_6C_4 = \dfrac{6!}{4!(6-4)!} = 15$ c. The answers are the same.

8. $_{12}C_8 = \dfrac{12!}{8!(12-8)!} = 495$ 9. $_{15}C_5 = \dfrac{15!}{5!(15-5)!} = 3003$

10. a. $_{10}C_3 = \dfrac{10!}{3!(10-3)!} = 120$ b. $_{10}C_7 = \dfrac{10!}{7!(10-7)!} = 120$

11. $_{12}C_3 = \dfrac{12!}{3!(12-3)!} = 220$ 12. $_8C_4 \cdot {_7}C_3 = \dfrac{8!}{4!(8-4)!} \cdot \dfrac{7!}{3!(7-3)!} = 2450$

13. a. $\dfrac{_8C_2 \cdot {_6}C_0}{_{14}C_2} = \dfrac{28}{91} = 0.31$ b. $\dfrac{_8C_0 \cdot {_6}C_2}{_{14}C_2} = \dfrac{15}{91} = 0.16$ c. $\dfrac{_8C_1 \cdot {_6}C_1}{_{14}C_2} = \dfrac{48}{91} = 0.53$

EXERCISES, SECTION 5.5 (Pages 139–140)

1. $-9/20, or \$-0.45   4. \$3,520   7. 97.5:2.5, or 39:1   10. 15:3, or 5:1
2. \$0   5. \$34/8 = \$4.25   8. 3:7   11. 9:1
3. \$160,000   6. 5:3   9. 26:26, or 1:1   12. 4:13

MASTERY TESTS

Form *A*

1. True
2. 1/365, assuming no leap year.
3. No

4. Unlikely, but not impossible.
5. $\dfrac{1}{2^{48}}$

6.

First Spin	Second Spin		First Spin	Second Spin
1	1		3	1
1	2		3	2
1	3		3	3
1	4		3	4
2	1		4	1
2	2		4	2
2	3		4	3
2	4		4	4

7. 13/16
8. $(14+16)/(14+16+8+12) = 30/50 = 0.6$

9. $22/50 = 0.44$
10. $1/(6!/2!) = 1/360$

Form *B*

1. $_{12}C_2 = \dfrac{12!}{2!(12-2)!} = 66$ 2. $_8P_2 = \dfrac{8!}{(8-2)!} = 56$

ANSWERS

3. Scheme *a* will result in $26 \cdot 26 \cdot 26 \cdot 10 \cdot 10 = 1{,}757{,}600$ possible tags. Scheme *b* will result in $26 \cdot 26 \cdot 10 \cdot 10 \cdot 10 = 676{,}000$ possible tags. Scheme *a* will result in more tags.

4.

Toss 1	Toss 2	Toss 3	Toss 4	Outcome

Only the first six outcomes are favorable.

5. Possible committees are

3 Republicans 3 Democrats		4 Republicans 2 Democrats		5 Republicans 1 Democrat		6 Republicans 1 Democrat	
$_8C_3 \cdot _6C_3$	$+$	$_8C_4 \cdot _6C_2$	$+$	$_8C_5 \cdot _6C_1$	$+$	$_8C_6 \cdot _6C_1 = 2534$ possible committees	

6. $21 \cdot 5 \cdot 20 = 2100$ possible words

7. $_{12}P_3 = \dfrac{12!}{(12-3)!} = 1320$

8. No, the probability is 0.51.

9. No, there are only $10 \cdot 10 \cdot 10$, or 1000, possible numbers available.

10. Yes, since there are $26 \cdot 26 \cdot 26$, or 17,576, possible codes available for the 13,475 bikes.

11. $\dfrac{1}{_4P_4} = \dfrac{1}{4!} = \dfrac{1}{24}$

12. The probability is greater.

13. 0, since it never occurred.

Chapter 6

EXERCISES, SECTION 6.1 (Pages 155–156)

1. *a.* Not mutually exclusive   *b.* Not mutually exclusive   *c.* Not mutually exclusive
   d. Mutually exclusive   *e.* Not mutually exclusive   *f.* Not mutually exclusive
   g. Not mutually exclusive   *h.* Not mutually exclusive   *i.* Mutually exclusive
   j. Not mutually exclusive.

2. $\dfrac{5}{14} + \dfrac{3}{7} - \dfrac{1}{7} = \dfrac{9}{14}$ 3. $\dfrac{7}{10} + \dfrac{3}{5} - \dfrac{4}{10} = \dfrac{9}{10}$ 4. $\dfrac{1}{3} + \dfrac{2}{9} = \dfrac{5}{9}$

5. 1/11   6. 14/99   7. 0.51   8. 0.4   9. 0.1   10. 0.47   11. 1   12. 11/60

EXERCISES, SECTION 6.2 (Pages 163–164)

1. 1/3   3. 2/3   5. 23/53   7. 1/2   9. 6/11   11. 14/43
2. 1/3   4. 3/4   6. 7/16   8. 7/11   10. 8/61

EXERCISES, SECTION 6.3 (Pages 168–169)

1. 0.2976   2. 0.2664   3. 0.2184   4. 0.1638   5. 0.5518   6. 0.2688   7. 21/50   8. 0.2242
9. $(0.55)(0.43)(0.72) = 0.1703$ 10. $(0.55)(0.43)(0.28) = 0.0662$ 11. $(1 - 0.62)(1 - 0.56) = 0.1672$
12. $(1 - 0.96)(1 - 0.52) = 0.0192$ 13. $(0.19)(0.19)(0.19)(0.19) = 0.0013$
14. $(1 - 0.7)(1 - 0.7)(1 - 0.7)(1 - 0.7)(1 - 0.7)(1 - 0.7)(1 - 0.7) = 0.0002$

EXERCISES, SECTION 6.4 (Pages 176–177)

1. $\dfrac{\left(\dfrac{3}{7}\right)\left(\dfrac{1}{2}\right)}{\left(\dfrac{6}{16}\right)\left(\dfrac{1}{2}\right) + \left(\dfrac{3}{7}\right)\left(\dfrac{1}{2}\right)} = \dfrac{8}{15}$

2. *a.* $\dfrac{\left(\dfrac{1}{7}\right)\left(\dfrac{1}{5}\right)}{\left(\dfrac{1}{7}\right)\left(\dfrac{1}{5}\right) + \left(\dfrac{1}{8}\right)\left(\dfrac{2}{5}\right) + \left(\dfrac{1}{9}\right)\left(\dfrac{2}{5}\right)} = 0.2325$

3. $\dfrac{\left(\dfrac{1}{9}\right)\left(\dfrac{2}{5}\right)}{\left(\dfrac{1}{9}\right)\left(\dfrac{2}{5}\right) + \left(\dfrac{1}{15}\right)\left(\dfrac{3}{5}\right)} = 0.5261$

b. $\dfrac{\left(\dfrac{1}{8}\right)\left(\dfrac{2}{5}\right)}{\left(\dfrac{1}{7}\right)\left(\dfrac{1}{5}\right) + \left(\dfrac{1}{8}\right)\left(\dfrac{2}{5}\right) + \left(\dfrac{1}{9}\right)\left(\dfrac{2}{5}\right)} = 0.4065$

4. 1/3   5. $\dfrac{(0.3)(0.6)}{(0.3)(0.6) + (0.7)(0.4)} = 0.3913$

c. $\dfrac{\left(\dfrac{1}{9}\right)\left(\dfrac{2}{5}\right)}{\left(\dfrac{1}{7}\right)\left(\dfrac{1}{5}\right) + \left(\dfrac{1}{8}\right)\left(\dfrac{2}{5}\right) + \left(\dfrac{1}{9}\right)\left(\dfrac{2}{5}\right)} = 0.3610$

6. $\dfrac{(0.5)(0.6)}{(0.5)(0.6) + (0.4)(0.3) + (0.1)(0.1)} = 0.6977$

7. $\dfrac{\left(\dfrac{4}{15}\right)\left(\dfrac{1}{2}\right)}{\left(\dfrac{6}{35}\right)\left(\dfrac{1}{2}\right) + \left(\dfrac{4}{15}\right)\left(\dfrac{1}{2}\right)} = 0.6087$

MASTERY TESTS

Form *A*

2. *a.* Not mutually exclusive   *b.* Not mutually exclusive
3. *a.* Not independent   *b.* Not independent
4. 0.37   5. False   6. True   7. False
8. No, not necessarily independent events.   9. (0.40)(0.45) = 0.18
10. (0.40)(0.40) + (0.12)(0.12) + (0.3)(0.03) + (0.45)(0.45) = 0.3778

Form *B*

1. *a.* (0.7)(0.8)(0.9) = 0.504
   b. (0.3)(0.2)(0.1) = 0.006
   c. (0.3)(0.8)(0.9) + (0.7)(0.2)(0.9) + (0.7)(0.8)(0.1) = 0.398
   d. p(1 succeeds) + p(2 succeed) + p(all succeed) = 0.398 + 0.092 + 0.006 = 0.496

2. $1 \cdot \dfrac{364}{365} = \dfrac{364}{365}$ 3. $1 \cdot \dfrac{364}{365} \cdot \dfrac{363}{365} = 0.99$

4. (0.79)(0.86)(0.53) = 0.36   5. (0.79)(0.86)(0.47) = 0.3193

6. *a.* $\dfrac{42}{50} \cdot \dfrac{41}{49} = 0.7029$ *b.* $\dfrac{8}{50} \cdot \dfrac{7}{49} = 0.0229$ *c.* $\left(\dfrac{8}{50}\right)\left(\dfrac{42}{49}\right) + \left(\dfrac{42}{50}\right)\left(\dfrac{8}{49}\right) = 0.2743$

7. (0.7)(0.7)(0.7)(0.7)(0.7)(0.7)(0.7) = 0.0824   8. 1/3
9. No. The probability is 1/2.

Chapter 7
EXERCISES, SECTION 7.1 (Pages 188–189)

1. *a.* 0, 1, 2, . . . , 227   *d.* 1, 2, 3, . . . , 29
   b. 0, 1, 2, . . . , 26   *e.* 0, 1, 2, 3, 4
   c. 0, 1, 2, 3, 4, 5   *f.* 0, 1, 2, . . . , 22
   g. 1, 2, ... It actually depends on the country.

2. No, because the sum of the probabilities is 1.03.   3. 0.16

4.

x	p (x)
0	1/8
1	3/8
2	3/8
3	1/8
	1

5.

x	p(x)
0	1/4
1	1/2
2	1/4
	1

6.

x	p(x)
0	25/36
1	10/36
2	1/36
	1

7.

x (in cents)	p (x)
2	1/36
6	2/36
10	1/36
11	2/36
15	2/36
20	1/36
26	2/36
30	2/36
35	2/36
50	1/36

x (in cents)	p (x)
51	2/36
55	2/36
60	2/36
75	2/36
100	1/36
101	2/36
105	2/36
110	2/36
125	2/36
150	2/36
200	1/36

EXERCISES, SECTION 7.3 (Pages 196–198)

1. Mean = 3.10; variance = 4.25, and standard deviation = $\sqrt{4.25}$, or 2.06
2. Mean = 2.73; variance = 2.32; and standard deviation = $\sqrt{2.32}$, or 1.52
3. Mean = 7.99
4. $\mu = 7.33$; $\sigma^2 = 4.49$; $\sigma = \sqrt{4.49}$, or 2.12
5. Mean = 1.84; variance = 1.51; and standard deviation = $\sqrt{1.51}$, or 1.23
6. Mean = 2.64; variance = 1.93; and standard deviation = $\sqrt{1.93}$, or 1.39
7. Mean = 5.44   8. Mean = 2.97   9. Mean = 14.01   10. Mean = 7.86

EXERCISES, SECTION 7.4 (Pages 211–212)

1. a. 0.27648 b. 0.0041 c. 0.0467
2. 0.3364   3. 0.3110   4. 0.0287   5. Approximately 0
6. p(at least 2 escape) = p(2 escape) + p(3 escape) + p(4 escape)
   = 0.375 + 0.250 + 0.0625 = 0.6875
7. a. 0.3164 b. 0.2109 c. 0.0039
8. a. p(at least 3 heads) = p(3 heads) + p(4 heads) + p(5 heads)
   = 0.3087 + 0.3602 + 0.1681 = 0.837
   b. p(at most 3 heads) = p(0 heads) + p(1 head) + p(2 heads) + p(3 heads)
   = 0.002 + 0.028 + 0.132 + 0.309 = 0.471
9. a. p(at least 4 cash) = p(4 cash) + p(5 cash) + p(6 cash) + p(7 cash) + p(8 cash)
   = 0.232 + 0.279 + 0.209 + 0.090 + 0.017 = 0.827
   b. p(at most 4 cash) = p(0 cash) + p(1 cash) + p(2 cash) + p(3 cash) + p(4 cash)
   = 0.001 + 0.008 + 0.041 + 0.124 + 0.232 = 0.406
10. 0.9993   11. 0.833   12. 0.312   13. 0.9972   14. 0.7463

ANSWERS

EXERCISES, SECTION 7.5 (Pages 214–215)

1. Mean = 60; variance = 48; standard deviation = $\sqrt{48}$, or 6.93.
2. Mean = 276; standard deviation = $\sqrt{22.08}$, or 4.70.
3. Mean = 20; standard deviation = $\sqrt{19}$, or 4.36.
4. Mean = 30; standard deviation = $\sqrt{21}$, or 4.58.
5. Mean 150; standard deviation = $\sqrt{37.5}$, or 6.12.
6. Mean = 1700; standard deviation = $\sqrt{1411}$, or 37.56.
7. 6480 students
8. Mean = 24; standard deviation = $\sqrt{21.12}$, or 4.60.
9. 6500
10. 1000

MASTERY TESTS

Form *A*

1. 1/2   2. 50   3. 5   4. A, B, C, D, or F in most colleges.
5. 0, 1, 2, . . . , 32   6. Choice *b*   7. Choice *a*   8. 17/40   9. Choice *b*
10. No, sum of the probabilities is larger than 1.   11. Choice *a*

Form *B*

1.

SAMPLE SPACE

Card 1	Card 2		*x (sum)*	Probability
5H	5H		4	1/16
3C	3C		5	2/16
2S	2S		6	1/16
7D	7D		7	2/16
5H	3C		8	2/16
3C	2S		9	2/16
2S	7D		10	3/16
7D	5H		12	2/16
5H	2S		14	1/16
3C	7D			1
2S	5H			
7D	3C			
5H	7D			
3C	5H			
2S	3C			
7D	2S			

2. Approximately 1
3. *a.* p(at most 1 cavity) $= p(0$ cavities) $+ p(1$ cavity)
$$= 0.0424 + 0.1413 = 0.1837$$
  b. $p(0$ cavities) $= 0.0424$
5. Mean $= 3.5$, variance $= 2.92$.
6. $0.0012 + 0.0093 + 0.0337 + 0.0785 + 0.1325 = 0.2552$
  No, the events are not necessarily independent.
7. The probabilities were not all the same.
8. *a.* 0.125 *b.* 0.25 *c.* 0.3125 *d.* 0.3125
9. *a.* 0.0046 *b.* 0.9619
10. 0.382

Chapter 8

EXERCISES, SECTION 8.2 (Pages 234–236)

1. *a.* 0.4292   *b.* 0.2123   *c.* 0.0268   *d.* 0.2177
   e. 0.0104   *f.* 0.8951   *g.* 0.9604   *h.* 0.1303
2. *a.* 0.9931   *b.* 0.8051   *c.* 0.0892   *d.* 0.0060
   e. Approximately 0 *f.* 0.0984   *g.* 0.7783   *h.* 0.0326
3. *a.* $z = -0.58$ *b.* $z = -0.92$ *c.* $z = -1.88$ *d.* $z = -2.33$
4. *a.* $z = +2.38$ or -2.38 *b.* $z = 2.04$ *c.* $z = -0.55$ *d.* $z = 2.08$ *e.* $z = 2.18$ *f.* $z = 2.93$
5. *a.* 0.6247 *b.* 0.2119 *c.* 0.0919 *d.* 0.1587
6. *a.* 89.44th percentile *b.* 26.43rd percentile *c.* 94.84th percentile *d.* 8.38th percentile
7. *a.* 121.56 lbs *b.* 126.24 lbs *c.* 139.74 lbs

8.

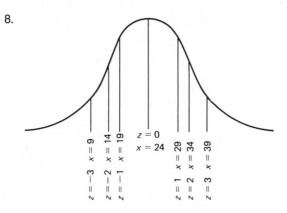

9. $\sigma = 80$ 10. $\mu = 48.912$

ANSWERS

EXERCISES, SECTION 8.3 (Pages 240–241)

1. 0.3085   2. 0.3594   3. 0.8472   4. 0.0764   5. 0.1335
6. *a.* 0.1056 *b.* 0.6217   7. 0.3085   8. 0.9236
9. *a.* 20.33% *b.* 51.92 *c.* 78.92
10. *a.* Approximately 1 *b.* 0.7910 *c.* 0.9985
11. Up to 67.68 minutes
12. 0.6369   13. 3.22 years   14. 0.9066   15. 0.5987

EXERCISES, SECTION 8.4 (Pages 247–248)

1. 0.9693   4. 0.7671   7. 0.1610   10. 0.1492
2. Approximately 0   5. 0.0401   8. 0.8315   11. 0.0350
3. 0.5438   6. 0.0073   9. 0.0071   12. 0.0582

MASTERY TESTS

Form *A*

1. Mean, median, and mode   6. Choice *c*
2. 100 percent   7. Choice *c*
3. Choice *c*   8. Choice *c*
4. Choice *a*   9. Choice *a*
5. Choice *c*   10. Choice *d*

Form *B*

1. *a.* 0.4129 *b.* 0.1335
2. Approximately 0
3. 87.8
4. 0.558
5. 0.3400
6. 0.0618
7. 0.1587
8. 0.5636
9. Approximately 1
10. 0.0019
11. 83.84

Chapter 9

EXERCISES, SECTION 9.2 (Pages 264–267)

1.

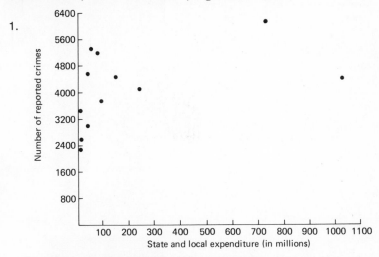

No particular type of relationship is suggested.

2. *a.* Negative correlation
   c. Negative correlation
   e. Probably some positive correlation
   g. Positive correlation
   i. Positive correlation
   k. Positive correlation

   b. Probably positive correlation
   d. Zero correlation
   f. Zero correlation
   h. Positive correlation
   j. Positive correlation

Depending upon your interpretation, your answers to several parts of this exercise may vary.

3.

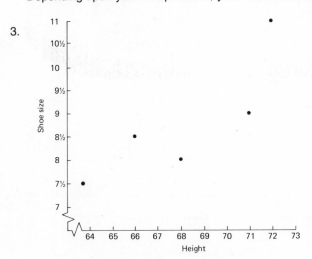

Coefficient of correlation = 0.84.

4. *a.*

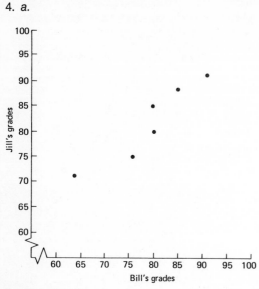

Coefficient of correlation = 0.93

5.

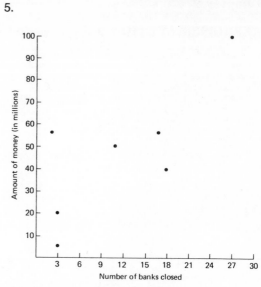

Coefficient of correlation = 0.76.

6. *a.*

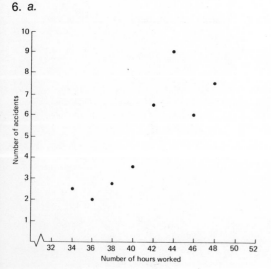

Coefficient of correlation = 0.85.

7.

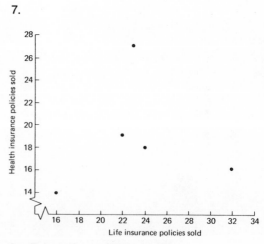

Coefficient of correlation = 0.06.

8.

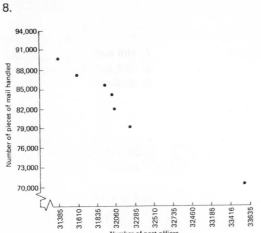

Coefficient of correlation = −0.96.

9.

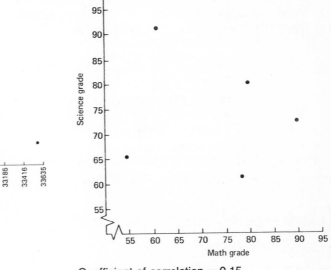

Coefficient of correlation = 0.15.

10. *a.*

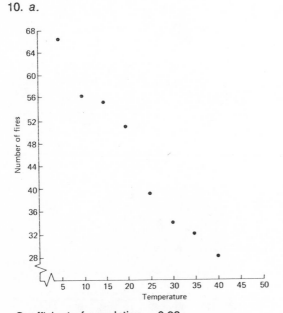

Coefficient of correlation − 0.98.

11. *a.*

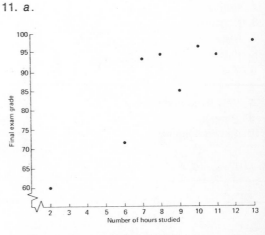

Coefficient of correlation = 0.86.

ANSWERS

EXERCISES, SECTION 9.3 (Page 268)

1. Not statistically significant
2. Statistically significant
3. Statistically significant
4. Statistically significant
5. Not statistically significant
6. Statistically significant
7. Not statistically significant
8. Not statistically significant
9. Statistically significant

EXERCISES, SECTION 9.5 (Pages 276–279)

1. a. $Y_{predicted} = 69.14 + 1.5(X - 69.71)$ b. 75.58
2. a. $Y_{predicted} = 62.5 + 3.6(X - 13.67)$ b. 49.29
3. $Y_{predicted} = 15 - 0.07(X - 88.33)$
4. $Y_{predicted} = 15.86 - 0.76(X - 15)$
5. $Y_{predicted} = 62.5 + 3.6(X - 13.67)$ b. 85.29
6. a. $Y_{predicted} = 8.2 + 0.93(X - 6.2)$ b. 9.87
7. $Y_{predicted} = 72.57 + 4.65(X - 9.14)$
8. a. $Y_{predicted} = 150.67 + 26.37(X - 13.83)$
   b. 260.63
9. $Y_{predicted} = 45.6 - 1.69(X - 58.4)$
10. $Y_{predicted} = 27 + 2.38(X - 7)$

EXERCISES, SECTION 9.6 (Page 282)

1. 1.32
2. 1.173
3. 0.5719
4. 1.263
5. 1.173
6. 0.8026
7. 3.462
8. 29.433
9. 1.995
10. 0.968

MASTERY TESTS

Form A

1. Positive
2. Negative
3. Negative
4. False
5. False
6. Yes
7. 0
8. Very large number
9. Strength
10. Strong negative

Form B

1. a.

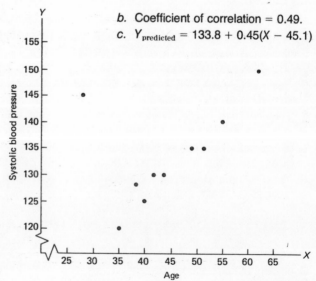

Age

b. Coefficient of correlation = 0.49.
c. $Y_{predicted} = 133.8 + 0.45(X - 45.1)$

2. The purpose of a scatter diagram is to determine if a relationship exists between the two variables.

4. $Y_{predicted} = 12.67 - 0.28(X - 106.83)$

5. a.

b. Based upon the given data, we conclude that a higher education will mean a higher salary.

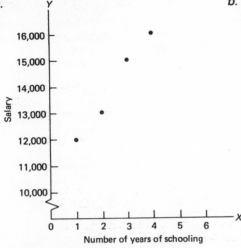

Coefficient of correlation = 0.99.

6. a. Coefficient of correlation = 0.26.   b. $Y_{predicted} = 143.57 + 0.60(X - 324.29)$

8. It would appear that there is a negative correlation between age and arithmetic skills and a positive correlation between age and reasoning and verbal skills.

Chapter 10

EXERCISES SECTION 10.1 (Pages 293–294)

1. Select those recipients whose case numbers are 126, 1063, 558, 1859, 2997, 2855, 594, 2562, 1829, 784, 1277, 2056, 10, 654, 2372, 1637, 1295, 1851, 2528, and 354.

2. Those prisoners whose numbers are 816, 309, 763, 78, 61, 277, 188, 174, 530, 709, 496, 889, 482, 772, 774, 893, 312, and 232.

3. Select pages 104, 223, 241, 421, 375, 289, 94, 103, 71, 23, 10, 70, 486, 326, 293, 24, 296, 7, 53, 5, 259, 97, 179, 465, and 145.

4. Those officers whose badge numbers are 191, 196, 69, 210, 114, 202, 19, 177, 195, 152, 76, 92, 212, 201, 41, 142, 147, 12, 106, and 55.

5. Those calculators whose serial numbers are 4934, 4968, 4401, 2533, 815, 5218, 3001, 151, 4944, 118, 2349, 5185, 3580, 4655, 5000, 4576, 4080, 599, 5553, and 1805.

6. Those returns whose last 4 digits are 8164, 3099, 7639, 785, 612, 2775, 9887, 1887, 1745, 5306, 7099, 4962, 8897, 4823, and 7723.

EXERCISES, SECTION 10.3 (Page 300)

1. Mean = 17, standard deviation = 7.29.

2. Mean = 263, standard deviation = 100.73.

3. Mean = 13, standard deviation = 5.94.

4. Mean = 59, standard deviation = 20.26.

ANSWERS

5. Mean = 55, standard deviation = 36.98.   7. Mean = 4, standard deviation = 1.15.
6. Mean = 24, standard deviation = 6.27.   8. Mean = 7, standard deviation = 1.09.

EXERCISES, SECTION 10.6 (Pages 309–310)

1. 0.0099   3. 0.0110   5. 0.0089   7. 0.9925
2. 0.0125   4. Between 4.94 and 5.06 ounces   6. 0.0013   8. 0.9985

MASTERY TESTS

Form A

1. No   3. True   5. Choice b   7. False   9. True
2. No   4. True   6. False   8. True   10. True

Form B

1. Call the auto repair shops which are 69th, 19th, 76th, 92nd, and 41st on the list.
2. Inspect sets whose numbers are 1637, 1810, 1391, 1630, 1988, 414, 1451, 669, 25, 1831, 1703, 454, 1419, 1873, 585, 1761, 1360, 473, 1539, and 862.
3. Mean = 11, standard deviation = 5.40.
4. Mean = 12,274; standard deviation = 2809.01.
5. Mean = 41, standard deviation = 12.91.
6. Mean = 4.1, standard deviation = 0.95.
7. Mean = 33,400; standard deviation = 9736.53.

Chapter 11
EXERCISES, SECTION 11.2 (Pages 321–322)

1. Interval $40.57 to $43.43   4. Interval $163.86 to $168.22   7. Interval $195.58 to $204.42
2. Interval $340.92 to $351.08   5. Interval 1.93 to 2.07 years   8. Interval 52.48 to 57.52
3. Interval $45.31 to $50.69   6. Interval 53.94 to 56.06 cents

EXERCISES, SECTION 11.3 (Pages 326–327)

1. 9.78 to 14.22   6. 4.71 to 11.29
2. 7.52 to 8.48   7. 1.71 to 4.29
3. −5.80 to 3.18   8. 8.85 to 11.15
4. 52.22 to 89.78   9. 7.26 to 8.74
5. 0.97 to 9.03   10. 26.59 to 29.41

EXERCISES, SECTION 11.5 (Pages 331–332)

1. 158   3. 53   5. 17   7. 35   9. $747.90 to $910.41
2. 12   4. 26   6. 44   8. $1609.34 to $2512.60   10. 1.33 to 2.13

EXERCISES, SECTION 11.6 (Pages 337–338)

1. 0.0228   3. 0.9792   5. 0.41 to 0.49   7. 0.62 to 0.70   9. 0.6847
2. 0.0062   4. 0.4090   6. 0.80 to 0.90   8. 0.0548   10. 0.0985

MASTERY TESTS

Form A

1. Confidence limits   3. Choice a 5. Choice a 7. Choice b 9. $\dfrac{p - \pi}{\sigma_p}$

2. False   4. Choice d 6. True   8. True   10. Choice a

Form B

1. 348.97 to 515.03   4. 0.9656   7. 17.28 to 18.72   10. 23
2. 0.53 to 0.61   5. 37.49 to 82.51 8. 177   11. 0.1922
3. $5839.93 to $6160.07   6. 2.14 to 15.86   9. 0.9980   12. 3.54 to 3.86

Chapter 12

EXERCISES, SECTION 12.3 (Pages 360–362)

1. Yes ($z = -8.37$, one-tailed test)   5. No ($z = -2.097$)   9. No ($z = -2.67$)
2. Yes ($z = 5.47$, one-tailed test)   6. Yes ($z = -23.33$)   10. No ($z = -1.18$)
3. No ($z = 1.70$)   7. Yes ($z = -7.62$)   11. No ($z = 1.39$)
4. Yes ($z = 3.80$)   8. No ($z = -2.00$)

EXERCISES, SECTION 12.4 (Pages 367–421)

1. Yes. There is a significant difference. $z = -2.93$
2. Yes. There is a significant difference. $z = -4.5$
3. No. There is no significant difference. $z = 1.31$
4. Yes. There is a significant difference. $z = 3.697$
5. Yes. There is a significant difference. $z = -2.71$

ANSWERS

6. No. There is no significant difference. $z = -2.195$
7. Yes. There is a significant difference. $z = -187.5$
8. No. There is no significant difference. $z = 1.84$ (two-tailed test)
9. No. There is no significant difference. $z = -1.14$
10. Yes. There is a significant difference. $z = 9.95$

EXERCISES, SECTION 12.5 (Pages 373–374)

1. Yes. Accept null hypothesis. $z = 0.44$
2. Yes. Accept null hypothesis. $z = 0.71$
3. No. Accept null hypothesis. $z = -0.56$
4. Yes. Reject null hypothesis. $z = -2.08$
5. No. Accept null hypothesis. $z = 0.68$
6. No. Accept null hypothesis. $z = -1.45$
7. No. We reject null hypothesis. $z = 2.00$ (two-tailed test)
8. Yes. Accept null hypothesis. $z = 0.67$
9. Yes. Accept null hypothesis. $z = 0.82$
10. Yes. Accept null hypothesis. $z = -0.54$
11. No. Reject null hypothesis. $z = -3.70$

MASTERY TESTS

Form A

1. False
2. Choice a
3. Choice b
4. Test statistic
5. False
6. Critical
7. One-sided or one-tailed
8. True
9. Critical value
10. Choice b

Form B

1. Yes. There is a significant difference. $z = 2.30$ (two-tailed test)
2. No. Reject the null hypothesis. $z = -3.54$
3. Yes. Reject the null hypothesis. $z = 2.15$
4. Yes. Accept the null hypothesis. $z = 1.04$
5. Yes. The difference is significant. $z = 3.9$
6. Yes. Reject newspaper claim. $z = -4.35$
7. Yes. Accept the claim. $z = 1.82$ (two-tailed test)
8. Yes. The difference is significant. $z = -2.61$
9. No. Claim is justified. $z = 0.69$
10. Yes. Difference is significant. $z = 3.26$

Chapter 13

EXERCISES, SECTION 13.1 (Pages 388–390)

1. Accept null hypothesis ($\chi^2 = 2.161$)
2. Reject null hypothesis ($\chi^2 = 4.609$).
3. Reject null hypothesis ($\chi^2 = 11.1649$).
4. Accept null hypothesis ($\chi^2 = 1.6444$).
5. Accept null hypothesis ($\chi^2 = 0.5275$).
6. Accept null hypothesis ($\chi^2 = 1.6096$).
7. Accept null hypothesis ($\chi^2 = 3.844$).
8. Accept null hypothesis ($\chi^2 = 1.3558$).
9. Reject null hypothesis ($\chi^2 = 26.0617$).
10. Accept null hypothesis ($\chi^2 = 1.6350$).
11. Reject null hypothesis ($\chi^2 = 25.1463$).

EXERCISES, SECTION 13.2 (Pages 394–397)

1. Reject null hypothesis ($\chi^2 = 8.1770$).
2. Reject null hypothesis ($\chi^2 = 52.9825$).
3. Reject null hypothesis ($\chi^2 = 17.157$).
4. Reject null hypothesis ($\chi^2 = 230.3946$).
5. Reject null hypothesis ($\chi^2 = 667.987$).
6. Accept null hypothesis ($\chi^2 = 0.0699$).
7. Accept null hypothesis ($\chi^2 = 6.5097$).
8. Reject null hypothesis ($\chi^2 = 677.3732$).
9. Accept null hypothesis ($\chi^2 = 1.9508$).
10. Reject null hypothesis ($\chi^2 = 68.7268$).

EXERCISES, SECTION 13.3 (Page 401)

1. Reject null hypothesis ($\chi^2 = 8.7586$).
2. Accept null hypothesis ($\chi^2 = 4.8842$).
3. Reject null hypothesis ($\chi^2 = 9.7292$).
4. Reject null hypothesis ($\chi^2 = 16.8945$).
5. Reject null hypothesis ($\chi^2 = 11.96$).
6. Reject null hypothesis ($\chi^2 = 24.2667$).

MASTERY TESTS

Form *A*

1. 0
2. Contingency table
3. Choice *c*
4. Choice *b*
5. $(r - 1)(c - 1)$
6. Choice *a*
7. Choice *e*
8. Choice *c*
9. Choice *a*
10. Choice *a*

Form *B*

1. Accept null hypothesis ($\chi^2 = 0.0256$).
2. Accept null hypothesis ($\chi^2 = 5.5804$).
3. Reject null hypothesis ($\chi^2 = 19.9836$).
4. Reject null hypothesis ($\chi^2 = 18.5696$).
5. Reject null hypothesis ($\chi^2 = 29.145$).
6. Accept null hypothesis ($\chi^2 = 3.014$).
7. Accept null hypothesis ($\chi^2 = 2.9964$).
8. Reject null hypothesis ($\chi^2 = 94.3454$).
9. Accept null hypothesis ($\chi^2 = 0.3121$).
10. Accept null hypothesis ($\chi^2 = 7,000$).

Chapter 14
EXERCISES, SECTION 14.1 (Pages 420–421)

1. Accept null hypothesis ($F = 0.024$).
2. Accept null hypothesis ($F = 0.37$).
3. Accept null hypothesis ($F = 0.44$).
4. Accept null hypothesis ($F = 0.84$).
5. Accept null hypothesis ($F = 0.63$).
6. Reject null hypothesis ($F = 3.74$).

MASTERY TESTS

Form *A*

1. True   2. Analysis of variance   3. True   4. Variances   5. *MS* (error)
6. $r(c - 1)$ where r is the number of levels and c is the number of repetitions of each.
7. False   8. False   9. False   10. $n - 1$ where n is the total number of samples

Chapter 15

EXERCISES, SECTION 15.1 (Pages 432–434)

1. Reject null hypothesis.
2. Using two-tailed test accept null hypothesis.
   Using one-tailed test reject null hypothesis.
3. Accept null hypothesis.
4. Accept null hypothesis.
5. Reject null hypothesis.

EXERCISES, SECTION 15.2 (Pages 438–440)

1. Accept null hypothesis.
2. Accept null hypothesis.
3. Accept null hypothesis.
4. Accept null hypothesis.
5. Accept null hypothesis.
6. Accept null hypothesis.

EXERCISES, SECTION 15.3 (Pages 445–447)

1. Random   3. Random   5. Random   7. Random   9. Random
2. Random   4. Random   6. Random   8. Random   10. Random

MASTERY TESTS
Form *A*

1. Distribution free
2. Samples are not independent.
3. True
4. Sign test

5. Difference between two sample means
6. Rank–sum
7. Runs
8. Each of the tied numbers is assigned mean of the ranks they occupy.
9. False
10. False

Form *B*

1. Reject null hypothesis.
2. Accept null hypothesis.
3. Random
4. Random
5. Accept null hypothesis.
6. Reject null hypothesis.
7. Random
8. Random
9. Accept null hypothesis.
10. Reject null hypothesis.
11. Random
12. Not random
13. Accept null hypothesis.
14. Random
15. Accept null hypothesis.

Index

FREQUENTLY USED FORMULAS

Relative frequency $\dfrac{f_i}{n}$

Mean $\dfrac{\Sigma x}{n} = \dfrac{x_1 + x_2 + \cdots + x_n}{n}$

Weighted mean $\bar{x}_w = \dfrac{\Sigma xw}{\Sigma w}$

Variance $\sigma^2 = \dfrac{\Sigma(x - \mu)^2}{n}$ or $\dfrac{\Sigma x^2}{n} - \dfrac{(\Sigma x)^2}{n^2}$

Standard deviation $\sigma = \sqrt{\dfrac{\Sigma(x - \mu)^2}{n}}$

Average deviation $\dfrac{\Sigma|x - \mu|}{n}$

Sample standard deviation $\sqrt{\dfrac{\Sigma(x - \bar{x})^2}{n - 1}}$

Percentile rank of X $\dfrac{B + \frac{1}{2}E}{n} \cdot 100$

z-score $z = \dfrac{x - \mu}{\sigma}$ **Original score** $x = \mu + z\sigma$

Probability, p $p = \dfrac{f}{n}$

$_nP_r = \dfrac{n!}{(n - r)!}$ $\qquad$ $_nP_n = n!$ $\qquad$ $_nC_r = \dfrac{n!}{r!(n - r)!}$

Number of permutations with repetitions $\dfrac{n!}{p!q!r! \cdots}$

Mathematical expectation $m_1p_1 + m_2p_2 + m_3p_3 + \cdots$

Addition rule (for mutually exclusive events) $p(A \text{ or } B) = p(A) + p(B)$

Addition rule (general case) $p(A \text{ or } B) = p(A) + p(B) - p(A \text{ and } B)$

Complement of event A $p(A') = 1 - p(A)$

Conditional probability formula $p(A|B) = \dfrac{p(A \text{ and } B)}{p(B)}$

Multiplication rule $p(A \text{ and } B) = p(A|B) \cdot p(B)$

Multiplication rule (for independent events) $p(A \text{ and } B) = p(A) \cdot p(B)$

Bayes' rule $\dfrac{p(B|A_n)p(A_n)}{p(B|A_1)p(A_1) + p(B|A_2)p(A_2) + \cdots + p(B|A_n)p(A_n)}$